T0134943

Jaya: An Advanced Optimization Algorithm and its Engineering Applications

Ravipudi Venkata Rao

Jaya: An Advanced Optimization Algorithm and its Engineering Applications

Ravipudi Venkata Rao, Ph.D., D.Sc.
Department of Mechanical Engineering
S.V. National Institute of Technology
Surat, Gujarat
India

ISBN 978-3-030-07680-1 ISBN 978-3-319-78922-4 (eBook)
https://doi.org/10.1007/978-3-319-78922-4

Softcover re-print of the Hardcover 1st edition 2019

Printed on acid-free paper

This Springer imprint is published by the registered company Springer International Publishing AG part of Springer Nature
The registered company address is: Gewerbestrasse 11, 6330 Cham, Switzerland

Dedicated to my parents (Lakshmi Narayana and Jayamma), dearest wife (Sujatha Rao) and beloved daughter (Jaya Lakshmi)

Foreword

It is a well-known fact that the traditional or classical optimization techniques impose some limitations on solving complex optimization problems. These limitations are mainly interrelated to their inherent search mechanisms. In order to overcome some of the deficiencies of the classical optimization procedures, metaheuristic optimization techniques (also called the advanced optimization techniques), mainly originated from artificial intelligence research, have been developed by the researchers. These algorithms are problem- and model-independent, and most of them are efficient and flexible. Research on these techniques is very active, and many new metaheuristics and improved versions of the older ones are continually appearing in the scientific literature.

In recent years, the field of optimization is witnessing a number of intelligent optimization algorithms, almost all of them based on a metaphor of some natural or man-made process. However, in this book, Prof. Ravipudi Venkata Rao describes a new optimization algorithm named as "Jaya" which is not based on any metaphor. The algorithm always tries to get closer to success (i.e., reaching the best solution) and tries to avoid failure (i.e., moving away from the worst solution). The algorithm strives to become victorious by reaching the best solution, and hence it is named as Jaya (a Sanskrit word meaning triumph or victory). The algorithm has been developed for the global optimization problems and can be used for solving the continuous and discrete optimization problems involving single objective or multiple or many objectives.

In addition to the basic Jaya algorithm, different variants of Jaya algorithm are described in this book. The applications of Jaya algorithm and its variants to different fields of engineering and sciences are also described. The readers may find that, in addition to being simple and powerful, the Jaya algorithm does not need any algorithm-specific parameters for its working, and hence it eliminates the disadvantages of many existing advanced optimization algorithms which face the burden of tuning of algorithm-specific parameters. Improper tuning of the algorithm-specific parameters may lead to local optima or inferior results.

The Jaya algorithm is gaining good reputation among the optimization research community. I believe that the researchers, scientists, engineers, and practitioners belonging to different disciplines of engineering and sciences (physical, life, and social) will find Jaya algorithm as a powerful tool to optimize the systems and processes. I hope the book will be a delight to the readers.

http://www.imiue.mech.pk.edu.pl

Kraków, Poland

Professor Jan Taler, Ph.D., D.Sc.
Institute of Thermal Power Engineering
Politechnika Krakowska
(Cracow University of Technology)

Preface

Keeping in view of the limitations of the traditional optimization techniques, the researchers have developed a number of advanced optimization algorithms popularly known as metaheuristics. The population-based metaheuristic algorithms have two important groups: Evolutionary Algorithms (EA) and swarm intelligence (SI)-based algorithms. Some of the recognized evolutionary algorithms are Genetic Algorithm (GA), Evolutionary Strategy (ES), Evolutionary Programming (EP), Differential Evolution (DE), etc. Some of the well-known swarm intelligence based algorithms are Particle Swarm Optimization (PSO), Ant Colony Optimization (ACO), Artificial Bee Colony (ABC), Firefly (FF) algorithm, Cuckoo search algorithm (CSA), etc. Besides the evolutionary and swarm intelligence based algorithms, there are some other algorithms which work on the principles of different natural phenomena and some of them are Harmony Search (HS) algorithm, Gravitational Search Algorithm (GSA), Biogeography-Based Optimization (BBO) algorithm, League championship algorithm (LCA), etc.

All the abovementioned algorithms are population-based optimization methods and have some limitations in one or the other aspect. The main limitation of all the algorithms is that different parameters (i.e., algorithm-specific parameters) are required for proper working of these algorithms. Proper tuning of these parameters is essential for the searching of the optimum solution by these algorithms. A change in the algorithm-specific parameters changes the effectiveness of the algorithm. The improper tuning of algorithm-specific parameters either increases the computational effort or yields the local optimal solution. Considering this fact, in the year 2011, I had introduced the teaching–learning-based optimization (TLBO) algorithm which does not require any algorithm-specific parameters. The TLBO algorithm requires only common controlling parameters like population size and number of generations for its working. The TLBO algorithm has gained wide acceptance among the optimization researchers.

Keeping in view of the success of the TLBO algorithm, another algorithm-specific parameter-less algorithm was proposed by me in 2016 and was named as Jaya algorithm. However, unlike two phases (i.e., teacher phase and the learner phase) of the TLBO algorithm, the Jaya algorithm has only one phase. The Jaya algorithm is simple in concept and has shown better performance as compared to other optimization algorithms. This algorithm can be used to obtain global solutions for continuous as well as discrete optimization problems with less computational effort and high consistency. In this book, *a posteriori* multi-objective versions of TLBO algorithm and Jaya algorithm are presented and applied to solve the single- and multi-objective optimization problems. In addition, improved versions of Jaya algorithm named as *Elitist Jaya*, *Quasi-Oppositional Jaya*, *Self-Adaptive Jaya*, *Self-Adaptive Multi-Population Jaya*, *Self-Adaptive Multi-Population Elitist Jaya*, *Multi-objective Jaya*, and *Multi-objective Quasi-Oppositional Jaya* are developed and applied to solve the engineering optimization problems.

The Jaya algorithm is gaining wide acceptance in the optimization research community in different fields of science and engineering. The major applications, as of March 2018, are found in the fields of electrical engineering, mechanical design, thermal engineering, manufacturing engineering, civil engineering, structural engineering, computer engineering, electronics engineering, biotechnology, and economics. Many research papers have been published in various reputed international journals of Elsevier, Springer-Verlag, Taylor & Francis and IEEE Transactions in addition to those published in the proceedings of international conferences. The number of research papers is continuously increasing at a faster rate. The algorithm has carved a niche for itself in the field of advanced optimization, and many more researchers may find this as a potential optimization algorithm.

This book provides a detailed understanding of the Jaya algorithm and its versions. Also, it provides the applications of Jaya algorithm and its versions in different fields of engineering. The computer codes of Jaya and its versions are also included in the book and these will be useful to the readers. The book is expected to be useful to various engineering professionals as it presents the powerful Jaya algorithm to make their tasks easier, logical, efficient, and effective. The book is intended for engineers, practitioners, managers, institutes involved in the optimization related projects, applied research workers, academics and graduate students in mechanical, manufacturing, electrical, computer, civil, and structural engineering. As such, this book is expected to become a valuable reference for those wishing to do research by making use of advanced optimization techniques for solving single- or multi-objective combinatorial design optimization problems.

I am grateful to Anthony Doyle and his team of Springer-Verlag, London, for their support and help in producing this book. I profusely thank Prof. Jan Taler of Cracow University of Technology, Poland for writing a nice foreword. I wish to thank various publishers of international journals for giving me the permission to reproduce certain portions of the published research works. I gratefully acknowledge the support of my past and present Ph.D. students (particularly, Kiran More, Dhiraj Rai, Ankit Saroj, and Gajanan Waghmare). My special thanks are due to the Director and my colleagues at S.V. National Institute of Technology, Surat, India.

While every attempt has been made to ensure that no errors (printing or otherwise) enter the book, the possibility of these creeping into the book is always there. I will be grateful to the readers if these errors are pointed out. Suggestions for further improvement of the book will be thankfully acknowledged.

Bangkok, Thailand
March 2018

Ravipudi Venkata Rao, Ph.D., D.Sc.
ravipudirao@gmail.com

Contents

Chapter 1
Introduction

Abstract This chapter presents an introduction to the single objective and multi-objective optimization problems and the optimization techniques to solve the same. The a priori and a posteriori approaches of solving the multi-objective optimization problems are explained. The importance of algorithm-specific parameter-less concept is emphasized.

1.1 Introduction to Optimization

Optimization may be defined as finding of solution of a problem where it is necessary to maximize or minimize a single or set of objective functions within a domain which contains the acceptable values of variables while some restrictions are to be satisfied. There might be a large number of sets of variables in the domain that maximize or minimize the objective function(s) while satisfying the described restrictions. They are called as the acceptable solutions and the solution which is the best among them is called the optimum solution of the problem. An objective function expresses the main aim of the model which is either to be minimized or maximized. For example, in a manufacturing process, the aim may be to maximize the profit or minimize the cost. In designing a structure, the aim may be to maximize the strength or minimize the deflection or a combination of many objectives. The use of optimization techniques for design of engineering systems helps the designers in improving the system's performance, utilization, reliability and cost. A typical engineering system is defined by a number of parameters and it is the designer's task to specify appropriate values for these variables. Skilled designers utilize their knowledge, experience, and judgment to specify these variables and design effective engineering systems. Because of the size and complexity of the typical design task, however, even the most skilled designers are unable to take into account all of the variables simultaneously. Optimization methodologies can be applied during the product development stage to ensure that the finished design will have high performance, high reliability, low weight, and/or low cost. Alternatively,

R. Venkata Rao, *Jaya: An Advanced Optimization Algorithm and its Engineering Applications*, https://doi.org/10.1007/978-3-319-78922-4_1

optimization methods can be applied to existing products to identify potential design improvements.

Thermal system design includes an optimization process in which the designer considers certain objectives such as heat transfer rate, thermal resistance, total cost, friction factor, pressure amplitude, effectiveness, cooling capacity, pressure drop, etc. depending on the requirements. While achieving the mentioned objectives, it is also desirable to minimize the total cost of the system. However, the design optimization for a whole thermal system assembly may be having a complex objective function with a large number of design variables. So, it is a good practice to apply optimization techniques for individual component or intermediate system than to a whole system. For example, in a thermal power plant, individual optimization of cooling tower, heat pipe and heat sink is computationally and mathematically simpler than the optimization of whole power plant.

In the case of manufacturing of products, in order to survive in a competitive market scenario, industries are required to manufacture products with highest quality at a lowest possible cost, fulfill the fast-changing customer desires, consider significance of aesthetics and conform to environmental norms, adopt flexibility in the production system and minimize time-to market of their products. Industries employ a number of manufacturing processes to transform raw materials into finishing goods and efficient utilization of these manufacturing processes is important to achieve the above goals. Manufacturing processes are characterized by a number of input parameters which significantly influence their performances. Therefore, determining best combination of the input parameters is an important task for the success of any manufacturing process.

A human process planner may select proper combination of process parameters for a manufacturing process using his own experience or machine catalog. The selected parameters are usually conservative and far from optimum. Selecting a proper combination of process parameters through experimentation is costly, time consuming and requires a number of experimental trials. Therefore researchers have opted to use optimization techniques for process parameter optimization of manufacturing processes. Researchers have used a number of traditional optimization techniques like geometric programming, nonlinear programming, sequential programming, goal programming, dynamic programming, etc. for process parameter optimization of manufacturing processes. Although the traditional optimization methods had performed well in many practical cases, but they have certain limitations which are mainly related to their inherent search mechanisms. Search strategies of these traditional optimization methods are generally depended on the type of objective and constraint functions (linear, non-linear, etc.) and the type of variables used in the problem modeling (integer, binary, continuous, etc.), their efficiency is also very much dependent on the size of the solution space, number of variables and constraints used in the problem modeling and the structure of the solution space (convex, non-convex, etc.), the solutions obtained largely depends upon the selected initial solution. They also do not provide generic solution approaches that can be used to solve problems where different types of variables, objective and constraint functions are used. In real world optimization problems the number of variables can be very large and the influence on objective function to be optimized can be very complicated. The objective function may have many local optima, whereas the designer or process planner (or the decision maker) is interested in the global optimum. Such problems cannot be effectively handled by classical methods or traditional methods that compute only local optima. So there remains a need for efficient and effective optimization methods for design problems. Continuous research is being conducted in this field and nature inspired heuristic optimization methods are proving to be better than deterministic methods and are widely used.

Therefore, keeping in view of the limitations of the traditional optimization techniques, the researchers have developed a number of advanced optimization algorithms popularly known as metaheuristics. The population-based metaheuristic algorithms have two important groups: evolutionary algorithms (EA) and swarm intelligence (SI) based algorithms. Some of the recognized evolutionary algorithms are: Genetic Algorithm (GA), Evolutionary Strategy (ES), Evolutionary Programming (EP), Differential Evolution (DE), etc. Some of the well-known swarm intelligence based algorithms are: Particle Swarm Optimization (PSO), Ant Colony Optimization (ACO), Artificial Bee Colony (ABC), Fire Fly (FF) algorithm, Cuckoo search algorithm (CSA), etc. Besides the evolutionary and swarm intelligence based algorithms, there are some other algorithms which work on the principles of different natural phenomena and some of them are: Harmony Search (HS) algorithm, Gravitational Search Algorithm (GSA), Biogeography-Based Optimization (BBO) algorithm, etc.

The optimization problem may contain only a single objective function or a number of objective functions. A problem containing only one objective function is called the single objective optimization problem. A problem containing more than one objective function is called the multiple or multi-objective optimization problem. An example of a constrained single objective optimization problem is given below.

Maximize Profit,

$$Y_1 = f(x_1, x_2, x_3, x_4) = 2x_1 + 3x_2 + 1.5x_3 + 0.8x_4 \tag{1.1}$$

Subject to the constraints:

$$x_1 + 2x_2 + x_3 \leq 150 \tag{1.2}$$

$$x_1 + x_2 + x_3 + 0.3x_4 \leq 170 \tag{1.3}$$

$$3x_1 + 1.5x_2 \leq 128 \tag{1.4}$$

The ranges of the variables:

$$0 \leq x_1 \leq 50 \tag{1.5}$$

$$0 \leq x_2 \leq 32 \tag{1.6}$$

$$0 \leq x_3 \leq 44 \tag{1.7}$$

$$0 \leq x_4 \leq 18 \tag{1.8}$$

In this optimization problem the objective function and the constraints are expressed as linear equations. However, in practical applications, these may be in the form of non-linear equations. From the considered ranges of the variables given by Eqs. (1.5)–(1.8), it can be observed that the variables can assume any value within the given ranges. For example, x_1 can assume any value between 0 and 50 (including 0 and 50), x_2 can assume any value between 0 and 32 (including 0 and 32), x_3 can assume any value between 0 and 44 (including 0 and 44) and x_4 can assume any value between 0 and 18 (including 0 and 18). The problem is to find out the best combination of x_1, x_2, x_3 and x_4 to obtain the maximum value of Y_1. We can solve the above single objective optimization problem by using traditional (such as simplex method, dynamic programming, separable programming, etc.) and advanced optimization methods such as genetic algorithm (GA), simulated annealing (SA), particle swarm optimization (PSO), ant colony optimization (ACO), artificial bee colony algorithm (ABC), etc. Then which method will give us the best solution (i.e. maximum value of Y_1)? We do not know! We can know only after applying these methods to the *same* problem and whichever method gives the best solution is called the best optimization method *for the given problem.*

The multi-objective optimization problems require the simultaneous optimization of multiple (often conflicting) objectives over a given space of candidate solutions. These problems occur in many practical applications, rather often as bi-objective problems, with typical pairs of objectives as quality vs cost, strength vs weight, or accuracy vs complexity. Suppose, we include another objective function of minimizing the cost to the above single objective optimization problem.

Minimize Cost,

$$Y_2 = f(x_1, x_2, x_3, x_4) = 0.75x_1 + 8x_2 + x_3 + 1.2x_4 \tag{1.9}$$

Then the problem is to find out the best combination of x_1, x_2, x_3 and x_4 to obtain the maximum value of Y_1 and minimum value of Y_2. Now let us assume a set of values (0, 0, 0, 0) of x_1, x_2, x_3 and x_4. These values satisfy the constraints and hence substituting these values in the objective function Y_2 leads to the value of cost of 0. This value is highly appreciable but at the same time if the same set of values (0, 0, 0, 0) is substituted in the first objective function Y_1 then it leads to the value of profit of 0. This is highly undesirable and hence we can say that the set of values (0, 0, 0, 0) of x_1, x_2, x_3 and x_4 do not give the optimum solution. Now let us assume another set of values (0, 5, 40, 10) of x_1, x_2, x_3 and x_4. These values also satisfy the constraints and hence substituting these values in the objective functions Y_1 and Y_2 lead to the values of 342 and 83 respectively. The set of values (0, 5, 40, 10) of x_1, x_2, x_3 and x_4 can be said as a feasible solution but not the optimum solution. Let us further assume another set of values (15, 18, 7, 24) of x_1, x_2, x_3 and x_4. These values also satisfy the constraints and hence substituting these values in the objective functions Y_1 and Y_2 lead to the values of 191.05 and 113.7 respectively. Thus, the set of values (15, 18, 7, 24) of x_1, x_2, x_3 and x_4 can be said as another feasible solution but not the optimum solution. Thus, how many feasible solutions are possible for the considered problem? Various combinations of x_1, x_2, x_3 and x_4 are possible and hence we can say that a large number (or almost infinite number) of feasible solutions may be possible for the considered problem. The two objectives are of conflicting type and optimal solution of one objective does not meet the optimal solution of the other and there exist a large number (or almost infinite number) of solutions (as variables can take any value within their bounds). In general, the multi-objective optimization problems have decision variable values which are determined in a continuous or integer domain with either an infinite or a large number of solutions, the best of which should satisfy the designer or decision maker's constraints and preference priorities.

Here also we can apply different optimization methods to solve the *same* bi-objective optimization problem and whichever method gives the best combination of solution (i.e. most maximum value of Y_1 and most minimum value of Y_2) is called the best optimization method *for the given problem*. In fact, the solution to the multi-objective optimization problem involves finding not one, but a set of solutions that represent the best possible trade-offs among the objective functions being optimized. Such trade-offs constitute the so called Pareto optimal set, and their corresponding objective function values form the so called Pareto front. The multi-objective optimization problem may contain any number of objectives more than one. The example described by Eqs. (1.1)–(1.9) is only for giving an idea to the readers about the concepts of single objective and multi-objective optimization problems.

1.2 A Priori and a Posteriori Approaches of Solving the Multi-objective Optimization Problems

There are basically two approaches to solve a multi-objective optimization problem and these are: a priori approach and a posteriori approach. In a priori approach, multi-objective optimization problem is transformed into a single objective optimization problem by assigning an appropriate weight to each objective. This ultimately leads to a unique optimum solution. In the a priori approach, the preferences of the decision maker are asked and the best solution according to the given preferences is found. The preferences of the decision maker are in the form of weights assigned to the objective functions. The weights may be assigned through any method like direct assignment, eigenvector method (Rao 2007), empty method, minimal information method, etc. Once the weights are decided by the decision maker, the multiple objectives are combined into a scalar objective via the weight vector. However, if the objective functions are simply weighted and added to produce a single fitness, the function with the largest range would dominate the evolution. A poor input value for the objective with the larger range makes the overall value much worse than a poor value for the objective with smaller range. To avoid this, all objective functions are normalized to have same range. For example, if $f_1(x)$ and $f_2(x)$ are the two objective functions to be minimized, then the combined objective function can be written as,

$$\min f(x) = \left\{ w_1 \left[\left(\frac{f_{1(x)}}{f_1^*} \right) \right] + w_2 \left[\left(\frac{f_{2(x)}}{f_2^*} \right) \right] \right\} \tag{1.10}$$

where, $f(x)$ is the combined objective function and f_1^* is the minimum value of the objective function $f_1(x)$ when solved it independently without considering $f_2(x)$ (i.e. solving the multi-objective problem as a single objective problem and considering only $f_1(x)$ and ignoring $f_2(x)$). And f_2^* is the minimum value of the objective function $f_2(x)$ when solved it independently without considering $f_1(x)$ (i.e. solving the multi-objective problem as a single objective problem considering only $f_2(x)$ and ignoring $f_1(x)$). w_1 and w_2 are the weights assigned by the decision maker to the objective functions $f_1(x)$ and $f_2(x)$ respectively.

Suppose $f_1(x)$ and $f_2(x)$ are not of the same type (i.e. minimization or maximization) but one is a minimization function (say $f_1(x)$) and the other is a maximization function (say $f_2(x)$). In that case, the Eq. (1.10) is written as Eq. (1.11) and f_2^* is the maximum value of the objective function $f_2(x)$ when solved it independently without considering $f_1(x)$.

$$\min f(x) = \left\{ w_1 \left[\left(\frac{f_{1(x)}}{f_1^*} \right) \right] - w_2 \left[\left(\frac{f_{2(x)}}{f_2^*} \right) \right] \right\} \tag{1.11}$$

In general, the combined objective function can include any number of objectives and the summation of all weights is equal to 1. The solution obtained by this process depends largely on the weights assigned to the objective functions. This approach does not provide a set of Pareto points. Furthermore, in order to assign weights to each objective the process planner is required to precisely know the order of importance of each objective in advance which may be difficult when the scenario is volatile or involves uncertainty. This drawback of a priori approach is eliminated in a posteriori approach, wherein it is not required to assign the weights to the objective functions prior to the simulation run.

A posteriori approach provides multiple tradeoff (Pareto-optimal) solutions for a multi-objective optimization problem in a single simulation run. The designer or process planner can then select one solution from the set of Pareto-optimal solutions based on the requirement or order of importance of objectives. On the other hand, as a priori approach provides only a single solution at the end of one simulation run, in order to achieve multiple trade-off solutions using a priori approach the algorithm has to be run multiple times with different combination of weights. Thus, a posteriori approach is very suitable for solving multi-objective optimization problems in machining processes wherein taking into account frequent change in customer desires is of paramount importance and determining the weights to be assigned to the objectives in advance is difficult. Evolutionary algorithms are popular approaches for generating the Pareto optimal solutions to a multi-objective optimization problem. Currently, most evolutionary multi-objective optimization algorithms apply Pareto-based ranking schemes (Simon 2013). Evolutionary algorithms such as the Non-dominated Sorting Genetic Algorithm-II (NSGA-II) and Strength Pareto Evolutionary Algorithm 2 (SPEA-2) have become standard approaches. The main advantage of evolutionary algorithms, when applied to solve multi-objective optimization problems, is the fact that they typically generate sets of solutions, allowing computation of an approximation of the entire Pareto front. The main disadvantage of evolutionary algorithms is their lower speed and the Pareto optimality of the solutions cannot be guaranteed. It is only known that none of the generated solutions dominates the others.

1.3 Algorithm-Specific Parameter-Less Concept

All the evolutionary and swarm intelligence based algorithms used for solving the single and multi-objective optimization problems are probabilistic algorithms and require common controlling parameters like population size, number of generations, elite size, etc. Besides the common control parameters, different algorithms require their own algorithm-specific control parameters. For example, GA uses mutation probability, crossover probability, selection operator; PSO algorithm uses inertia weight, social and cognitive parameters; ABC uses number of onlooker bees, employed bees, scout bees and limit; HS algorithm uses harmony memory consideration rate, pitch adjusting rate, and the number of improvisations. Similarly, the other algorithms such as evolutionary programming (EP), differential evolution (DE), ant colony optimization (ACO), fire fly (FF), cuckoo search algorithm (CSA), gravitational search algorithm (GSA), bio-geography based optimization (BBO), flower pollination algorithm (FPA), ant lion optimization (ALO), invasive weed optimization (IWO), etc. need the tuning of respective algorithm-specific parameters. The proper tuning of the algorithm-specific parameters is a very crucial factor which affects the performance of the above mentioned algorithms. The improper tuning of algorithm-specific parameters either increases the computational effort or yields the local optimal solution. Considering this fact, Rao et al. (2011) introduced the teaching-learning-based optimization (TLBO) algorithm which does not require any algorithm-specific parameters. The TLBO algorithm requires only common controlling parameters like population size and number of generations for its working. The TLBO algorithm has gained wide acceptance among the optimization researchers (Rao 2016a).

Keeping in view of the success of the TLBO algorithm, another algorithm-specific parameter-less algorithm was proposed by Rao (2016b) and was named as Jaya algorithm. However, unlike two phases (i.e. teacher phase and the learner phase) of the TLBO algorithm, the Jaya algorithm has only one phase. The Jaya algorithm is simple in concept and has shown better performance as compared to other optimization algorithms. In this book a posteriori multi-objective versions of TLBO algorithm and Jaya algorithm are presented and applied to solve the single and multi-objective optimization problems. In addition, improved versions of Jaya algorithm named as *Elitist Jaya*, *Quasi-Oppositional Jaya*, *Self-Adaptive Jaya*, *Self-Adaptive Multi-Population Jaya*, and *Self-Adaptive Multi-Population Elitist Jaya*, and *Multi-Objective Quasi-Oppositional Jaya* are developed and applied to solve the single and multi-objective engineering optimization problems.

The details of the working of TLBO algorithm, NSTLBO algorithm, Jaya algorithm and its variants are given in the next chapter.

References

Rao, R. V. (2007). *Decision making in the manufacturing environment using graph theory and fuzzy multiple attribute decision making methods*. London: Springer Verlag.

Rao, R. V., Savsani, V. J., & Vakharia, D. P. (2011). Teaching–learning-based optimization: A novel method for constrained mechanical design optimization problems. *Computer-Aided Design, 43,* 303–315.

Rao, R. V. (2016a). *Teaching learning based optimization algorithm and its engineering applications*. Switzerland: Springer International Publishing.

Rao, R. V. (2016b). Jaya: A simple and new optimization algorithm for solving constrained and unconstrained optimization problems. *International Journal of Industrial Engineering Computations, 7,* 19–34.

Simon, D. (2013). *Evolutionary optimization algorithms*. New York: Wiley.

Chapter 2
Jaya Optimization Algorithm and Its Variants

Abstract This chapter presents the details of TLBO algorithm, NSTLBO algorithm, Jaya algorithm and its variants named as *Self-Adaptive Jaya*, *Quasi-Oppositional Jaya*, *Self-Adaptive Multi-Population Jaya*, *Self-Adaptive Multi-Population Elitist Jaya, Chaos Jaya, Multi-Objective Jaya,* and *Multi-Objective Quasi-Oppositional Jaya.* Suitable examples are included to demonstrate the working of Jaya algorithm and its variants for the unconstrained and constrained single and multi-objective optimization problems. Three performance measures of coverage, spacing and hypervolume are also described to assess the performance of the multi-objective optimization algorithms.

2.1 Teaching-Learning-Based Optimization (TLBO) Algorithm

Rao et al. (2011) developed the TLBO algorithm which does not require tuning of any algorithm-specific parameters for its working. The algorithm describes two basic modes of the learning: (i) through teacher (known as teacher phase) and (ii) through interaction with the other learners (known as learner phase). In this optimization algorithm a group of learners is considered as population and different subjects offered to the learners are considered as different design variables of the optimization problem and a learner's result is analogous to the 'fitness' value of the optimization problem. The best solution in the entire population is considered as the teacher. The design variables are actually the parameters involved in the objective function of the given optimization problem and the best solution is the best value of the objective function.

R. Venkata Rao, *Jaya: An Advanced Optimization Algorithm and its Engineering Applications*, https://doi.org/10.1007/978-3-319-78922-4_2

The working of TLBO is divided into two parts, 'Teacher phase' and 'Learner phase'. Working of both these phases is explained below (Rao 2016a).

2.1.1 Teacher Phase

It is the first part of the algorithm where learners learn through the teacher. During this phase a teacher tries to increase the mean result of the class in the subject taught by him or her depending on his or her capability. At any iteration *i*, assume that there are '*m*' number of subjects (i.e. design variables), '*n*' number of learners (i.e. population size, $k = 1, 2, \ldots, n$) and $M_{j,i}$ be the mean result of the learners in a particular subject '*j*' ($j = 1, 2, \ldots, m$). The best overall result $X_{total\text{-}kbest,i}$ considering all the subjects together obtained in the entire population of learners can be considered as the result of best learner *kbest*. However, as the teacher is usually considered as a highly learned person who trains learners so that they can have better results, the best learner identified is considered by the algorithm as the teacher. The difference between the existing mean result of each subject and the corresponding result of the teacher for each subject is given by,

$$Difference\_Mean_{j,k,i} = r_i \left(X_{j,kbest,i} - T_F M_{j,i} \right) \tag{2.1}$$

where, $X_{j,kbest,i}$ is the result of the best learner in subject *j*. T_F is the teaching factor which decides the value of mean to be changed, and r_i is the random number in the range [0, 1]. Value of T_F can be either 1 or 2. The value of T_F is decided randomly with equal probability as,

$$T_F = round[1 + rand(0, 1)\{2 - 1\}] \tag{2.2}$$

T_F is not a parameter of the TLBO algorithm. The value of T_F is not given as an input to the algorithm and its value is randomly decided by the algorithm using Eq. (2.2). After conducting a number of experiments on many benchmark functions it is concluded that the algorithm performs better if the value of T_F is between 1 and 2. However, the algorithm is found to perform much better if the value of TF is either 1 or 2 and hence to simplify the algorithm, the teaching factor is suggested to take either 1 or 2 depending on the rounding up criteria given by Eq. (2.2). Based on the $Difference\_Mean_{j,k,i}$, the existing solution is updated in the teacher phase according to the following expression.

$$X'_{j,k,i} = X_{j,k,i} + Difference\_Mean_{j,k,i} \tag{2.3}$$

where, $X'_{j,k,i}$ is the updated value of $X_{j,k,i}$. $X'_{j,k,i}$ is accepted if it gives better function value. All the accepted function values at the end of the teacher phase are maintained and these values become the input to the learner phase. The learner phase depends upon the teacher phase.

2.1.2 *Learner Phase*

It is the second part of the algorithm where learners increase their knowledge by interaction among themselves. A learner interacts randomly with other learners for enhancing his or her knowledge. A learner learns new things if the other learner has more knowledge than him or her. Considering a population size of '*n*', the learning phenomenon of this phase is explained below.

Randomly select two learners P and Q such that $X'_{total\text{-}P,i} \neq X'_{total\text{-}Q,i}$ (where, $X'_{total\text{-}P,i}$ and $X'_{total\text{-}Q,i}$ are the updated function values of $X_{total\text{-}P,i}$ and $X_{total\text{-}Q,i}$ of P and Q respectively at the end of teacher phase)

$$X''_{j,P,i} = X'_{j,P,i} + r_i\left(X'_{j,P,i} - X'_{j,Q,i}\right),\ \text{If}\ X'_{total-P,i} < X'_{total-Q,i} \tag{2.4}$$

$$X''_{j,P,i} = X'_{j,P,i} + r_i\left(X'_{j,Q,i} - X'_{j,P,i}\right),\ \text{If}\ X'_{total-Q,I} < X'_{total-P,i} \tag{2.5}$$

$X''_{j,P,I}$ is accepted if it gives a better function value.

Equations (2.4) and (2.5) are for minimization problems. In the case of maximization problems, Eqs. (2.6) and (2.7) are used.

$$X''_{j,P,i} = X'_{j,P,i} + r_i\left(X'_{j,P,i} - X'_{j,Q,i}\right),\ \text{If}\ X'_{total-Q,i} < X'_{total-P,i} \tag{2.6}$$

$$X''_{j,P,i} = X'_{j,P,i} + r_i\left(X'_{j,Q,i} - X'_{j,P,i}\right),\ \text{If}\ X'_{total-P,i} < X'_{total-Q,i} \tag{2.7}$$

Figure 2.1 gives the flowchart for the TLBO algorithm. For the TLBO algorithm the maximum number of function evaluations = 2 × population size × no. of iterations.

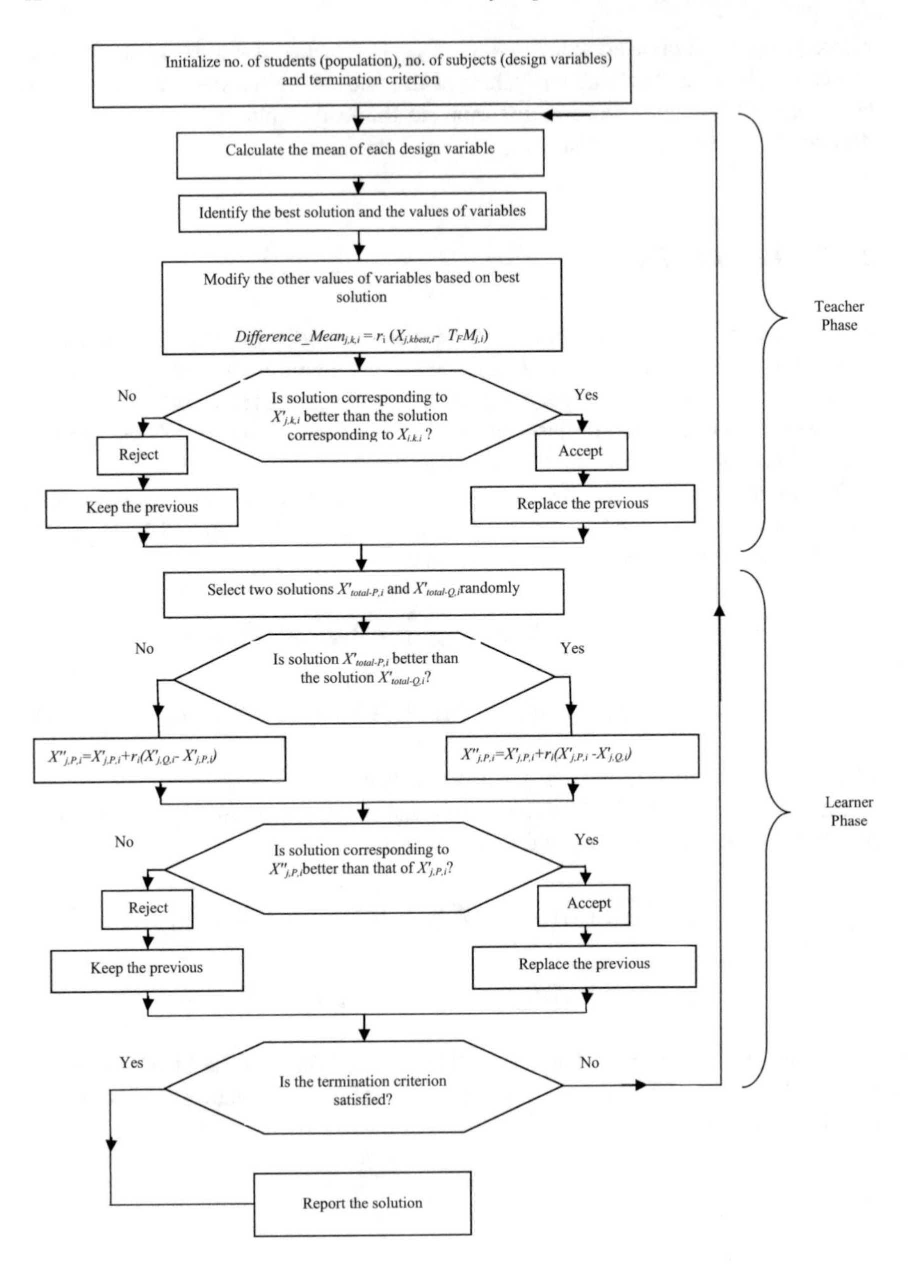

Fig. 2.1 Flowchart of TLBO algorithm

2.2 Non-dominated Sorting Teaching-Learning-Based Optimization (NSTLBO) Algorithm

The NSTLBO algorithm is developed for solving the multi-objective optimization problems. The NSTLBO algorithm is an extension of the TLBO algorithm. It is a posteriori approach for solving multi-objective optimization problems and maintains a diverse set of solutions. NSTLBO algorithm consists of teacher phase and learner phase similar to the TLBO algorithm. However, in order to handle multiple objectives, effectively and efficiently the NSTLBO algorithm is incorporated with non-dominated sorting approach and crowding distance computation

In the beginning, an initial population is randomly generated with *NP* number of solutions. This initial population is then sorted and ranked based on constraint-dominance and non-dominance concept. The superiority among the solutions is first determined based on the constraint-dominance concept and then on the non-dominance concept and then on the crowding distance value of the solutions. The learner with the highest rank (rank = 1) is selected as the teacher of the class. In case, there exists more than one learner with the same rank then the learner with the highest value of crowding distance is selected as the teacher of the class. This ensures that the teacher is selected from the sparse region of the search space.

Once the teacher is selected, learners are updated based on the teacher phase of the TLBO algorithm i.e. according to Eqs. (2.1)–(2.3). After the teacher phase, the set of updated learners (new learners) is concatenated to the initial population to obtain a set of 2*NP* solutions (learners). These learners are again sorted and ranked based on the constraint-dominance concept, non-dominance concept and the crowding distance value for each learner is computed. A learner with a higher rank is regarded as superior to the other learner. If both the learners hold the same rank, then the learner with higher crowding distance value is seen as superior to the other. Based on the new ranking and crowding distance value *NP* number of best learners are selected. These learners are further updated according to the learner phase of the TLBO algorithm i.e. according to Eqs. (2.4) and (2.5) or (2.6) and (2.7)

The superiority among the learners is determined based on the constraint-dominance, non-dominance rank and the crowding distance value of the learners. A learner with a higher rank is regarded as superior to the other learner. If both the learners hold the same rank, then the learner with higher crowding distance value is seen as superior to the other. After the end of the learner phase, the new learners are combined with the old learners and again sorted and ranked. Based on the new ranking and crowding distance value *NP* number of best learners are selected, and these learners are directly updated based on the teacher phase in the next iteration (Rao 2016a; Rao et al. 2016). Figure 2.2 shows the flowchart of the NSTLBO algorithm.

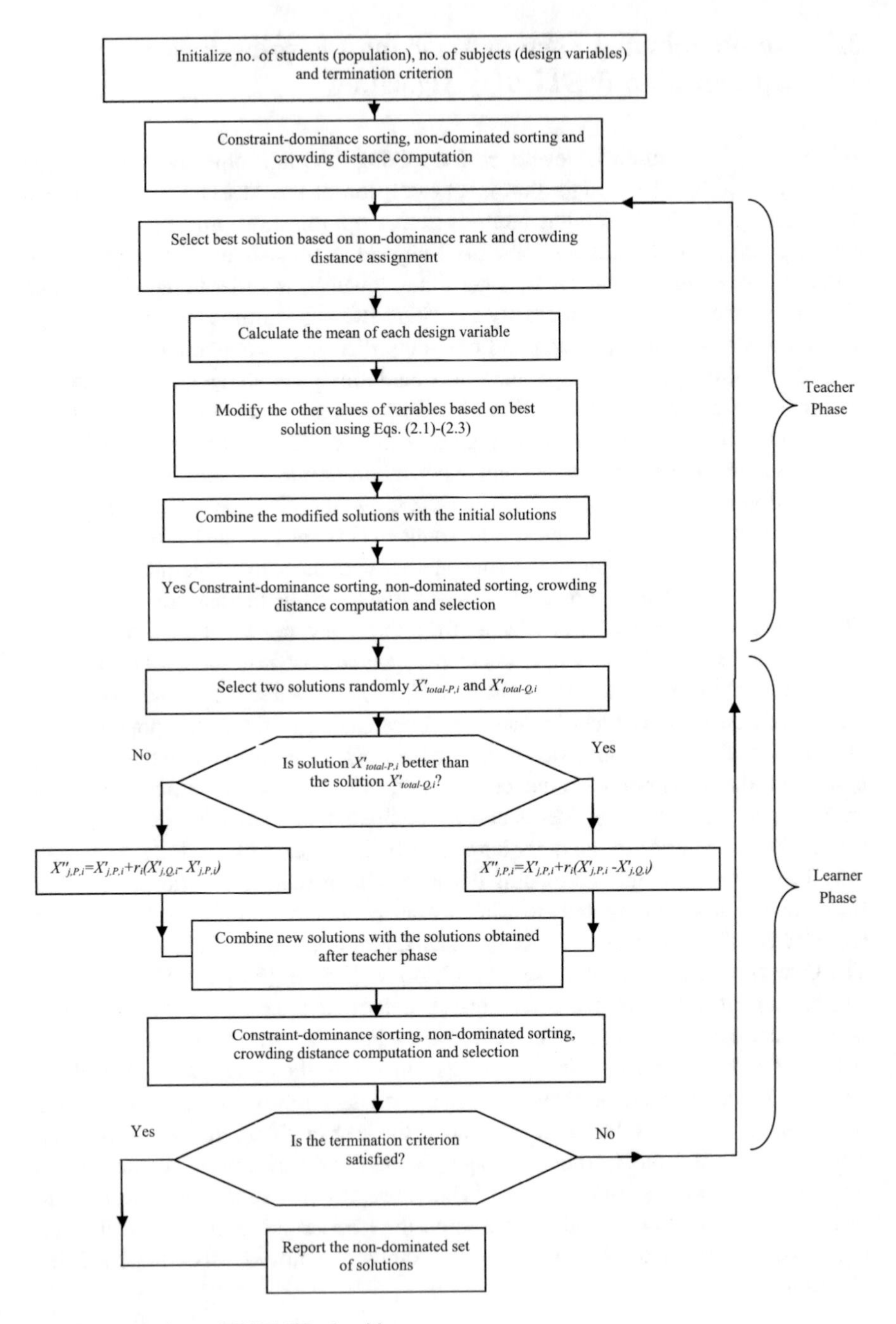

Fig. 2.2 Flowchart of NSTLBO algorithm

2.2.1 Non-dominated Sorting of the Population

In this approach the population is sorted into several ranks (fronts) based on the dominance concept as follows: a solution x_i is said to dominate other solution x_j if and only if solution x_i is no worse than solution x_j with respect to all the objectives and the solution x_i is strictly better than solution x_j in at least one objective. If any of the two conditions are violated, then solution x_i does not dominate solution x_j.

Among a set of solutions P, the non-dominated solutions are those that are not dominated by any solution in the set P. All such non-dominated solutions which are identified in the first sorting run are assigned rank one (first front) and are deleted from the set P. The remaining solutions in set P are again sorted and the procedure is repeated until all the solutions in the set P are sorted and ranked. In the case of constrained multi-objective optimization problems constrained-dominance concept may be used. For the NSTLBO algorithm the maximum number of function evaluations = 2 × population size × no. of iterations.

2.2.2 Crowding Distance Computation

The crowding distance is assigned to each solution in the population with an aim to estimate the density of solutions surrounding a particular solution i. Thus, average distance of two solutions on either side of solution i is measured along each of the M objectives. This quantity is called as the crowding distance (CD_i). The following steps may be followed to compute the CD_i for each solution i in the front F.

Step 1: Determine the number of solutions in front F as $l = |F|$. For each solution i in the set assign $CD_i = 0$.

Step 2: For each objective function $m = 1, 2, \ldots, M$, sort the set in the worst order of f_m.

Step 3: For $m = 1, 2, \ldots, M$, assign largest crowding distance to boundary solutions in the sorted list ($CD_1 = CD_l = \infty$), and for all the other solutions in the sorted list $j = 2$ to $(l - 1)$, crowding distance is as follows:

$$CD_j = CD_j + \frac{f_m^{j+1} - f_m^{j-1}}{f_m^{\max} - f_m^{\min}} \tag{2.8}$$

where, j is a solution in the sorted list, f_m is the objective function value of mth objective, $f_m^{\max}$ and $f_m^{\min}$ are the population-maximum and population-minimum values of the mth objective function.

2.2.3 Crowding-Comparison Operator

Crowding-comparison operator is used to identify the superior solution among two solutions under comparison, based on the two important attributes possessed by every individual i in the population i.e. non-domination rank ($Rank_i$) and crowding distance (CD_i). Thus, the crowded-comparison operator $(\prec_n)$ is defined as follows: $i \prec_n j$ if ($Rank_i < Rank_j$) or (($Rank_i = Rank_j$) and ($CD_i > CD_j$)). That is, between two solutions (i and j) with differing non-domination ranks, the solution with lower or better rank is preferred. Otherwise, if both solutions belong to the same front ($Rank_i = Rank_j$), then the solution located in the lesser crowded region ($CD_i > CD_j$) is preferred.

2.2.4 Constraint-Dominance Concept

A solution i is said to constrained-dominate a solution j, if any of the following conditions is true. Solution i is feasible and solution j is not. Solutions i and j are both infeasible, but solution i has a smaller overall constraint violation. Solutions i and j are feasible and solution i dominates Solution j. The constraint-dominance concept ensures that the feasible solutions get a higher rank as compared to the infeasible solutions. Among the feasible solutions, the solutions which are superior (non-dominated) are ranked higher than the dominated feasible solutions. Among infeasible solutions, higher rank is assigned to the solutions with less overall constraint violation.

2.3 Jaya Algorithm

Rao (2016b) developed the Jaya algorithm which is simple to implement and does not require tuning of any algorithm-specific parameters. In the Jaya algorithm P initial solutions are randomly generated obeying the upper and lower bounds of the process variables. Thereafter, each variable of every solution is stochastically updated using Eq. (2.9). Let f is the objective function to be minimized (or maximized). Let there be 'd' number of design variables. Let the objective function value corresponding to the best solution be *f_best* and the objective function value corresponding to the worst solution be *f_worst*.

$$\begin{aligned} A(i+1,j,k) = & A(i,j,k) + r(i,j,1)(A(i,j,b) - |A(i,j,k)|) \\ & - r(i,j,2)(A(i,j,w) - |A(i,j,k)|) \end{aligned} \tag{2.9}$$

Here b and w represent the index of the best and worst solutions among the current population. i, j, k are the index of iteration, variable, and candidate solution.

$A(i, j, k)$ means the jth variable of kth candidate solution in ith iteration. $r(i, j, 1)$ and $r(i, j, 2)$ are numbers generated randomly in the range of [0, 1]. The random numbers $r(i, j, 1)$ and $r(i, j, 2)$ act as scaling factors and ensure good diversification. The main aim of the Jaya algorithm is to improve the fitness (objective function value) of each candidate solution in the population. Therefore, the Jaya algorithm tries to move the objective function value of each solution towards the best solution by updating the values of the variables. Once the values of the variables are updated, the updated (new) solutions are compared with the corresponding old solutions and only the good solutions (i.e. solutions having better objective function value) are considered for the next generation. In the Jaya algorithm, a solution moves closer to the best solution in every generation, but at the same time a candidate solutions moves away from the worst solution. Thereby, a good intensification and diversification of the search process is achieved. The algorithm always tries to get closer to success (i.e. reaching the best solution) and tries to avoid failure (i.e. moving away from the worst solution). The algorithm strives to become victorious by reaching the best solution and hence it is named as **Jaya** (a Sanskrit word meaning **victory** or **triumph**).

The flowchart for Jaya algorithm is shown in Fig. 2.3 (Rao et al. 2017). For every candidate solution the Jaya algorithm evaluates the objective function only

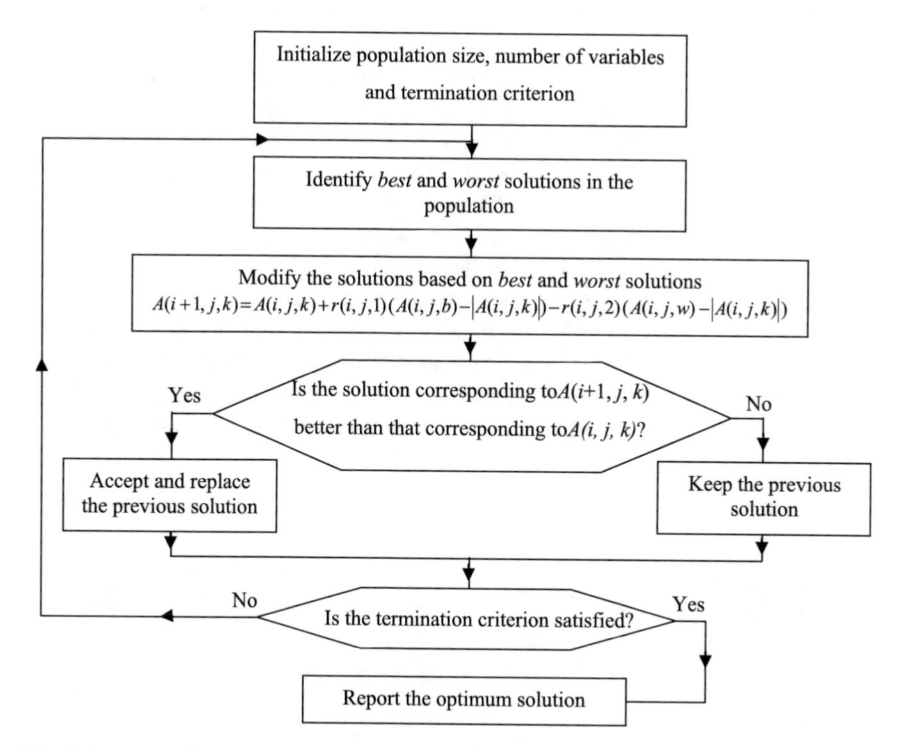

Fig. 2.3 Flowchart of Jaya algorithm (Rao et al. 2017; Reprinted with permission from Elsevier)

once in each generation. Therefore, the total number of function evaluations required by Jaya algorithm = population size × no. of generations for a single simulation run.

2.3.1 Demonstration of Jaya Algorithm on Unconstrained Optimization Problems

To demonstrate the working of Jaya algorithm, an unconstrained benchmark function of Sphere is considered. The objective function is to find out the values of x_i that minimize the value of the Sphere function.

$$\text{Minimize, } f(x_i) = \sum_{i=1}^{n} x_1^2 \tag{2.10}$$

Range of variables: $-100 \leq x_i \leq 100$.

The known solution to this benchmark function is 0 for all x_i values of 0. Now to demonstrate the Jaya algorithm, let us assume a population size of 5 (i.e. candidate solutions), two design variables x_1 and x_2 and two iterations as the termination criterion. The initial population is randomly generated within the ranges of the variables and the corresponding values of the objective function are shown in Table 2.1. As it is a minimization function, the lowest value of $f(x)$ is considered as the best solution and the highest value of $f(x)$ is considered as the worst solution.

From Table 2.1 it can be seen that the best solution is corresponding the 4th candidate and the worst solution is corresponding to the 3rd candidate. Now assuming random numbers $r_1 = 0.58$ and $r_2 = 0.81$ for x_1 and $r_1 = 0.92$ and $r_2 = 0.49$ for x_2, the new values of the variables for x_1 and x_2 are calculated using Eq. (2.9) and are placed in Table 2.2. For example, for the 1st candidate, the new values of x_1 during the first iteration is calculated as, $-5 + 0.58\,(-8-|-5|) - 0.81\,(70-|-5|) = -65.19$. The new value of x_2 during the first iteration is calculated as, $18 + 0.92\,(7-|18|) - 0.49\,(-6-|18|) = 19.64$. Similarly, the new values of x_1 and x_2 for the other candidates are calculated. Table 2.2 shows the new values of x_1 and x_2 and the corresponding values of the objective function.

Table 2.1 Initial population (Sphere function)

Candidate	x_1	x_2	$f(x)$	Status
1	−5	18	349	
2	14	63	4165	
3	70	−6	4936	*Worst*
4	−8	7	113	*Best*
5	−12	−18	468	

Table 2.2 New values of the variables and the objective function during first iteration (Sphere function)

Candidate	x_1	x_2	$f(x)$
1	−65.19	19.64	4635
2	−44.12	45.29	3997.7
3	24.76	0.8	613.697
4	−67.5	13.37	4735
5	−70.58	−16.36	5249.18

Now, the values of $f(x)$ of Tables 2.1 and 2.2 are compared and the best values of $f(x)$ are considered and placed in Table 2.3. This completes the first iteration of the Jaya algorithm.

From Table 2.3 it can be seen that the best solution is corresponding to the 4th candidate and the worst solution is corresponding to the 2nd candidate. Now, during the second iteration, assuming random numbers $r_1 = 0.27$ and $r_2 = 0.23$ for x_1 and $r_1 = 0.38$ and $r_2 = 0.51$ for x_2, the new values of the variables for x_1 and x_2 are calculated using Eq. (2.9). Table 2.4 shows the new values of x_1 and x_2 and the corresponding values of the objective function during the second iteration.

Now, the values of $f(x)$ of Tables 2.3 and 2.4 are compared and the best values of $f(x)$ are considered and placed in Table 2.5. This completes the second iteration of the Jaya algorithm.

Table 2.3 Updated values of the variables and the objective function based on fitness comparison at the end of first iteration (Sphere function)

Candidate	x_1	x_2	$f(x)$	Status
1	−5	18	349	
2	−44.12	45.29	3997.7	*Worst*
3	24.76	0.8	613.697	
4	−8	7	113	*Best*
5	−12	−18	468	

Table 2.4 New values of the variables and the objective function during second iteration (Sphere function)

Candidate	x_1	x_2	$f(x)$
1	2.7876	−0.0979	7.7803
2	−37.897	30.74	2381.13
3	31.757	−19.534	1390.08
4	−0.3324	−12.528	157.06
5	−4.4924	−36.098	1323.247

Table 2.5 Updated values of the variables and the objective function based on fitness comparison at the end of second iteration (Sphere function)

Candidate	x_1	x_2	$f(x)$	Status
1	2.7876	−0.0979	7.7803	*Best*
2	−37.897	30.74	2381.13	*Worst*
3	24.76	0.8	613.697	
4	−8	7	113	
5	−12	−18	468	

From Table 2.5 it can be seen that the best solution is corresponding to the 1st candidate and the worst solution is corresponding to the 2nd candidate. It can also be observed that the value of the objective function is reduced from 113 to 7.7803 in just two iterations. If we increase the number of iterations then the known value of the objective function (i.e. 0) can be obtained within next few iterations. Also, it is to be noted that in the case of maximization function problems, the best value means the maximum value of the objective function and the calculations are to be proceeded accordingly. Thus, the proposed method can deal with both minimization and maximization problems.

To further demonstrate the working of Jaya algorithm, another benchmark function of Rastrigin is considered (Rao et al. 2017). The objective function is to find out the values of x_i that minimize the value of the Rastrigin function.

$$f_{\min} = \sum_{i=1}^{D} \left[x_i^2 - 10\cos(2\pi x_i) + 10\right] \quad (2.11)$$

Range of variables: $-5.12 \leq x_i \leq 5.12$.

The known solution to this benchmark function is 0 for all x_i values of 0. Now to demonstrate the Jaya algorithm, let us assume a population size of 5 (i.e. candidate solutions), two design variables x_1 and x_2 and three generations as the termination criterion. The initial population is randomly generated within the ranges of the variables and the corresponding values of the objective function are shown in Table 2.6. As it is a minimization function, the lowest value of the Rastrigin function is considered as the best solution and the highest value is considered as the worst solution.

From Table 2.6 it can be seen that the best solution is corresponding to the 4th candidate and the worst solution is corresponding to the 3rd candidate. Now assuming random numbers $r_1 = 0.38$ and $r_2 = 0.81$ for x_1 and $r_1 = 0.92$ and $r_2 = 0.49$ for x_2, the new values of the variables for x_1 and x_2 are calculated using Eq. (2.9) and are placed in Table 2.7. For example, for the 1st candidate, the new values of x_1 and x_2 during the first generation are calculated as shown below.

Table 2.6 Initial population (Rastrigin function)

Candidate	x_1	x_2	$f(x)$	Status
1	−4.570261872	0.045197073	40.33111401	
2	3.574220009	1.823157605	40.59470054	
3	−2.304524513	4.442417134	57.75775652	*Worst*
4	−1.062187325	−0.767182961	11.39297485	*Best*
5	−0.84373426	3.348170112	32.15169977	

Table 2.7 New values of the variables and the objective function during first generation (Rastrigin function)

Candidate	x_1	x_2	$f(x)$
1	−1.1420155950	−2.8568303882	16.96999238
2	6.574168285* (5.12)	−1.84339288493	36.785779377
3	0.1494454699	−0.350414953	20.138516097
4	0.857586897	−3.979664354	20.398015196
5	0.982105144	−0.974135756	2.108371360

*The value of the variable x_1 exceeds the upper bound value. Therefore, the upper bound value of the variable x_1 is considered here

$$\begin{aligned} A_{2,1,1} &= A_{1,1,1} + r(1,\,1,\,1)\left(A_{1,1,b} - \left|A_{1,1,1}\right|\right) - r(1,\,1,\,2)\left(A_{1,1,w} - \left|A_{1,1,1}\right|\right) \\ &= -4.570261872 + 0.38(-1.062187325 - |-4.570261872|) \\ &\quad - 0.81(-2.304524513 - |-4.570261872|) \\ &= -1.142020267 \end{aligned}$$

$$\begin{aligned} A_{2,2,1} &= A_{1,2,1} + r(1,\,2,\,1)\left(A_{1,2,b} - \left|A_{1,2,1}\right|\right) - r(1,\,2,\,2)\left(A_{1,2,w} - \left|A_{1,2,1}\right|\right) \\ &= 0.045197073 + 0.92(-0.767182961 - |0.045197073|) \\ &\quad -0.49(-4.442417134 - |0.045197073|) \\ &= -2.856829068 \end{aligned}$$

Similarly, the new values of x_1 and x_2 for the other candidates are calculated. Table 2.7 shows the new values of x_1 and x_2 and the corresponding values of the objective function.

Now, the values of $f(x)$ of Tables 2.6 and 2.7 are compared and the best values of $f(x)$ are considered and placed in Table 2.8. This completes the first generation of the Jaya algorithm.

From Table 2.8 it can be seen that the best solution is corresponding to the 5th candidate and the worst solution is corresponding to the 2nd candidate. Now, during the second generation, assuming random numbers $r_1 = 0.65$ and $r_2 = 0.23$ for x_1 and $r_1 = 0.38$ and $r_2 = 0.51$ for x_2, the new values of the variables for x_1 and x_2 are calculated using Eq. (2.9). Table 2.9 shows the new values of x_1 and x_2 and the corresponding values of the objective function during the second generation.

Table 2.8 Updated values of the variables and the objective function based on fitness comparison at the end of first generation (Rastrigin function)

Candidate	x_1	x_2	$f(x)$	Status
1	−1.1420155950	−2.8568303882	16.96999238	
2	5.12	−1.84339288493	36.785779377	*Worst*
3	0.1494454699	−0.350414953	20.138516097	
4	−1.062187325	−0.767182961	11.39297485	
5	0.982105144	−0.974135756	2.108371360	*Best*

Table 2.9 New values of the variables and the objective function during second generation (Rastrigin function)

Candidate	x_1	x_2	$f(x)$
1	−2.160893801	−1.9154836537	14.4049230292
2	2.4303683436	−1.0337930259	26.25808867
3	−0.452553284	0.265097775	30.7811944122
4	−2.0475376579	−0.0974903420	6.4628364079
5	0.030389327	−0.2775393237	11.98141715291

Now, the values of $f(x)$ of Tables 2.8 and 2.9 are compared and the best values of $f(x)$ are considered and placed in Table 2.10. This completes the second generation of the Jaya algorithm.

From Table 2.10 it can be seen that the best solution is corresponding to the 5th candidate and the worst solution is corresponding to the 2nd candidate. Now, during the third generation, assuming random numbers $r_1 = 0.01$ and $r_2 = 0.7$ for x_1 and $r_1 = 0.02$ and $r_2 = 0.5$ for x_2, the new values of the variables for x_1 and x_2 are calculated using Eq. (2.9). Table 2.11 shows the new values of x_1 and x_2 and the corresponding values of the objective function during the third generation.

Now, the values of $f(x)$ of Tables 2.10 and 2.11 are compared and the best values of $f(x)$ are considered and placed in Table 2.12. This completes the third generation of the Jaya algorithm.

From Table 2.12 it can be seen that the best solution is corresponding the 5th candidate and the worst solution is corresponding to the 2nd candidate. It can also

Table 2.10 Updated values of the variables and the objective function based on fitness comparison at the end of second generation (Rastrigin function)

Candidate	x_1	x_2	$f(x)$	Status
1	−2.160893801	−1.9154836537	14.4049230292	
2	2.4303683436	−1.0337930259	26.25808867	*Worst*
3	0.1494454699	−0.350414953	20.138516097	
4	−2.0475376579	−0.0974903420	6.4628364079	
5	0.982105144	−0.974135756	2.108371360	*Best*

Table 2.11 New values of the variables and the objective function during third generation (Rastrigin function)

Candidate	x_1	x_2	$f(x)$
1	−2.361313867	−0.498637656	42.261754812
2	2.415885712	−0.040158576	24.790382977
3	−1.438873945	0.315198022	35.424089418
4	−2.326173463	0.446718820	39.660957216
5	−0.0316790957	−0.0091367953	0.2150036577

Table 2.12 Updated values of the variables and the objective function based on fitness comparison at the end of third generation (Rastrigin function)

Candidate	x_1	x_2	f(x)	Status
1	−2.160893801	−1.9154836537	14.4049230292	
2	2.415885712	−0.040158576	24.790382977	*Worst*
3	0.1494454699	−0.350414953	20.138516097	
4	−2.0475376579	−0.0974903420	6.4628364079	
5	−0.0316790957	−0.0091367953	0.2150036577	*Best*

be observed that the value of the objective function is reduced from 11.39297485 to 0.2150036577 in just three generations. If we increase the number of generations then the known value of the objective function (i.e. 0) can be obtained within next few generations. In the case of maximization problems, the best value means the maximum value of the objective function and the calculations are to be proceeded accordingly. Thus, the proposed method can deal with both minimization and maximization problems. Figure 2.4 shows the convergence graph of Jaya algorithm for Rastrigin function for three generations.

2.3.2 Demonstration of Jaya Algorithm on Constrained Optimization Problems

To demonstrate the working of Jaya algorithm, a constrained benchmark function of Himmelblau is considered. The objective function is to find out the values of x_1 and x_2 that minimize the Himmelblau function.

Benchmark function: Himmelblau

$$\text{Minimize}, f(x_i) = \left(x_1^2 + x_2 - 11\right)^2 + \left(x_1 + x_2^2 - 7\right)^2 \tag{2.12}$$

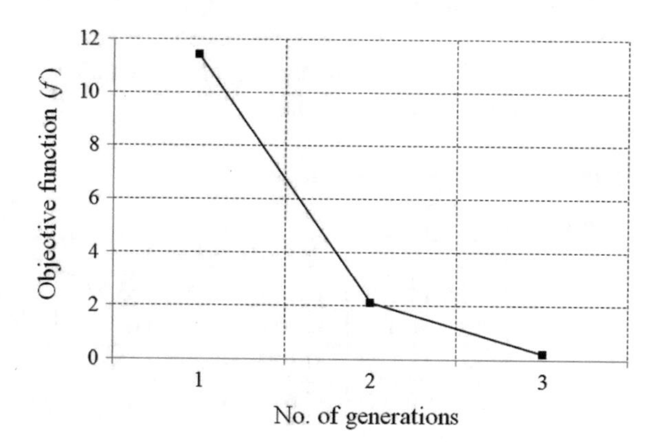

Fig. 2.4 Convergence graph of Jaya algorithm for Rastrigin function for three generations (Rao et al. 2017; Reprinted with permission from Elsevier)

Constraints:

$$g_1(x) = 26 - (x_1 - 5)^2 - x_2^2 \geq 0 \quad (2.13)$$

$$g_2(x) = 20 - 4x_1 - x_2 \geq 0 \quad (2.14)$$

Ranges of variables: $-5 \leq x_1, x_2 \leq 5$.

The known solution to this benchmark function is 0 for $x_1 = 3$ and $x_2 = 2$ and $g_1(x) = 18$ and $g_2(x) = 6$. Now to demonstrate the Jaya algorithm, let us assume a population size of 5, two design variables x_1 and x_2 (i.e. number of subjects) and two iterations as the termination criterion. The initial population is randomly generated within the ranges of the variables and the corresponding values of the objective function are shown in Table 2.13. As it is a minimization function, the lowest value of $f(x)$ is considered as the best value. If the constraints are violated then penalties are assigned to the objective function. There are many ways to assign the penalties and in this example a static penalty approach is used. A penalty p_1 for violation of $g_1(x)$ is considered as $10*(g_1(x))^2$ and the penalty p_2 for violation of $g_2(x)$ is considered as $10*(g_2(x))^2$. As it is a minimization problem, the values of penalties are added to the value of the objective function $f(x)$ and the fitness function is $f'(x) = f(x) + 10*(g_1(x))^2 + 10*(g_2(x))^2$. The function $f'(x)$ is called the pseudo-objective function.

It may be noted that $10*(g_1(x))^2$ is used as the penalty function in this example for the violation in constraint $g_1(x)$ and $10*(g_2(x))^2$ is used as the penalty function for the violation in constraint $g_2(x)$. Higher penalties are desirable and one may use $50*(g_1(x))^2$ or $100*(g_1(x))^2$ or $500*(g_1(x))^2$ or any such penalty for violation of $g_1(x)$ and $50*(g_2(x))^2$ or $100*(g_2(x))^2$ or $500*(g_2(x))^2$ or any such penalty for violation of $g_2(x)$. The assignment of penalties for violations depends upon the designer/decision maker/user based on the optimization problem. Sometimes, the penalty functions assigned may be different for different constraints depending upon the application as decided by the designer/decision maker/user.

Table 2.14 is prepared assuming random numbers $r_1 = 0.25$ and $r_2 = 0.43$ for x_1 and $r_1 = 0.47$ and $r_2 = 0.33$ for x_2.

Now, the values of $f'(x)$ of Tables 2.13 and 2.14 are compared and the best values of $f'(x)$ are considered and placed in Table 2.15.

Table 2.16 is prepared assuming random numbers $r_1 = 0.15$ and $r_2 = 0.50$ for x_1 and $r_1 = 0.32$ and $r_2 = 0.09$ for x_2.

Table 2.13 Initial population (Himmelblau function)

Candidate	x_1	x_2	$f(x)$	$g_1(x)$	p_1	$g_2(x)$	p_2	$f'(x)$	Status
1	3.22	0.403	13.139	22.669	0	6.717	0	13.13922	*Best*
2	0.191	2.289	77.710	−2.366	55.97	16.947	0	133.6902	
3	3.182	0.335	14.024	22.582	0	6.937	0	14.02423	
4	1.66	4.593	261.573	−6.251	390.78	8.767	0	652.3543	*Worst*
5	2.214	0.867	43.641	17.486	0	10.277	0	43.64116	

Table 2.14 New values of the variables, objective function and the penalties (Himmelblau function)

Candidate	x_1	x_2	$f(x)$	$g_1(x)$	p_1	$g_2(x)$	p_2	$f'(x)$
1	3.890	−0.979	14.596	23.809	0	5.4165	0	14.596
2	0.316	0.642	144.541	3.6530	0	18.091	0	144.541
3	3.845	−1.038	11.890	23.590	0	5.654	0	11.890
4	2.05	2.623	21.160	10.413	0	9.1763	0	21.160
5	2.703	−0.580	33.912	20.389	0	9.765	0	33.912

Table 2.15 Updated values of the variables, objective function and the penalties based on fitness comparison (Himmelblau function)

Candidate	x_1	x_2	$f(x)$	$g_1(x)$	p_1	$g_2(x)$	p_2	$f'(x)$	Status
1	3.22	0.403	13.139	22.669	0	6.717	0	13.139	
2	0.191	2.289	77.710	−2.366	55.979	16.947	0	133.690	Worst
3	3.845	−1.038	11.890	23.590	0	5.6543	0	11.890	Best
4	2.05	2.623	21.160	10.413	0	9.176	0	21.160	
5	2.703	−0.580	33.912	20.389	0	9.765	0	33.912	

Table 2.16 New values of the variables, objective function and the penalties based on fitness comparison during second iteration (Himmelblau function)

Candidate	x_1	x_2	$f(x)$	$g_1(x)$	p_1	$g_2(x)$	p_2	$f'(x)$
1	4.828	−0.227	150.551	25.918	0	0.914	0	150.551
2	0.739	1.224	107.853	6.347	0	15.818	0	107.853
3	5.673	−1.815	379.177	22.251	0	−0.878	7.718	386.896
4	3.248	1.482	3.493	20.737	0	5.522	0	3.493
5	4.131	−1.252	24.885	23.676	0	4.726	0	24.885

Now, the values of $f'(x)$ of Tables 2.15 and 2.16 are compared and the best values of $f'(x)$ are considered and are placed in Table 2.17. This completes the second iteration of the Jaya algorithm.

Table 2.17 Updated values of the variables, objective function and the penalties based on fitness comparison at the end of second iteration (Himmelblau function)

Candidate	x_1	x_2	$f(x)$	$g_1(x)$	p_1	$g_2(x)$	p_2	$f'(x)$	Status
1	3.22	0.403	13.139	22.6691	0	6.717	0	13.1392	
2	0.739	1.224	107.853	6.34704	0	15.818	0	107.8535	Worst
3	3.845	−1.038	11.8909	23.590	0	5.654	0	11.8909	
4	3.248	1.482	3.493	20.737	0	5.5224	0	3.4932	Best
5	4.131	−1.252	24.885	23.6767	0	4.726	0	24.885	

It can be noted that the minimum value of the objective function in the randomly generated initial population is −13.13922 and it has been reduced to 3.4932 at the end of the second iteration. If we increase the number of iterations then the known value of the objective function (i.e. 0) can be obtained within next few iterations.

It is to be noted that in the case of maximization problems, the best value means the maximum value of the objective function and the calculations are to be proceeded accordingly. The values of penalties are to be subtracted from the objective function in the case of maximization problems (i.e. $f'(x) = f(x) - 10*(g_1(x))^2 - 10*(g_2(x))^2$).

It may be mentioned here that the absolute value $|A(i,j,k)|$ is used in Eq. (2.9), instead of the actual value, for improving the exploration capability of the algorithm. One may try without using the absolute value in Eq. (2.9) but that may not give better results while solving different types of problems.

2.4 Self-adaptive Jaya Algorithm

Just like all population-based algorithms, the basic Jaya algorithm requires the common control parameters of population size and no. of generations (but it does not require the algorithm-specific parameters) and thus no tuning of such parameters is required. The choice of a particular population size for appropriate case studies is a difficult task (Teo 2006). However, not much work has been conducted yet on self-adaptive population sizes; hence this aspect is taken up and the proposed algorithm is named as self-adaptive Jaya algorithm. The key feature is that the self-adaptive Jaya algorithm determines the population size automatically (Rao and More 2017a, b). Hence, the user need not concentrate on choosing the population size. Let the random initial population is (10*d), where d is the number of design variables then the new population is formulated as,

$$n_{new} = round(n_{old} + r^* n_{old}). \tag{2.15}$$

Where, r is a random value between [− 0.5, 0.5]; it acts as a comparative population development rate. The population size may decrease or increase due to negative or positive random value of r. For example, if the population size n_{old} is 20 and random number generated r is 0.23 then the new population size for the next iteration n_{new} will be 25 (as per Eq. 2.15). If the random number generated r is −0.16 then the new population size for the next iteration n_{new} will be 17.

The flowchart of the proposed self-adaptive Jaya algorithm is presented in Fig. 2.5. Elitism is implemented when the population size of the next generations is larger than the population size of the present generation ($n_{new} > n_{old}$). Then all of the existing population will go into the next generation and the best optimal solutions in the current population are assigned to the remaining $n_{new} - n_{old}$ solutions. When the population size of the next generation is less than the population size of the current generation ($n_{new} < n_{old}$), then, only the best population of the current generation will go to the next generation. No change takes place if the

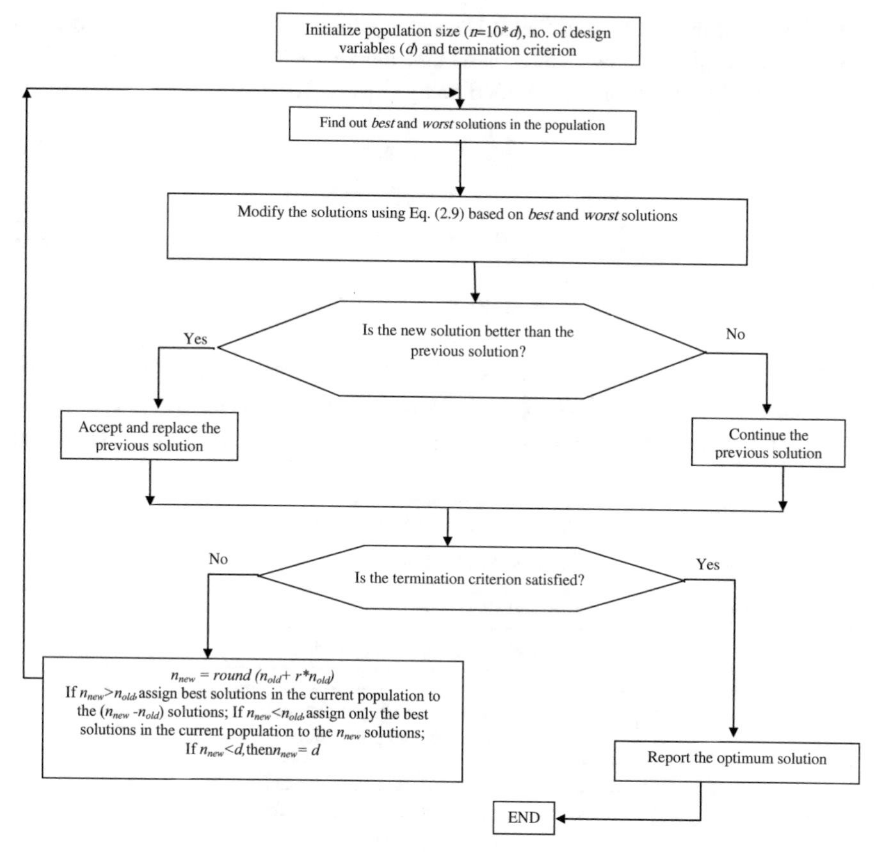

Fig. 2.5 Flowchart for self-adaptive Jaya algorithm (Rao and More 2017a; Reprinted with permission from Elsevier)

population sizes of the next and current generations are equal ($n_{new} = n_{old}$). If the population size decreases and becomes less than the number of design variables (d), then the population size is considered equal to the number of design variables ($if\ n_{new} < d,\ then\ n_{new} = d$). Thus, the solutions will not be stuck-up in local optima.

2.5 Quasi-oppositional Based Jaya (QO-Jaya) Algorithm

In order to further diversify the population and improve the convergence rate of Jaya algorithm, the concept of opposition based learning is introduced in the Jaya algorithm. To achieve better approximation, a population opposite to the current population is generated and both are considered at the same time. However, in order to maintain the stochastic nature of Jaya algorithm quasi-opposite population is generated. A quasi-opposite value of a variable of a candidate solution is not a mirror point of the variable; rather it is a value which is randomly chosen between the center of the search space and the mirror point of the variable. The quasi-opposite population is generated using Eqs. (2.16)–(2.18).

$$A^{q}_{i,j,k} = \mathrm{rand}(a, b) \tag{2.16}$$

$$a = \frac{A_k^L + A_k^U}{2} \tag{2.17}$$

$$b = A_k^L + A_k^U - A_{i,j,k} \tag{2.18}$$

$i = 1, 2, 3, \ldots, g$; $j = 1, 2, 3, \ldots, P$; $k = 1, 2, 3, \ldots, d$ where, A_k^L and A_k^U are the lower and upper bound values of the *k*th variable; $A^q_{i,j,k}$ is the quasi-opposite value of $A_{i,j,k}$. For example, if the lower and upper bounds of a variable are 10 and 60

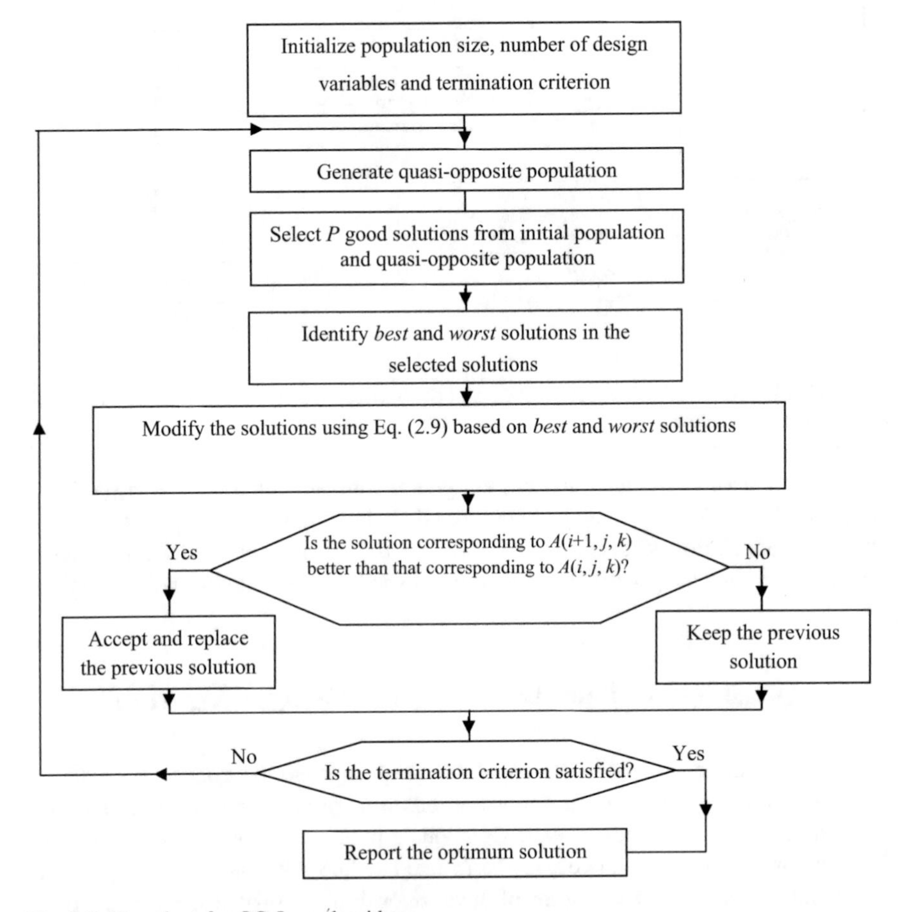

Fig. 2.6 Flowchart for QO-Jaya algorithm

respectively and the value of a variable during the current iteration, $A_{i,j,k}$, is 23 then a = 35 (the middle value of the given bounds), b = 10 + 60 − 23 = 47. In fact, the exact opposite value (mirror point) will be 47. However, the quasi-opposite value of the variable, $A^{q}_{i,j,k}$ = *rand* (35, 47). That means, any value between 35 and 47 will be generated. Figure 2.6 gives the flowchart for QO-Jaya algorithm (Rao and Rai 2017a, b). The total number of function evaluations required by QO-Jaya algorithm = population size × no. of generations for a single simulation run.

2.6 Self-adaptive Multi-population (SAMP) Jaya Algorithm

Multi-population based advanced optimization methods are used for improving the diversity of search by splitting the entire population into groups (sub-populations) and allocating these throughout the search space so that the problem changes can be detected effectively. This basic idea is used for keeping the diversity of the search process by allocating different sub-populations to different areas. Each population is subjected to either diversifying or intensifying the search processes of the algorithm. The interaction between the sub-populations takes place by means of a merge and divide process whenever there is a change in the solution is observed. The multi-population approaches are found effective while dealing with various problems and these have outperformed the existing fixed population size methods for different problems. The multi-population approaches are useful for maintaining the population diversity. Overall diversity of the search process can be maintained by allocating the entire population into groups, because various sub-populations can be situated in different regions of the problem search space and these are having the ability of search in various regions simultaneously.

The selection of number of sub-populations is a critical issue in algorithm's performance. It is related with the complexity of the problem. The size of sub-populations continuously changes during the search process. The solutions in the sub-populations may also not be enough for enough diversity. In order to address these issues, the present work proposes a self-adaptive multi-population (SAMP) Jaya algorithm for the engineering optimization problems. In order to effectively monitor the problem landscape changes, the SAMP-Jaya algorithm adaptively changes the number of sub-populations based on the change strength of the problem landscape. The following modifications are made to the basic Jaya algorithm:

- The proposed SAMP-Jaya algorithm uses a number of sub-populations by dividing it into number of groups based on the quality of the solutions. Use of the number of sub-populations distributes the solution over the search space

rather than concentrating on a particular region. Therefore, the proposed algorithm is expected to produce optimum solution.
- During the search process, the SAMP-Jaya algorithm modifies the number of sub-populations based on change strength of the problem for monitoring the landscape changes. It means that the number of sub-populations will be increased or decreased. In the SAMP-Jaya algorithm, the number of sub-populations is modified adaptively based on the strength of the solution change. This feature supports the search process for tracing the optimum solution and for improving the diversification of the search process. Furthermore, the duplicate solutions are replaced by newly generated solutions for maintaining the diversity and for enhancing the exploration procedure.

Figure 2.7 presents the flowchart of the proposed SAMP-Jaya algorithm (Rao and Saroj 2017). The basic steps of the SAMP-Jaya algorithm are as follows:

Step 1: It starts with the setting of the number of design variables (*d*), number of populations (*P*) and termination criterion (termination criterion may be the no. of iterations or maximum number of function evaluations or prescribed time or accuracy required, etc.).

Step 2: The next step is to calculate the initial solutions based on the defined fitness function for the problem.

Step 3: The entire population is grouped into *m* number of groups based on the quality of the solutions (Initially m = 2 is considered).

Step 4: Each sub-population uses Jaya algorithm for modifying the solutions in each group independently. Each modified solution is accepted *if and only if* it is better than the old solution.

Step 5: Merge the entire sub-population together. Check whether *f*(*best_before*) is better than *f*(*best_after*). Here, *f*(*best_before*) is the previous best solution of the entire population and *f*(*best_after*) is the current best solution in the entire population. If the value of *f*(*best_after*) is better than the value of *f* (*best_before*), *m* is increased by 1 (i.e. $m = m + 1$) with the aim of the increasing the exploration feature of the search process. Otherwise, *m* is decreased by 1 (i.e. $m = m - 1$) as the algorithm needs to be more exploitive than explorative.

Step 6: Check the stopping condition(s). If the search process has reached the termination criterion, then terminate the loop and report the best optimum solution. Otherwise, (a) replace the duplicate solutions with randomly generated solutions and (b) re-divide the population as per step 3 and follow the subsequent steps.

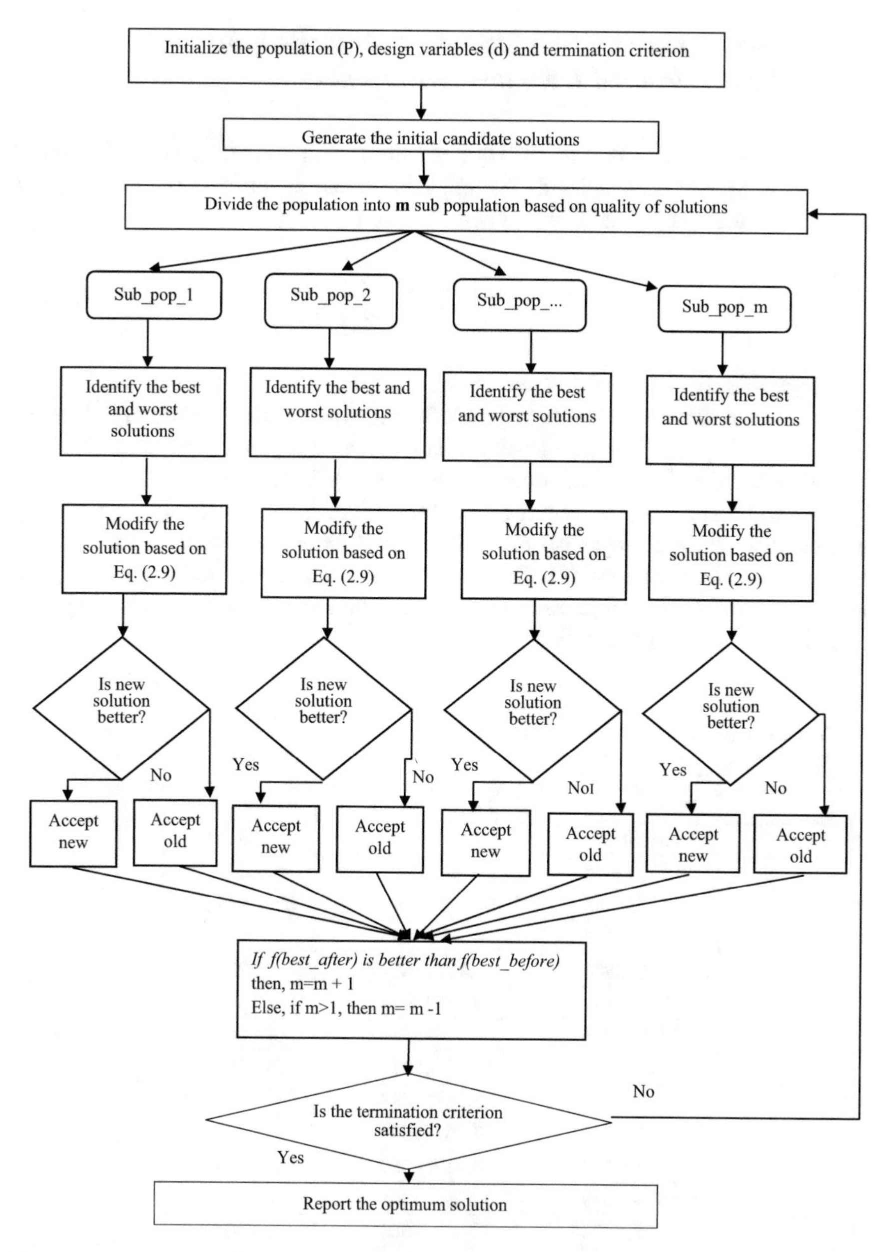

Fig. 2.7 Flowchart of the SAMP-Jaya algorithm (Rao and Saroj 2017; Reprinted with permission from Elsevier)

2.6.1 Demonstration of SAMP-Jaya Algorithm on Unconstrained Optimization Problem

To demonstrate the working of Jaya algorithm, an unconstrained benchmark function of Sphere is considered. The objective function is to find out thevalues of x_i that minimize the value of the Sphere function.

$$\text{Minimize, } f(x_i) = \sum_{i=1}^{m} x_1^2 \tag{2.19}$$

Range of variables: $-100 \le x_i \le 100$.

The known solution to this benchmark function is 0 for all x_i values of 0. Now to demonstrate the SAMP-Jaya algorithm, let us assume a population size of 20 (i.e. candidate solutions), initial no. of sub-populations as 2, two design variables x_1 and x_2 and two iterations as the termination criterion. The initial population is randomly generated within the ranges of the variables and the corresponding values of the objective function are shown in Table 2.18. As it is a minimization function, the lowest value of $f(x)$ is considered as the best solution and the highest value of $f(x)$ is considered as the worst solution.

Table 2.18 Initial solutions for the first iteration

Candidate	x_1	x_2	$f(x)$	Status
1	−73.27921791	27.70914736	6137.640625	
2	−23.01138101	53.13962513	3353.343415	
3	30.58313643	−23.70218694	1497.1219	
4	−39.99628345	−31.9720889	2621.917159	
5	83.78530734	−8.746637339	7096.481391	
6	**−11.50060948**	**−9.162836077**	**216.2215834**	**Best**
7	89.05640184	−56.17620949	11086.80922	
8	**76.48055919**	**−96.02491544**	**15070.06032**	**Worst**
9	−31.64702335	53.20548914	3832.358162	
10	−31.4392597	23.76128083	1553.025517	
11	−9.395767582	−97.96747947	9685.907483	
12	19.81623349	20.31366967	805.3282854	
13	29.88349019	−31.45587348	1882.494963	
14	−1.340148885	40.35478905	1630.304998	
15	77.56052029	−88.98843077	13934.57512	
16	−80.32760989	29.95655709	7349.920224	
17	52.81417825	97.59178832	12313.49457	
18	−74.93504172	−27.10456496	6349.917919	
19	35.24601806	−24.84836576	1859.72307	
20	72.69169829	−41.6046498	7015.029886	

Now, various steps of SAMP-Jaya algorithm are given below:

Step 1: No. of design variables (d) = 2; Populations size (P) = 20; Sub-population (initially) = 2 and termination criteria = 2 iterations.

Step 2: The next step is to calculate the initial solutions for the problem.

Step 3: The entire population is grouped into m number of groups based on the quality of the solutions (Initially m = 2 is considered).

Step 4: Each sub-population uses Jaya algorithm for modifying the solutions in each group independently. Each modified solution is accepted *if and only if* it is better than the old solution. Table 2.20a, b show the random numbers used for sub-populations 1 and 2 respectively.

Table 2.19 **a** Sub-population 1 for the first iteration. **b** Sub-population 2 for the first iteration

Candidate	x_1	x_2	$f(x)$	Status
(a)				
1	−11.50060948	−9.162836077	216.2215834	**Best**
2	19.81623349	20.31366967	805.3282854	
3	30.58313643	−23.70218694	1497.1219	
4	−31.4392597	23.76128083	1553.025517	
5	−1.340148885	40.35478905	1630.304998	
6	35.24601806	−24.84836576	1859.72307	
7	29.88349019	−31.45587348	1882.494963	
8	−39.99628345	−31.9720889	2621.917159	
9	−23.01138101	53.13962513	3353.343415	
10	−31.64702335	53.20548914	3832.358162	**Worst**
(b)				
11	−73.2792	27.70915	6137.641	**Best**
12	−74.935	−27.1046	6349.918	
13	72.6917	−41.6046	7015.03	
14	83.78531	−8.74664	7096.481	
15	−80.3276	29.95656	7349.92	
16	−9.39577	−97.9675	9685.907	
17	89.0564	−56.1762	11086.81	
18	52.81418	97.59179	12313.49	
19	77.56052	−88.9884	13934.58	
20	76.48056	−96.0249	15070.06	**Worst**

Table 2.20 **a** Random numbers for sub-population 1 during first iteration. **b** Random numbers for sub-population 2 during second iteration

	x_1	x_2
(a)		
r_1	0.133474990584297	0.672650514427429
r_2	0.202584980871927	0.868515198201092
(b)		
r_1	0.751157321899609	0.419379649992969
r_2	0.000231065199416736	0.149463732367703

Now, Tables 2.19a and 2.21a are compared and the updated values (i.e. better values) are kept in Table 2.22a. Similarly, Tables 2.19b and 2.21b are compared and the updated values (i.e. better values) are kept in Table 2.22b.

Step 5: Merge the entire sub-population together. Table 2.23 shows the merged population at the end of first iteration.
Check whether *f*(*best_before*) is better than *f*(*best_after*). Here, *f* (*best_before*) is the previous best solution of the entire population and *f* (*best_after*) is the current best solution in the entire population. At the end of the first iteration it is found that the value of *f*(*best_after*) is better than the value of *f*(*best_before*) and hence m is increased by 1 (i.e. $m = 2 + 1 = 3$) with the aim of the increasing the exploration feature of the search process.

Step 6: Check the stopping condition(s). If the search process has reached the termination criterion, then terminate the loop and report the best optimum solution. However, as the termination criterion is decided as 2 iterations in this demonstrative example, the process is repeated by going to step 3 again (Table 2.24).

Table 2.21 **a** Modified solutions of Table 2.19a. **b** Modified solutions of Table 2.19b

Candidate	x_1	x_2	$f(x)$
1	−5.829634592	−59.74132243	3603.010245
2	26.06190108	−28.08076219	1467.751893
3	37.57290457	−71.43292794	6514.386352
4	−24.39032489	−23.95788576	1168.868238
5	3.62863667	−4.114295299	30.09442989
6	42.5580379	−72.3546108	7046.376294
7	36.82490579	−77.66794112	7388.382764
8	−32.35597281	−78.08304816	7143.871387
9	−16.54489681	11.17463866	398.6061597
10	−24.58373	11.2534031	730.9988619
(b)			
11	−100* (−183.36839)	46.2029	12134.71
12	−100* (−186.26762)	−8.44762	10071.36
13	−36.9563	−26.8615	2087.309
14	−34.1932	14.8654	1390.153
15	−100* (−195.70961)	47.8437	12289.02
16	−71.5132	−98.4376	14804.09
17	−32.8803	−45.3662	3139.202
18	−41.9073	97.22311	11208.55
19	−35.7436	−87.0349	8852.684
20	−36.0126	−95.9707	10507.28

*The value of the variable x_1 exceeds the lower bound value. Therefore, the lower bound value of the variable x_1 is considered here

Table 2.22 **a** Updated values of the variables and the objective function corresponding to sub-population 1. **b** Updated values of the variables and the objective function corresponding to sub-population 2

Candidate	x_1	x_2	$f(x)$
(a)			
1	−11.50060948	−9.162836077	216.2215834
2	19.81623349	20.31366967	805.3282854
3	30.58313643	−23.70218694	1497.1219
4	−24.39032489	−23.95788576	1168.868238
5	3.62863667	−4.114295299	30.09442989
6	35.24601806	−24.84836576	1859.72307
7	29.88349019	−31.45587348	1882.494963
8	−39.99628345	−31.9720889	2621.917159
9	−16.54489681	11.17463866	398.6061597
10	−24.58373	11.2534031	730.9988619
(b)			
11	−73.27921791	27.70914736	6137.640625
12	−74.93504172	−27.10456496	6349.917919
13	−36.95629967	−26.86151225	2087.308926
14	−34.19317294	14.8654008	1390.153217
15	−80.32760989	29.95655709	7349.920224
16	−9.395767582	−97.96747947	9685.907483
17	−32.8802817	−45.36616784	3139.202109
18	−41.907268	97.22310601	11208.55145
19	−35.74360395	−87.03492993	8852.684252
20	−36.01259391	−95.97067381	10507.27715

Table 2.23 Merged populations at the end of first iteration

Candidate	x_1	x_2	$f(x)$	Status
1	−11.50060948	−9.162836077	216.2215834	
2	19.81623349	20.31366967	805.3282854	
3	30.58313643	−23.70218694	1497.1219	
4	−24.39032489	−23.95788576	1168.868238	
5	**3.62863667**	**−4.114295299**	**30.09442989**	**Best**
6	35.24601806	−24.84836576	1859.72307	
7	29.88349019	−31.45587348	1882.494963	
8	−39.99628345	−31.9720889	2621.917159	
9	−16.54489681	11.17463866	398.6061597	
10	−24.58373	11.2534031	730.9988619	
11	−73.27921791	27.70914736	6137.640625	

(continued)

Table 2.23 (continued)

Candidate	x_1	x_2	$f(x)$	Status
12	−74.93504172	−27.10456496	6349.917919	
13	−36.95629967	−26.86151225	2087.308926	
14	−34.19317294	14.8654008	1390.153217	
15	−80.32760989	29.95655709	7349.920224	
16	−9.395767582	−97.96747947	9685.907483	
17	−32.8802817	−45.36616784	3139.202109	
18	**−41.907268**	**97.22310601**	**11208.55145**	**Worst**
19	−35.74360395	−87.03492993	8852.684252	
20	−36.01259391	−95.97067381	10507.27715	

Table 2.24 **a** Sub-population 1 for the second iteration. **b** Sub-population 2 for the second iteration. **c** Sub-population 3 for the second iteration

Candidate	x_1	x_2	$f(x)$	Status
(a)				
1	3.62863667	−4.114295299	30.09442989	**Best**
2	−11.50060948	−9.162836077	216.2215834	
3	−16.54489681	11.17463866	398.6061597	
4	−24.58373	11.2534031	730.9988619	
5	19.81623349	20.31366967	805.3282854	
6	−24.39032489	−23.95788576	1168.868238	
7	−34.19317294	14.8654008	1390.153217	**Worst**
(b)				
8	30.58313643	−23.7021869	1497.1219	**Best**
9	35.24601806	−24.8483658	1859.72307	
10	29.88349019	−31.4558735	1882.494963	
11	−36.95629967	−26.8615122	2087.308926	
12	−39.99628345	−31.9720889	2621.917159	
13	−32.8802817	−45.3661678	3139.202109	**Worst**
(c)				
14	−73.27921791	27.70914736	6137.640625	**Best**
15	−74.93504172	−27.10456496	6349.917919	
16	−80.32760989	29.95655709	7349.920224	
17	−35.74360395	−87.03492993	8852.684252	
18	−9.395767582	−97.96747947	9685.907483	
19	−36.01259391	−95.97067381	10507.27715	
20	−41.907268	97.22310601	11208.55145	**Worst**

Iteration2:

Step 3: The entire population is grouped into 3 sub-groups based on the quality of the solutions.

Step 4: Each sub-population uses Jaya algorithm for modifying the solutions in each group independently. Each modified solution is accepted *if and only if* it is better than the old solution. Table 2.25a–c show the random numbers used for sub-populations 1, 2 and 3 respectively.
Now, Tables 2.24a and 2.26a are compared and the updated values (i.e. better values) are kept in Table 2.27a. Similarly, Tables 2.24b and 2.26b are compared and the updated values are kept in Table 2.27b. Similarly, Tables 2.24c and 2.26c are compared and the updated values are kept in Table 2.27c.

Table 2.25 **a** Random numbers for sub-population 1 during second iteration. **b** Random numbers for subpopulation 2 during second iteration. **c** Random numbers for subpopulation 3 during second iteration

	x_1	x_2
(a)		
r_1	0.273834301544268	0.87242501896288
r_2	0.601251143911812	0.321188452831265
(b)		
r_1	0.284293074456312	0.435315805784207
r_2	0.903759051181981	0.925105644987009
(c)		
r_1	0.505292452205831	0.62758180927003
r_2	0.71926392682241	0.0239128813596552

Table 2.26 **a** Modified solutions of Table 2.24a. **b** Modified solutions of Table 2.24b. **c** Modified solutions of Table 2.24c

Candidate	x_1	x_2	$f(x)$
(a)			
1	26.36904296	−14.74625455	912.77845
2	13.81721329	−22.57773561	700.6695284
3	10.42451059	−3.34924002	119.8878298
4	5.017726783	−3.313893421	36.15947168
5	47.85673162	0.752022918	2290.8323
6	5.147807804	−45.52835767	2099.331278
7	−1.445422693	−1.692960929	4.955363468
(b)			
8	87.93877499	19.55746474	8115.722573
9	95.49015313	18.97267266	9478.331653
10	86.80572171	15.60145508	7778.638722

(continued)

Table 2.26 (continued)

Candidate	x_1	x_2	$f(x)$
11	24.34729668	17.94554487	914.8334361
12	23.19047942	15.33807673	773.0549335
13	25.89836019	8.504281563	743.0478656
(c)			
14	−64.48460517	26.04686832	4836.703653
15	−65.78612991	−28.40187639	5134.481471
16	−70.02484232	26.93758663	5629.112115
17	−34.98054188	−100* (−124.5103405)	11223.63831
18	−14.27039091	−100* (−142.0425305)	10203.64406
19	−35.19197565	−100* (−138.8403153)	11238.47515
20	−39.82535764	53.59741007	4458.741478

*The value of the variablen x_2 exceeds the lower bound value. Therefore, the lower bound value of the variable x_2 is considered here

Table 2.27 **a** Updated values of the variables and the objective function corresponding to sub-population 1 (second iteration). **b** Updated values of the variables and the objective function corresponding to sub-population 2 (second iteration). **c** Updated values of the variables and the objective function corresponding to sub-population 3 (second iteration)

Candidate	x_1	x_2	$f(x)$
(a)			
1	3.62863667	−4.114295299	30.09442989
2	−11.50060948	−9.162836077	216.2215834
3	10.42451059	−3.34924002	119.8878298
4	5.017726783	−3.313893421	36.15947168
5	19.81623349	20.31366967	805.3282854
6	−24.39032489	−23.95788576	1168.868238
7	−1.445422693	−1.692960929	4.955363468
(b)			
8	30.58313643	−23.70218694	1497.1219
9	35.24601806	−24.84836576	1859.72307
10	29.88349019	−31.45587348	1882.494963
11	24.34729668	17.94554487	914.8334361
12	23.19047942	15.33807673	773.0549335
13	25.89836019	8.504281563	743.0478656
(c)			
14	−64.48460517	26.04686832	4836.703653
15	−65.78612991	−28.40187639	5134.481471
16	−70.02484232	26.93758663	5629.112115
17	−35.74360395	−87.03492993	8852.684252
18	−9.395767582	−97.96747947	9685.907483
19	−36.01259391	−95.97067381	10507.27715
20	−39.82535764	53.59741007	4458.741478

Step 5: Merge the entire sub-population together. Table 2.28 shows the merged population at the end of second iteration.
Check whether *f*(*best_before*) is better than *f*(*best_after*). Here, *f* (*best_before*) is the previous best solution of the entire population and *f* (*best_after*) is the current best solution in the entire population. At the end of the second iteration it is found that the value of *f*(*best_after*) is better than the value of *f*(*best_before*) and hence m is increased by 1 (i.e. $m = 3 + 1 = 4$) with the aim of the increasing the exploration feature of the search process.

Step 6: Check the stopping condition(s). If the search process has reached the termination criterion, then terminate the loop and report the best optimum solution. The process is to be continued till the termination criterion is reached. However, the termination criterion is decided as 2 iterations in this demonstrative example and hence the process is stopped.

Table 2.28 Merged populations at the end of second iteration

Candidate	x_1	x_2	$f(x)$	Status
1	3.62863667	−4.114295299	30.09442989	
2	−11.50060948	−9.162836077	216.2215834	
3	10.42451059	−3.34924002	119.8878298	
4	5.017726783	−3.313893421	36.15947168	
5	19.81623349	20.31366967	805.3282854	
6	−24.39032489	−23.95788576	1168.868238	
7	**−1.445422693**	**−1.692960929**	**4.955363468**	**Best**
8	30.58313643	−23.70218694	1497.1219	
9	35.24601806	−24.84836576	1859.72307	
10	29.88349019	−31.45587348	1882.494963	
11	24.34729668	17.94554487	914.8334361	
12	23.19047942	15.33807673	773.0549335	
13	25.89836019	8.504281563	743.0478656	
14	−64.48460517	26.04686832	4836.703653	
15	−65.78612991	−28.40187639	5134.481471	
16	−70.02484232	26.93758663	5629.112115	
17	−35.74360395	−87.03492993	8852.684252	
18	−9.395767582	−97.96747947	9685.907483	
19	**−36.01259391**	**−95.97067381**	**10507.27715**	**Worst**
20	−39.82535764	53.59741007	4458.741478	

2.7 Self-adaptive Multi-population Elitist (SAMPE) Jaya Algorithm

In SAMPE-Jaya algorithm the following modifications are added to the basic Jaya algorithm:

- The proposed algorithm uses number of sub-populations by dividing it into number of groups based on the quality of the solutions (value of fitness function). Furthermore, the worst solutions of the inferior group (populations having poor fitness values) are replaced by the solutions of the superior group such as populations having higher fitness values (elite solutions). Use of the number of sub-populations distributes the solution over the search space rather than concentrating in a particular area. Therefore, the proposed algorithm is expected to produce optimum solution and to monitor the problem landscape changes.
- During the search process, SAMPE-Jaya algorithm modifies the number of sub-populations based on change strength of the problem for monitoring the landscape changes. It means that the number of sub-populations will be increased or decreased. In the SAMPE-Jaya algorithm, the number of sub-populations is modified adaptively based on the strength of the solution change (e.g. improvement in the fitness value). This feature supports the search process for tracing the optimum solution and improving the exploration and diversification of the search process. Furthermore, the duplicate solutions are replaced by newly generated solutions for maintaining the diversity and enhancing the exploration procedure.

The basic steps of the SAMPE-Jaya algorithm are as follows:

Step 1: It starts with the setting of the number of design variables (d), number of populations (P), elite size (ES) and termination criterion (termination criterion may be no. of iterations or maximum number of function evaluations, accuracy required, etc.).

Step 2: Next step is to calculate the initial solutions based on the defined fitness function for the defined problem.

Step 3: The entire population is grouped into m number of groups based on the quality of the solutions (Initially m = 2 is considered) and replace the worst solutions (equals to ES) of the inferior group with solutions of the superior group (elite solutions).

Step 4: Each sub-population uses Jaya algorithm for modifying the solutions in each group independently. Modified solutions are kept *if and only if* they are better than the old solutions.

Step 5: Combine the entire sub-population. Check whether *f*(*best_before*) is better than *f*(*best_after*). Here, *f*(*best_before*) is the previous best solution of the entire population and *f*(*best_after*) is the current best solution in the entire population. If the value of *f*(*best_after*) is better than the value of *f* (*best_before*), m is increased by 1 (i.e. $m = \mathrm{m} + 1$) with the aim of the increasing the exploration feature of the search process. Otherwise, m is decreased by 1 (i.e. $m = \mathrm{m} - 1$) as the algorithm needs to be more exploitive than explorative.

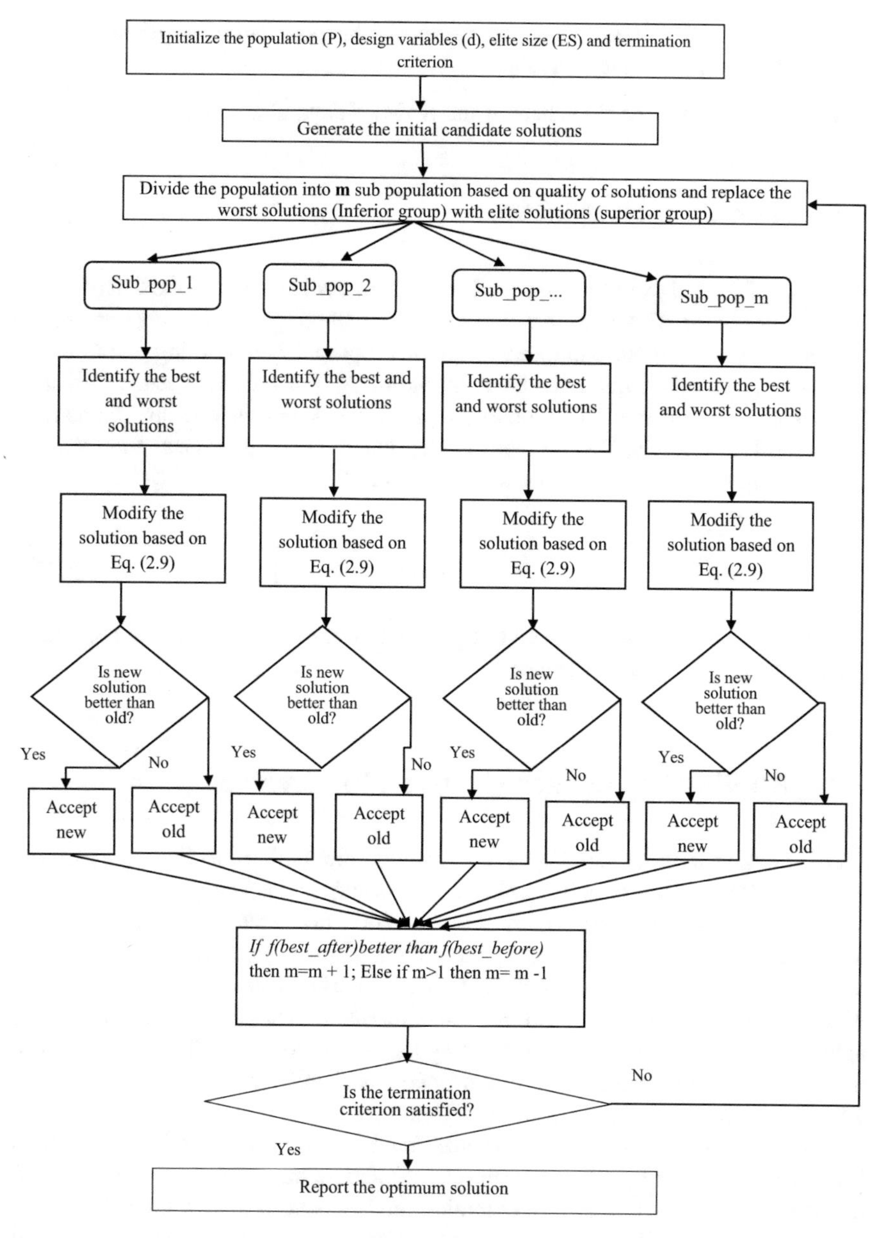

Fig. 2.8 Flowchart of the SAMPE-Jaya algorithm (Rao and Saroj 2018; Reprinted with permission from Springer Science + Business Media)

Step 6: Check the stopping condition(s). If the search process has reached the termination criterion, then terminate the loop and report the best optimum solution. Otherwise, (a). replace the duplicate solutions with randomly generated solutions and (b). re-divide the population as per step 3 and follow the subsequent steps.

Figure 2.8 shows the flowchart of the SAMPE-Jaya algorithm.

2.8 Chaotic Jaya Algorithm

A variant of Jaya algorithm based on the chaos theory is named as the Chaotic-Jaya (CJaya) algorithm. This is to improve the convergence speed and to better explore the search space without entrapment into local optima. The working of the CJaya algorithm is exactly similar to the Jaya algorithm, the only difference is that in CJaya algorithm, the random numbers are produced using a chaotic random number generator. For example, if a tent map chaotic function is used as the chaotic random number generator then it is expressed by Eq. (2.20).

$$x_{k+1} = \left\{ \begin{array}{ll} \frac{x_k}{0.7} & x_k < 0.7 \\ \frac{10}{3}(1 - x_k) & x_k \geq 0.7 \end{array} \right\} \tag{2.20}$$

where, x_{k+1} is the newly generated chaotic random number and x_k is the previously generated chaotic random number.

2.9 Multi-objective Jaya (MO-Jaya) Algorithm

MO-Jaya algorithm is developed for solving the multi-objective optimization problems (Rao et al. 2017). The MO-Jaya algorithm is a posteriori *version* of Jaya algorithm for solving multi-objective optimization problems. The solutions in the MO-Jaya algorithm are updated in the similar manner as in the Jaya algorithm based on Eq. (2.9). However, in order to handle multiple objectives effectively and efficiently the MO-Jaya algorithm is incorporated with non-dominated sorting approach and crowding distance computation mechanism. The concepts of non-dominated sorting and crowding distance mechanism are already explained in Sect. 2.2 in the case of NSTLBO algorithm.

In the case of single-objective optimization it is easy to decide which solution is better than the other based on the objective function value. But in the presence of multiple conflicting objectives determining the best and worst solutions from a set of solutions is difficult. In the MO-Jaya algorithm, the task of finding the best and worst solutions is accomplished by comparing the rank assigned to the solutions based on the constraints-dominance concept, non-dominance concept and the crowding distance value.

In the beginning, an initial population is randomly generated with *NP* number of solutions. This initial population is then sorted and ranked based on constraint-dominance and non-dominance concept. The superiority among the solutions is first determined based on the constraint-dominance concept and then on non-dominance concept and then on the crowding distance value of the solutions. A solution with a higher rank is regarded as superior to the other learner. If both the solutions hold the same rank, then the solution with higher crowding distance value is seen as superior to the other. This ensures that the solution is selected from the sparse region of the search space. The solution with the highest rank (rank = 1) is selected as the best solution. The solution with the lowest rank is selected as the worst solution. Once the best and worst solutions are selected, the solutions are updated based on the Eq. (2.9).

Once all the solutions are updated, the set of updated solutions (new solutions) is concatenated to the initial population to obtain a set of 2*NP* solutions. These solutions are again sorted and ranked based on the constraint-dominance concept, non-dominance concept and the crowding distance value for each solution is computed. Based on the new ranking and crowding distance value *NP* number of good solutions are selected. The superiority among the solutions is determined based on the non-dominance rank and the crowding distance value of the solutions. A solution with a higher rank is regarded as superior to the other solution. If both the solutions hold the same rank, then the solution with higher crowding distance value is seen as superior to the other. The flowchart of MO-Jaya algorithm is given in Fig. 2.9. For every candidate solution the MO-Jaya algorithm evaluates the objective function only once in each generation. Therefore, the total no. of function evaluations required by MO-Jaya algorithm = population size × no. of generations. However, when the algorithm is run more than once, then the number of function evaluations is to be calculated as: no. of function evaluations = no. of runs × population size × number of generations.

2.9.1 Demonstration of MO-Jaya Algorithm on a Constrained Bi-objective Optimization Problem

Let us consider the example of a bi-objective optimization problem of cutting parameters in turning process. Yang and Natarajan (2010) used differential evolution and non-dominated sorting genetic algorithm-II approaches for solving the problem. The same problem is considered here to demonstrate the working of the MO-Jaya algorithm. The problem has two objectives of minimizing the tool wear (T_w) and maximizing the metal removal rate (M_r). The objective functions, constraints and the ranges of the cutting parameters are given below.
Objective functions:

$$\text{Minimize}\ (T_w) = 0.33349\, v^{0.1480} f^{0.4912} d^{0.2898} \tag{2.21}$$

$$\text{Maximize}\ (M_r) = 1000\, v f d \tag{2.22}$$

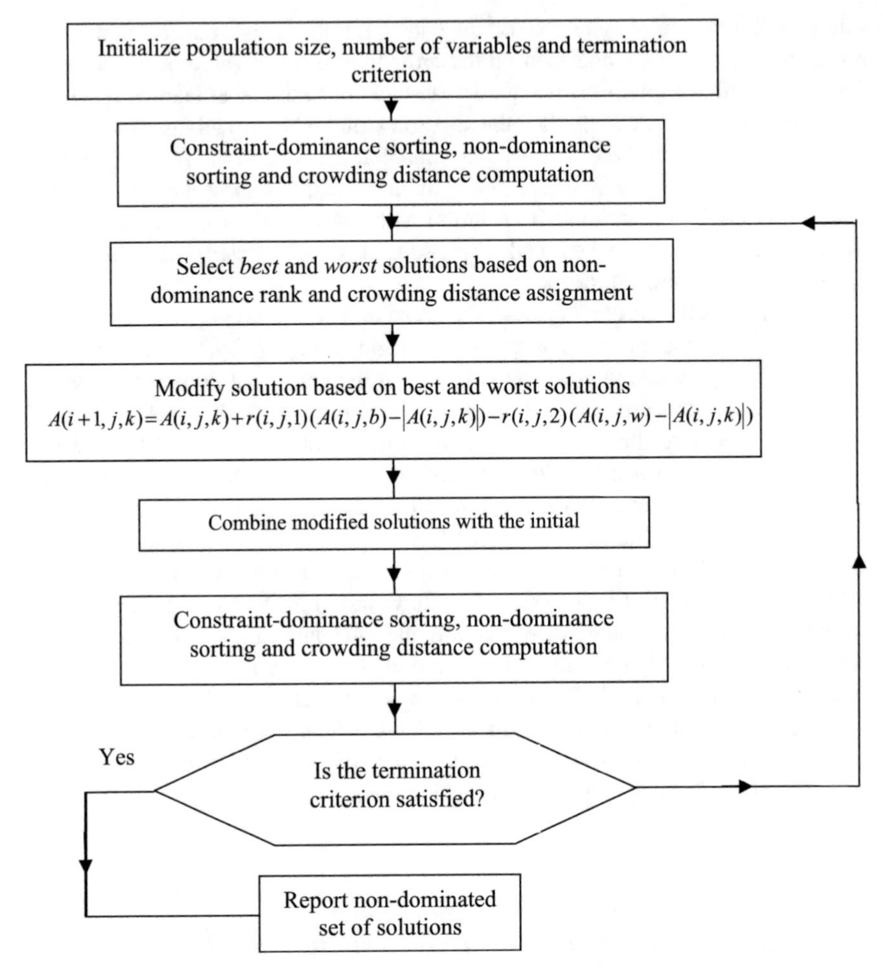

Fig. 2.9 Flowchart of MO-Jaya algorithm (Rao et al. 2017; Reprinted with permission from Elsevier)

Constraints:

Temperature constraint:

$$88.5168\, v^{0.3156} f^{0.2856} d^{0.2250} \leq 500 \tag{2.23}$$

Surface roughness constraint:

$$18.5167\, v^{-0.0757} f^{0.7593} d^{0.1912} \leq 2 \tag{2.24}$$

Parameter bounds:

$$\text{Cutting speed (m/min): } 42 \le v \le 201 \quad (2.25)$$

$$\text{Feed rate (mm/rev): } 0.05 \le f \le 0.33 \quad (2.26)$$

$$\text{Depth of cut (mm): } 0.5 \le d \le 2.5 \quad (2.27)$$

v = speed (m/min); f = feed (mm/rev); d = depth of cut (mm); M_r = metal removal rate (mm^3/min); T_w = tool wear (mm); T = tool-workpiece interface temperature (°C) and R_a = surface roughness (μm).

Now to demonstrate the MO-Jaya algorithm, let us assume a population size of 5, three design variables v, f and d and one iteration as the termination criterion. The initial population is randomly generated within the ranges of the variables and the corresponding values of the objective functions are shown in Table 2.29. The mean values of v, f and d are also shown.

$$(Z_T)_{max} = 0.000; \ (Z_{Ra})_{max} = 0.9371$$

Z' =overall constraint violation and it can be written as,

$$Z' = \frac{Z_T}{(Z_T)_{max}} + \frac{Z_{Ra}}{(Z_{Ra})_{max}} \quad (2.28)$$

In Table 2.29, the values under Z_T and Z_{Ra} represent the values by which these constraints are violated by the candidate solution and $(Z_T)_{max}$ and $(Z_{Ra})_{max}$ represent the maximum values of violations of the constraints of tool-workpiece interface temperature and surface roughness. For example, $(Z_{Ra})_{max}$ = 6.5973 − 2 = 4.5973. The crowding distance *CD* is 0. If there are more constraints (whether "$\geq$" or "$\leq$" or "=") then the (extent of constraint violation/$(Z_X)_{max}$) can be added to Z'. Here (Z_X) can be any constraint.

The random numbers r_1 and r_2 for v, f and d can be taken as 0.54 and 0.29; 0.48 and 0.37; and 0.62 and 0.14 respectively. Using Eq. (2.9) the new values of v, f and d are calculated. Table 2.30 shows the new values of v, f and d and the corresponding values of the objective functions and the values of constraints.

The following important points may be remembered in the cases of both NSTLBO and MO-Jaya algorithms: When Z' values are different for all the candidate solutions then the ranking may be done based only on Z' values and there is no need of ranking using non-dominance relations and the *CD* values. Non-dominance relations for candidate solutions come into play only when the candidate solutions have same value of Z'. If Z' values are same for more than 2 candidate solutions then non-dominance concept can be applied to those solutions to rank them. If non-dominance ranks are also same for the candidate solutions then the *CD* values can be calculated and then the ranking can be done. It is also to be remembered while using the non-dominance relations to rank the candidate

Table 2.29 Initial population

S. No.	v	f	d	M_r	T_w	T	R_a	Z_T	Z_{Ra}	Z'	Rank	Status
1	187.9435	0.05	1.5595	14654.894	0.189	217.0482	1.3946	0	0	0	1	Best
2	191.7079	0.0583	2.3379	26129.704	0.2299	249.9699	1.6907	0	0	0	1	
3	171.541	0.0941	1.811	29233.177	0.2657	261.2649	2.3355	0	0.3355	0.36015	2	
4	191.4187	0.1711	0.5	16375.87	0.2494	240.1621	2.8515	0	0.8515	0.90865	3	
5	180.9845	0.1315	1.6239	38647.946	0.3058	285.2846	2.9371	0	0.9371	1	4	Worst

Table 2.30 New values of the variables, objective functions, constraints, and violations

S. No.	v	f	d	M_r	T_w	T	R_a	Z_T	Z_{Ra}	Z'
A	189.9616	0.0198	1.5505	5831.8021	0.1199	166.9384	0.6889	0	0	0
B	192.7849	0.0272	1.9553	10253.103	0.1502	193.4776	0.9154	0	0	0
C	177.6597	0.0591	1.6813	17653.126	0.208	227.4739	1.6132	0	0	0
D	192.568	0.1276	0.9995	24559.391	0.2642	258.5964	2.6041	0	0.6041	0.64465
E	184.7424	0.0924	1.584	27039.193	0.2561	258.1636	2.2326	0	0.2326	0.24821

solutions, some solutions may get 1st rank, some may get 2nd rank, some may get 3rd rank, and so on. Then, *CD* measure will be used to distinguish the solutions that are having the same rank. After distinguishing the 1st ranked solutions using the *CD* measure, then the 2nd ranked solutions will be distinguished, and so on. After this exercise the solutions are arranged as per *CD*s of 1st ranked solutions, then as per *CD*s of 2nd ranked solutions, and so on.

Now the solutions obtained in Table 2.29 are combined with the randomly generated solutions of Table 2.29 and are shown in Table 2.31.

Calculation of the crowding distance

Step 1: Sort and rank the population of Table 2.31 based on constraint-dominance and the non-dominated sorting concept.

Step 2: Collect all rank 1 solutions

Step 3: Determine the minimum and maximum values of both the objective functions for the entire population from Table 2.31 and these are, $(M_r)_{\min} = 5831.8021; (M_r)_{\max} = 38647.946; (T_w)_{\min} = 0.1199; (T_w)_{\max} = 0.3058$.

Step 4: Consider only the first objective and sort all the values of the first objective function in the ascending order irrespective of the values of the second objective function.

Step 4a: Assign crowding distance as infinity to the first and last solutions (i.e., the best and the worst solutions) as shown in Table 2.31a.

Step 4b: The crowding distances d_{B}, d_1 and d_{C} are calculated as follows (please note that B is between A and 1; 1 is between B and C; C is between 1 and 2)

$$d_B^{(1)} = 0 + \frac{(M_r)_1 - (M_r)_A}{(M_r)_{\max} - (M_r)_{\min}} = 0 + \frac{14654.894 - 5831.8021}{38647.946 - 5831.8021} = 0.2688$$

$$d_1^{(1)} = 0 + \frac{(M_r)_C - (M_r)_B}{(M_r)_{\max} - (M_r)_{\min}} = 0 + \frac{17653.126 - 10253.103}{38647.946 - 5831.8021} = 0.2255$$

$$d_C^{(1)} = 0 + \frac{(M_r)_2 - (M_r)_1}{(M_r)_{\max} - (M_r)_{\min}} = 0 + \frac{26129.704 - 14654.894}{38647.946 - 5831.8021} = 0.34967$$

Table 2.31b shows the crowding distance assignment as per the first objective.

Step 5: Consider only the second objective and sort all the values of the second objective function in the ascending order irrespective of the values of the first objective function.

Step 5a: Assign crowding distance as infinity to the first and the last solutions (i.e. the best and the worst solutions) as shown in Table 2.31c.

Step 5b: The crowding distances d_{B}, d_1 and d_{C} are calculated as,

Table 2.31 Combined population

S. No.	V	f	d	M_r	T_w	T	R_a	Z_T	Z_{Ra}	Z'	Rank	CD
1	187.9435	0.05	1.5595	14654.894	0.189	217.0482	1.3946	0	0	0	1	0.53642
2	191.7079	0.0583	2.3379	26129.704	0.2299	249.9699	1.6907	0	0	0	1	∞
3	171.541	0.0941	1.811	29233.177	0.2657	261.2649	2.3355	0	0.3355	0.36015	3	∞
4	191.4187	0.1711	0.5	16375.87	0.2494	240.1621	2.8515	0	0.8515	0.90865	5	∞
5	180.9845	0.1315	1.6239	38647.946	0.3058	285.2846	2.9371	0	0.9371	1	6	∞
A	189.9616	0.0198	1.5505	5831.8021	0.1199	166.9384	0.6889	0	0	0	1	∞
B	192.7849	0.0272	1.9553	10253.103	0.1502	193.4776	0.9154	0	0	0	1	0.6405
C	177.6597	0.0591	1.6813	17653.126	0.208	227.4739	1.6132	0	0	0	1	0.56998
D	192.568	0.1276	0.9995	24559.391	0.2642	258.5964	2.6041	0	0.6041	0.64465	4	∞
E	184.7424	0.0924	1.584	27039.193	0.2561	258.1636	2.2326	0	0.2326	0.24821	2	∞

S. No.	Objective functions		Rank	Crowding distance
	M_r	T_w		
(a) Crowding distance assignment to the first and last solutions as per the first objective				
A	5831.8021	0.1199	1	∞
B	10253.103	0.1502	1	
1	14654.894	0.189	1	
C	17653.126	0.208	1	
2	26129.704	0.2299	1	∞
(b) Crowding distance assignment as per the first objective				
A	5831.8021	0.1199	1	∞
B	10253.103	0.1502	1	0.2688
1	14654.894	0.189	1	0.2255
C	17653.126	0.208	1	0.34967
2	26129.704	0.2299	1	∞

(continued)

Table 2.31 (continued)

S. No.	Objective functions		Rank	Crowding distance
	M_r	T_w		
(c) Crowding distance assignment to the first and last solutions as per the second objective				
A	5831.8021	0.1199	1	∞
B	10253.103	0.1502	1	
1	14654.894	0.189	1	
C	17653.126	0.208	1	
2	26129.704	0.2299	1	∞
(d) Crowding distance assignment as per the second objective				
A	5831.8021	0.1199	1	∞
B	10253.103	0.1502	1	0.6405
1	14654.894	0.189	1	0.53642
C	17653.126	0.208	1	0.56998
2	26129.704	0.2299	1	∞

$$d_B^{(2)} = d_B^{(1)} + \frac{(T_w)_1 - (T_w)_A}{(T_w)_{\max} - (T_w)_{\min}} = 0.2688 + \frac{0.189 - 0.1199}{0.3058 - 0.1199} = 0.6405$$

$$d_1^{(2)} = d_1^{(1)} + \frac{(T_w)_C - (T_w)_B}{(T_w)_{\max} - (T_w)_{\min}} = 0.2255 + \frac{0.208 - 0.1502}{0.3058 - 0.1199} = 0.53642$$

$$d_C^{(2)} = d_C^{(1)} + \frac{(T_w)_2 - (T_w)_1}{(T_w)_{\max} - (T_w)_{\min}} = 0.34967 + \frac{0.2299 - 0.189}{0.3058 - 0.1199} = 0.56968$$

Table 2.31d shows the crowding distance assignment as per the second objective.

Table 2.32 shows the arrangement of candidate solutions based on the values of constraint dominance, non-dominance rank and the crowing distance values at the end of first iteration.

At the end of the first iteration, two solutions (i.e. 2nd and Ath) are having the 1st rank based on Z' (and $Z' = 0$ for both the solutions). We may use the non-dominance concept to distinguish these two solutions. However, it can be seen from the values of the objectives of these two solutions it can be understood that both are of non-dominated type. Furthermore, their *CD* values are ∞. As both 2nd and Ath solutions are having the 1st rank any one of these two solutions can be taken as the best solution for the next iteration. The solutions at the end of first iteration become input to the second iteration.

2.10 Multi-objective Quasi-oppositional Jaya (MOQO-Jaya) Algorithm

In order to further diversify the population and improve the convergence rate of MO-Jaya algorithm, in this paper the concept of opposition based learning is introduced in the MO-Jaya algorithm. To achieve better approximation, a population opposite to the current population is generated according to Eqs. (2.13)–(2.15) and both are considered at the same time. The solutions in the MOQO-Jaya algorithm are updated in the similar manner as in the Jaya algorithm based on Eq. (2.9). However, in order to handle multiple objectives effectively and efficiently the MOQO-Jaya algorithm is incorporated with non-dominated sorting approach and crowding distance computation mechanism.

Similar to the MO-Jaya algorithm, in the MOQO-Jaya algorithm, the task of finding the best and worst solutions is accomplished by comparing the rank assigned to the solutions based on the non-dominance concept and the crowding distance value. In the beginning, an initial population is randomly generated with *P* number of solutions. Then a quasi-opposite population is generated which is combined to the initial population. This combined population is then sorted and ranked based on the non-dominance concept and crowding distance value is computed for each solution. Then *P* best solutions are selected from the combined population based on rank and crowding distance. Further, the solutions are modified

Table 2.32 Candidate solutions based on constraint-dominance, non-dominance rank and crowding distance

S. No.	V	f	d	M_r	T_w	T	R_a	Z_T	Z_{Ra}	Z'	Rank	CD
2	191.7079	0.0583	2.3379	26129.704	0.2299	249.9699	1.6907	0	0	0	1	∞
A	189.9616	0.0198	1.5505	5831.8021	0.1199	166.9384	0.6889	0	0	0	1	∞
B	192.7849	0.0272	1.9553	10253.103	0.1502	193.4776	0.9154	0	0	0	1	0.6405
C	177.6597	0.0591	1.6813	17653.126	0.208	227.4739	1.6132	0	0	0	1	0.56998
1	187.9435	0.05	1.5595	14654.894	0.189	217.0482	1.3946	0	0	0	1	0.53642

according to Eq. (2.9) of Jaya algorithm. For this purpose the best and worst solutions are to be identified. The solution with the highest rank (rank = 1) is selected as the best solution. The solution with the lowest rank is selected as the worst solution. In case, there exists more than one solution with the same rank then the solution with the highest value of crowding distance is selected as the best solution and vice versa. This ensures that the best solution is selected from the sparse region of the search space. Once all the solutions are updated, the set of updated solutions (new solutions) is combined to the initial population. These solutions are again sorted and ranked based on the non-dominance concept and the crowding distance value for each solution is computed. Based on the new ranking and crowding distance value *P* number of good solutions are selected. The superiority among the solutions is determined based on the non-dominance rank and the crowding distance value of the solutions. A solution with a higher rank is regarded as superior to the other solution. If both the solutions hold the same rank, then the solution with higher crowding distance value is seen as superior to the other. The number of function evaluations required by MOQO-Jaya algorithm = population size × no. of generations for a single simulation run. Figure 2.10 shows the flowchart of MOQO-Jaya algorithm.

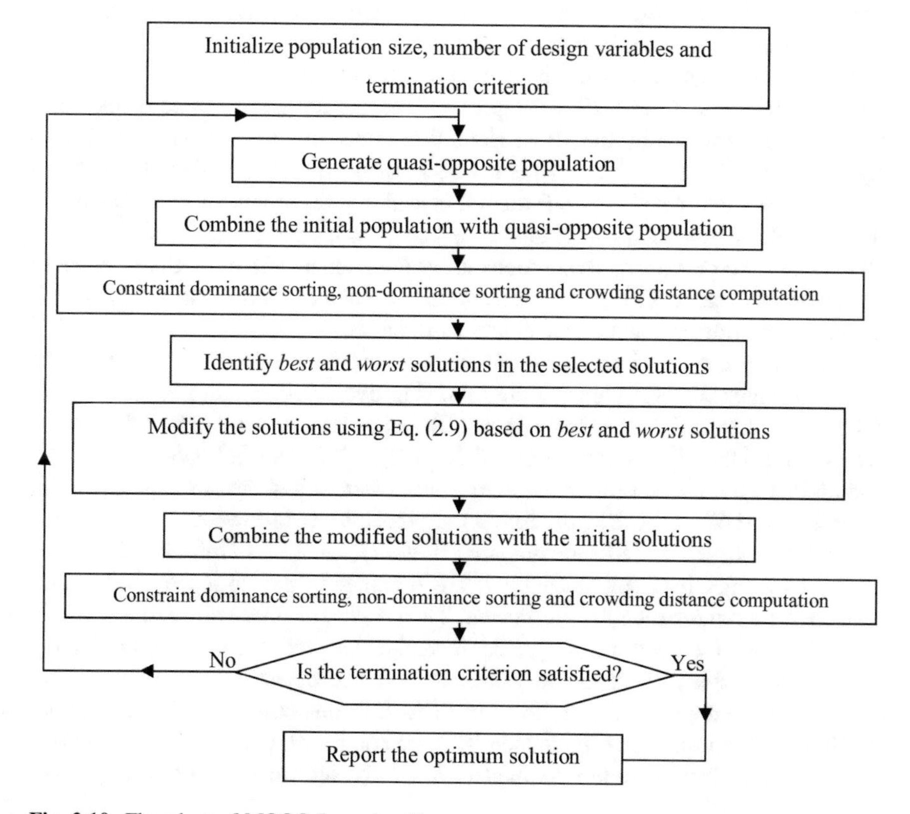

Fig. 2.10 Flowchart of MOQO-Jaya algorithm

2.11 Performance Measures

The main aim behind adopting a posteriori approach to solve multi-objective optimization problems is to obtain a diverse set of Pareto optimal solutions (Zhou et al. 2011). Thus, in order to assess the performance of any multi-objective optimization algorithm three performance measures can be adopted and they are as follows:

2.11.1 Coverage

This performance measure compares two sets of non-dominated solutions (*A*, *B*), and it gives the percentage of individuals of one set dominated by the individuals of the other set. It is defined as follows:

$$Cov(A,B) = \frac{|\{b \in B| \exists\ a \in A : a \prec= b\}|}{|B|} \quad (2.29)$$

where, *A* and *B* are the two non-dominated set of solutions under comparison; $a \prec= b$ means *a* dominates *b* or is equal to *b*.

The value *Cov*(*A*, *B*) = 1 means that all points in *B* are dominated by or equal to all points in *A* and *Cov*(*A*, *B*) = 0 represents the situation when none of the solutions in *B* are covered by the set *A*. Here, it is imperative to consider both *Cov*(*A*, *B*) and *Cov*(*B*, *A*), since *Cov*(*A*, *B*) is not necessarily equal to 1 − *Cov*(*B*, *A*). When *Cov*(*A*, *B*) = 1 and *Cov*(*B*, *A*) = 0 then, it is said that the solutions in *A* completely dominate the solutions in *B* (i.e. this is the best possible performance of *A*). *Cov*(*A*, *B*) represent the percentage of solutions in set *B* which are either inferior or equal to the solutions in set *A*; *Cov*(*B*, *A*) represent the percentage of solutions in set *A* which are either inferior or equal to the solutions in set *B*.

For example, for a bi-objective optimization problem of minimization type, let the non-dominated set *A* contains the following five solutions: *a* (1.2, 7.8), *b* (2.8, 5.1), *c* (4.0, 2.8), *d* (7.0, 2.2) and *e* (8.4, 1.2). Let us assume another non-dominated set *B* contains the following solutions: *f* (1.3, 8.2), *g*(2.7, 4.9), *h*(3.9, 3.0), *i*(7.3, 2.1) and *j*(8.2, 1.5). *A* and *B* form two Pareto fronts. Then to find the coverage of *A* with respect to *B* (i.e. *Cov*(*A*, *B*)), the following procedure is followed:

First we take *a*(1.2, 7.8) and compare this with *f*(1.3, 8.2). In this case the values of the two objectives contained by *a* are dominating the corresponding values contained by *f* and hence *f* is eliminated. Then *a* is compared with *g*, then with *h*, then with *i* and then with *j* in sequence. It can be observed that *a* is not dominating *g*, *h*, *i* and *j* with respect to the values of both the objectives. Then the values contained by *b* are compared with *g*, *h*, *i* and *j*. In this case, *g*, *h*, *i* and *j* are not dominated. Then the values contained by *c* are compared with *g*, *h*, *i* and *j*. In this case also, *g*, *h*, *i* and *j* are not dominated. Similarly, the values contained by *d* and

e are compared with g, h, i and j and it is observed that g, h, i and j are not dominated. The point f is not considered as it is already removed. Any point that is removed will vanish from the set. Thus, in the present example, the set A is dominating only f of B. Hence, $Cov(A, B)$ is calculated as the ratio of the no. of dominated points of B to the total no. of points of B. Thus, $Cov(A, B) = 1/5 = 0.2$ (or 20%). Similarly, $Cov(B, A)$ can be calculated. In this case only b is removed and hence $Cov(B, A) = 1/5 = 0.2$ (or 20%). In the present example, equal no. of points (i.e. 5 points) is assumed in each set of A and B. However, the total no. of points of A and B can be different but the procedure of calculation of coverage remains the same as described in this section.

2.11.2 Spacing

This performance measure quantifies the spread of solutions (i.e. how uniformly distributed the solutions are) along a Pareto front approximation. It is defined as follows:

$$S = \sqrt{\frac{1}{|n-1|}\sum\nolimits_{i=1}^{n}\left(\bar{d}-d_i\right)^2} \tag{2.30}$$

where, n is the number of non-dominated solutions.

$$d_i = \min_{i,i \neq j} \sum_{m=1}^{k} \left|f_m^i - f_m^j\right|, \quad i,j = 1,2,\ldots.,n \tag{2.31}$$

where, k denotes the number of objectives and f_m is the objective function value of the mth objective. Since different objective functions are added together in Eq. (2.18), normalizing the objectives is essential.

$$\bar{d} = \sum_{i=1}^{n} d_i/|n| \tag{2.32}$$

The spacing measure determines the smoothness of a Pareto-front and uniformity of the distribution of Pareto-points along the Pareto-front. For this purpose the distance variance of the neighboring non-dominated solutions is calculated. $S = 0$ indicates that all the non-dominated points on a Pareto front are equidistantly spaced. The spacing measure is a suitable metric to compare the quality of Pareto-fronts obtained by different algorithms, when the Pareto-fronts are regular and contain a set of distinct points with no solutions with same values of objectives (i.e. no duplicate solutions).

2.11.3 Hypervolume

The hypervolume (HV) is used to compare the quality of Pareto-fronts obtained by optimization algorithms in the case of multi-objective optimization problems. HV gives the volume of the search space which is dominated by a Pareto-front obtained by a particular algorithm with respect to a given reference point. Therefore for a particular algorithm a higher value of HV is desirable which indicates the quality of the Pareto-front obtained by the algorithm.

Mathematically, for a Pareto-front containing Q solutions, for each solution i belongs to Q, a hypervolume v_i is constructed with reference point W and the solution i as the diagonal corners of the hypercube. Thereafter the union of these hypercubes is found and its hypervolume is calculated as follows.

$$HV = volume\left(\bigcup_{i=1}^{|Q|} v_i\right) \tag{2.33}$$

The hypervolume (HV) concept can be understood by the following bi-objective optimization example. Figure 2.11 shows the two objectives f_1 and f_2 and the Pareto-front *a-b-c-d-e*.

In Fig. 2.11, the points *a*, *b*, *c*, *d* and *e* are the set of non-dominated points. The shaded region is the area of the search space dominated by the Pareto-front *a-b-c-d-e*. Considering N (11.0, 10.0) as the reference point (it can be chosen arbitrarily), the hypervolume is calculated as follows.

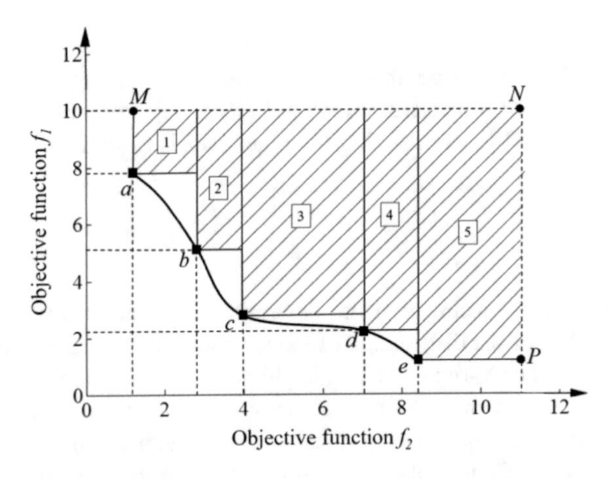

Fig. 2.11 Hypervolume obtained by the Pareto-optimal points *a* (1.2, 7.8), *b* (2.8, 5.1), *c* (4.0, 2.8), *d* (7.0, 2.2) and *e* (8.4, 1.2)

$$
\begin{aligned}
\mathrm{A1} &= (2.8 - 1.2) \times (10.0 - 7.8) = 3.52 \\
\mathrm{A2} &= (4.0 - 2.8) \times (10.0 - 5.1) = 5.88 \\
\mathrm{A3} &= (7.0 - 4.0) \times (10.0 - 2.8) = 21.60 \\
\mathrm{A4} &= (8.4 - 7.0) \times (10.0 - 2.2) = 10.92 \\
\mathrm{A5} &= (11 - 8.4) \times (10.0 - 1.2) = 22.88 \\
HV &= \mathrm{A1} + \mathrm{A2} + \mathrm{A3} + \mathrm{A4} + \mathrm{A5} = 64.80
\end{aligned}
$$

In fact, what we calculated above considering 2-dimensional graph (for 2 objectives) is the "area". If 3 objectives are considered then we have to calculate the "volume". If more than 3 objectives are considered then we have to calculate the "hypervolume" . Special algorithms are available to calculate the hypervolume (Beume et al. 2009; Jiang et al. 2015).

If any particular optimization algorithm generates a Pareto-front *f-g-h-i-j* (instead of *a-b-c-d-e*) then we calculate the hypervolume for the Pareto-front *f-g-h-i-j* and compare that value with the hypervolume obtained by the Pareto-front *a-b-c-d-e*. The optimization algorithm giving higher hypervolume is considered better. Hypervolume gives the volume of the search space dominated by a Pareto-front obtained by a particular optimization algorithm.

References

Beume, N., Fonseca, C. M., Manuel, L.-I., Paquete, L., & Vahrenhold, J. (2009). On the complexity of computing the hypervolume indicator. *IEEE Transactions on Evolutionary Computation, 13*(5), 1075–1082.

Jiang, S., Zhang, J., Ong, Y.-S., Zhang, A. N., & Tan, P. S. (2015). A simple and fast hypervolume indicator-based multiobjective evolutionary algorithm. *IEEE Transactions on Cybernetics, 45* (10), 2202–2213.

Rao, R. V. (2016a). *Teaching learning based optimization algorithm and its engineering applications*. Switzerland: Springer.

Rao, R. V. (2016b). Jaya: A simple and new optimization algorithm for solving constrained and unconstrained optimization problems. *International Journal of Industrial Engineering Computations, 7,* 19–34.

Rao, R. V., & More, K. (2017a). Design optimization and analysis of selected thermal devices using self-adaptive Jayaalgorithm. *Energy Conversion and Management, 140,* 24–35.

Rao, R. V., & More, K. (2017b). Optimal design and analysis of mechanical draft cooling tower using improved Jaya algorithm. *International Journal of Refrigeration*. https://doi.org/10.1016/j.ijrefrig.2017.06.024.

Rao, R. V., & Rai, D. P. (2017a). Optimization of welding processes using quasi oppositional based Jaya algorithm. *Journal of Experimental & Theoretical Artificial Intelligence, 29*(5), 1099–1117.

Rao, R. V., & Rai, D. P. (2017b). Optimization of submerged arc welding process using quasi-oppositional based Jaya algorithm. *Journal of Mechanical Science and Technology, 31*(5), 1–10.

Rao, R. V., Rai, D. P., Balic, J. (2016). Multi-objective optimization of machining and micro-machining processes using non-dominated sorting teaching–Learning-based optimization algorithm. *Journal of Intelligent Manufacturing*, 2016. https://doi.org/10.1007/s10845-016-1210-5.

Rao, R. V., Rai, D. P., & Balic, J. (2017). A multi-objective algorithm for optimization of modern machining processes. *Engineering Applications of Artificial Intelligence, 61,* 103–125.

Rao, R. V., & Saroj, A. (2017). A self-adaptive multi-population based Jaya algorithm for engineering optimization. *Swarm and Evolutionary Computation*. https://doi.org/10.1016/j.swevo.2017.04.008.

Rao, R. V., & Saroj, A. (2018). An elitism-based self-adaptive multi-population Jaya algorithm and its applications. *Soft Computing*. https://doi.org/10.1007/s00500-018-3095-z.

Rao, R. V., Savsani, V. J., & Vakharia, D. P. (2011). Teaching–learning-based optimization: A novel method for constrained mechanical design optimization problems. *Computer-Aided Design, 43,* 303–315.

Simon, D. (2013). *Evolutionary optimization algorithms*. New York: Wiley.

Teo, T. (2006). Exploring dynamic self-adaptive populations in differential evolution. *Soft Computing, 10,* 673–686.

Yang, S. H., & Natarajan, U. (2010). Multiobjective optimization of cutting parameters in turning process using differential evolution and non-dominated sorting genetic algorithm-II approaches. *International Journal of Advanced Manufacturing Technology, 49,* 773–784.

Zhou, A., Qu, B.-Y., Li, H., Zhao, S.-Z., Suganthan, P. N., & Zhang, Q. (2011). Multiobjective evolutionary algorithms: a survey of the state of the art. *Swarm and Evolutionary Computation, 1*(1), 32–49.

Chapter 3
Application of Jaya Algorithm and Its Variants on Constrained and Unconstrained Benchmark Functions

Abstract This chapter presents the results of application of Jaya algorithm and its variants like SAMP-Jaya and SAMPE-Jaya algorithms on 15 unconstrained benchmark functions given in CEC 2015 as well as 15 other unconstrained functions and 5 constrained benchmark functions. The results are compared with those given by the other well known optimization algorithms. The results have shown the satisfactory performance of Jaya algorithm and its variants for the considered CEC 2015 benchmark functions and the other constrained and unconstrained optimization problems. The statistical tests have also supported the performance supremacy of the variants of the Jaya algorithm.

3.1 Applications on Unconstrained Uni-modal and Multi-modal Benchmark Functions

This section presents the analysis of the performance of the proposed approaches on fifteen unconstrained (i.e. uni-modal and multi-modal) benchmark problems. The optimum values of all the considered benchmark problems are zero except for the function O_8 with the optimum value of −418.9829 * d. Table 3.1 presents the fifteen unconstrained benchmark problems (Ngo et al. 2016).

The results obtained by the Jaya algorithm and its variants SAMP-Jaya and SAMPE-Jaya for the uni-modal and multimodal problems are compared with GA and its variants, PSO and its variants. The comparisons of the results are presented in Tables 3.2 and 3.3. The optimization results obtained by SAMPE-Jaya algorithm for functions (O_1 to O_{15}) with 50,000 and 200,000 function evaluations over 30 independent runs for the 30 dimension problems are presented in Tables 3.2 and 3.3 respectively. The results are compared with the Jaya algorithm, extraordinariness particle swarm optimizer (EPSO) (Ngo et al. 2016), gravitational search algorithm (GSA) (Rashedi et al. 2009) real coded genetic algorithm (RGA) (Haupt and Haupt 2004), PSO (Eberhart and Kennedy 1995), cooperative PSO (CPSO) (Bergh and Engelbrecht 2004), comprehensive learning PSO (LPSO) (Liang and Qin 2006),

R. Venkata Rao, *Jaya: An Advanced Optimization Algorithm and its Engineering Applications*, https://doi.org/10.1007/978-3-319-78922-4_3

Table 3.1 Summary of uni-modal and multi modal problems (Ngo et al. 2016; Reprinted with permission from Elsevier)

Test function	Interval
$O_1(y) = \sum_{i=1}^{d} y_i^2$	$[-100, 100]^d$
$O_2(y) = \sum_{i=1}^{d} abs\|y_i\| + \prod_{i=1}^{d} abs\|y_i\|$	$[-10,10]^d$
$O_3(y) = \sum_{m=1}^{d} \left(\sum_{n=1}^{m} y_n\right)^2$	$[-100,100]^d$
$O_4(y) = \max\{\|y_i\|,\ 1 \le i \le d\}$	$[-100,100]^d$
$O_5(y) = \sum_{m=1}^{d} \left[100\left(y_{m+1} - y_m^2\right) + (y_i - 1)^2\right]$	$[-30,30]^d$
$O_6(y) = \sum_{i=1}^{d} \lfloor y_i + 0.5 \rfloor^2$	$[-100,100]^d$
$O_7(y) = \sum_{i=1}^{d} i y_i^4 + rand[0,1]$	$[-1.28,1.28]^d$
$O_8(y) = \sum_{m=1}^{d} -y_m \sin\left(\sqrt{\|y_m\|}\right)$	$[-500,500]^d$
$O_9(y) = \sum_{m=1}^{d} \left[y_m^2 - 10\cos(2\pi y_m) + 10\right]$	$[-5.12,5.12]^d$
$O_{10}(y) = 20\left(-0.2\sqrt{\frac{1}{d}\sum_{m=1}^{d} y_m^2}\right) - \exp\left(\frac{1}{d}\sum_{m=1}^{d}\cos(2\pi y_m)\right) + 20 + e$	$[-32,32]^d$
$O_{11}(y) = \frac{1}{400}\sum_{m=1}^{d} y_m^2 - \prod_{m=1}^{d}\cos\left(\frac{y_m}{\sqrt{m}}\right) + 1$	$[-50,50]^d$
$O_{12}(Y) = \frac{\pi}{d}\left\{10\sin^2(\pi x_1)\sum_{m=1}^{d-1}(x_m - 1)^2\left[1 + 10\sin^2(\pi x_{m+1})\right] + (x_d - 1)^2\right\} + \sum_{m=1}^{d} v(y_m, 10, 100, 4)$ $x_m = 1 + \frac{y_m + 1}{4}$ $v(y,a,k,l) = \begin{cases} k(y_m - a)^l & y_m > a \\ 0 & -a < y_i < a \\ k(-y_i - a)^m & y_i < a \end{cases}$	$[-50,50]^d$
$O_{13}(y) = 0.1\left\{\sin^2(3\pi y_1)\sum_{m=1}^{d}(y_i - 1)^2\left[1 + \sin^2(3\pi y_1 + 1)\right] + (y_d - 1)^2\left[1 + \sin^2(2\pi x_d + 1)\right]\right\}$ $+ \sum_{m=1}^{d} v(y_i, 5, 100, 4)$	$[-50,50]^d$
$O_{14}(y) = \sum_{m=1}^{d} \left[x_m - 10\cos(2\pi x_m) + 10\right]$ $x_m = \begin{cases} y_m & \|y_m\| < 0.5 \\ \frac{round(2y_m)}{2} & \|y_m\| \ge 0.5 \end{cases}$	$[-5.12,5.12]^d$
$O_{15}(y) = \sin^2(\pi y_1)$ $+ \sum_{m=1}^{d-1}\left\{(y_i - 1)^2\left[1 + 10\sin^2(\pi y_{m+1})\right]\right\}^2 + (y_d - 1)^2\left[1 + \sin^2(\pi y_d)\right]$	$[-10,10]^d$

Table 3.2 Comparative results for the uni-modal and multimodal problems with maximum function evaluations of 50,000 (Rao and Saroj 2018; Reprinted with permission from Springer Nature)

	RGA		GSA		EPSO		Jaya algorithm		SAMP–Jaya algorithm		SAMPE–Jaya	
	Mean	SD	Mean	SD	Mean	SD	Mean	SD	Mean	SD	Mean	SD
O_1	2.313E+01	1.215E+01	6.800E−17	2.180E−17	7.776E−18	6.010E−18	7.321e−19	2.820e−18	2.799e−19	8.270e−19	**3.951e−22**	**6.517e−22**
O_2	1.073E+00	2.666E−01	6.060E−08	1.190E−08	6.787E−12	3.008E−12	4.532910503e−11	1.05439e−10	1.343855e−12	2.163436e−11	**5.213e−15**	**7.931e−15**
O_3	5.617E+02	1.256E+02	9.427E+02	2.466E+02	2.121E−01	5.461E−01	5.623442438e−26	1.634578e−25	**1.546e−35**	**4.484e−35**	**1.546e−35**	**4.484e−35**
O_4	1.178E+01	1.576E+00	4.207E+00	1.122E+00	**9.941E−03**	**9.855E−03**	5.060152	7.352134	1.152804	2.808808	0.106004	0.049688
O_5	1.180E+03	5.481E+02	4.795E+01	3.956E+00	1.785E−02	2.136E−02	1.752707344e−06	1.75270734e−05	1.8006e−09	1.84426e−08	**5.273e−11**	**4.699e−09**
O_6	2.401E+01	1.017E+01	9.310E−01	2.510E+00	**0.00E+00**	**0.00E+00**	**0.00+00**	**0.00E+00**	**0.00E+00**	**0.00E 00**	**0.00E+00**	**0.00E 00**
O_7	6.750E−02	2.870E−02	7.820E−02	4.100E−02	6.470E−04	4.542E−04	7.4145641e−19	4.052875e−18	3.077026e−28	1.14118e−27	**8.639e−30**	**2.716e−29**
O_8	−1.248E+04	5.326E+01	−3.604E +03	5.641E+02	**−1.257E+04**	**3.851E−12**	−11466.336436	1288.140757	−11763.9563	838.950302	−12072.030	428.7556
O_9	5.902E+00	1.171E+00	2.940E+01	4.727E+00	**2.274E−**	**2.832E−14**	116.24670	58.29225	68.516505	32.543715	49.150865	11.442653
O_{10}	2.140E+00	4.014E−01	4.800E−09	5.420E−10	1.284E−09	7.280E−10	9.35277331e−11	1.4691025e−10	7.07370e−11	6.71682e−11	**9.681e−14**	**1.502e−13**
O_{11}	1.168E+00	7.950E−02	1.669E+01 1.669E+01	4.283E+00	2.533E−08	1.311E−07	2.71E−14	3.16095E−14	1.45E−14	1.9475E−14	**2.220E−17**	**4.965E−17**
O_{12}	5.100E−02	3.520E−02	5.049E−01	4.249E−01	**6.055E−20**	**8.970E−20**	1.6426611e−06	1.03669020 e−04	3.258634e−12	6.1328517e −10	3.2586e−12	6.13285e −10
O_{13}	8.170E−02	1.074E+01	3.405E+00	3.683E+00	**9.373E−16**	**1.802E−15**	9.13E−11	2.84099E−10	**2.80E−13**	1.19953E−12	**2.80E−13**	1.1995E−12
O_{14}	–	–	–	–	–	–	0.0000E+000	0.00000E+000	**0.00E+00**	**0.00E+000**	**0.00E+00**	**0.00E+000**
O_{15}	–	–	–	–	–	–	2.20E−25	2.74948E−26	8.00E−27	**2.74948E−27**	1.50E−32	**0.00E+000**

Note Bold value shows better solution

Table 3.3 Comparative results for the uni-modal and multimodal problems with maximum function evaluations of 200,000 (Ngo et al. 2016; Reprinted with permission from Elsevier; Rao and Saroj 2018; Reprinted with permission from Springer Nature)

Test function		PSO	CPSO	CLPSO	FFIPS	F-PSO	AIWPSO	EPSO	Jaya algorithm	SAMP-Jaya	SAMPE-Jaya
O_1	M	5.198E−70	5.146E−13	4.894E−39	4.588E−27	2.409E−16	**33.370E−134**	1.662E−74	7.057e−90	4.939e−89	1.36E−102
	SD	1.130E−74	7.759E−25	6.781E−39	1.958E−53	2.005E−31	**5.172E−267**	2.761E−75	3.445e−89	1.894726−88	3.61E−102
O_2	M	2.070E325	1.253E−07	8.868E−24	2.324E−16	1.580E−11	13653E−58	1.903E−47	6.924e−69	**9.500e−96**	**3.43E−81**
	SD	1.442E−49	1.179E−14	7.901E−49	1.141E−32	1.030E−22	**73735E−123**	2.152E−47	3.779e−68	4.3716557e−95	1.02E−79
O_3	M	1.458E+00	1.889E+03	1.922E+02	9.463E+00	1.732E+02	1.9587E−10	2.014E−03	1.16539e−132	**0.0000E+00**	**0.0000E+00**
	SD	1.78E+00	9.911E+06	3.843E+02	2.598E+01	9.158+03	1.201E−19	1.934E−3	2.860e−132	**0.0000E+00**	**0.0000E+00**
O_4	M	–	–	–	–	–	–	–	7.16E−05	**6.91E−08**	**6.91E−08**
	SD	–	–	–	–	–	–	–	0.0002	**6.91E−08**	**6.91E−08**
O_5	M	2.540E+01	8.265E−01	1.322E+01	2.671E+01	2.816+E01	2.500E+00	2.824E−05	1.765e−17	2.25200e−25	**1.24E−26**
	SD	5.903E+02	2.345E+00	2.148E+02	2.003E+02	2.313E+02	1.600E+01	3.650E−05	9.6721e−17	6.19346e−25	**3.01E−26**
O_6	M	**0.000E+000**	**0.000E+00**	**0.00E+00**	**0.000E+00**	**0.000E+00**	**0.000E+00**	**0.000E+00**	**0.00E+0**	**0.000E+00**	**0.000E+00**
	SD	**0.000E+000**	**0.000E+00**	**0.000E+0**	**0.00E+0**	**0.00E+00**	**0.000E+00**	**0.0000E+00**	**0.000E+00**	**0.000E+00**	**0.000E+00**
O_7	M	1.238E−02	1.076E−02	4.064E−03	3.305E−03	4.169E−03	5.524E−03	2.580E−04	1.0322e−106	2.1803727e−111	**4.12E−128**
	SD	2.311E−05	2.770E−05	9.618E−07	83668E−07	2.401E−06	1.536E−05	1.871E−04	5.653e−106	1.19423e−110	**1.84E−127**
O_8	M	−1.100E+04	−1.213E+04	−1.255E+04	−1.105E+04	−1.122E+04	**−1.257E+04**	**−1.257E+04**	−11224.9936	−12276.045275	−12246.045
	SD	1.375E+05	3.380E04	4.257E+03	9.442E+05	2.227+E05	**1.141E−25**	2.482E−12	1069.593223	569.593223	377.337923
O_9	M	3.476E+01	3.601E−13	0.0000E+00	5.850E01	7.384E+01	1.658E−01	**0.0000E+0**	76.981618	59.245665	29.848726
	SD	1.064E+02	1.540E−24	0.0000E+00	1.919E+02	3.706E+02	2.105E−01	**0.0000E+0**	26.329608	23.896502	11.512196
O_{10}	M	1.492E−14	1.609E−07	9.237E−15	1.386E−14	2.179E−09	6.987E−15	1.214E−14	5.15143e−15	**4.44089209e−15**	**4.44E−15**
	SD	1.863E−29	7.861E−14	6.616E−30	2.323E−29	1.719E−18	4.207E−31	**3.106E−15**	2.7040e−15	**0.0000E+00**	**0.00E+00**
O_{11}	M	2.162E−02	2.125E−02	0.000E+00.	2.478E−04	1.474E−03	2.852E−02	**0.0000E+0**	**0.000E+00**	**0.000E+00**	**0.000E+00**
	SD	4.502E−04	6.314E−04	0.000E+00	1.827E−06	1.285E−05	7.664E−04	0.0000E+0	0.000E+00	**0.000E+00**	**0.000E+00**

(continued)

Table 3.3 (continued)

Test function		PSO	CPSO	CLPSO	FFIPS	F-PSO	AIWPSO	EPSO	Jaya algorithm	SAMP-Jaya	SAMPE-Jaya
O_{12}	M	–	–	–	–	–	–	–	5.0968e−12	**1.570544e−18**	**1.500544e−18**
	SD	–	–	–	–	–	–	–	5.09684e−09	**6.494772e−11**	**6.5435e−12**
O_{13}	M	–	–	–	–	–	–	–	5.64E−32	**1.54E−33**	**1.54E−33**
	SD	–	–	–	–	–	–	–	2.739 E−31	**2.2504E−34**	**2.2504E−35**
O_{14}	M	2.096E+01	5.137E−13	5.944E−24	6.188E+01	7.035E+01	1.184E−16	**0.00E+0**	**0.00E+0**	**0.00E+0**	**0.00E+0**
	SD	1.8333E+02	5.944E−24	1.333E−01	1.401E+02	2.960E+02	4.207E−31	**0.000E+0**	**0.00E+0**	**0.00E+0**	**0.00E+0**
O_{15}	M	1.142E−29	2.091E−15	1.295E−29	1.027E−28	5.514E−18	**1.500E−32**	**1.500E−32**	**1.500E−32**	**1.500E−32**	**1.50E−32**
	SD	3.233E−57	1.295E−29	1.500E−32	1.005E−56	1.450E−34	**1.240E−94**	1.113E−47	2.7837E−45	4.658019e−48	3.780E−46

Note Bold value shows better solution

fully informed particle swarm (FIPS) (Mendes et al. 2004), Frankenstein's PSO (F-PSO) (Oca and Stutzle 2009) and adaptive inertia weight PSO (AIWPSO) (Nickabadi et al. 2011).

An observation can be made from Table 3.2 that the mean function value obtained by the SAMPE-Jaya algorithm is better or equal in 10 cases out of 15 cases in comparison to the other algorithms. The SAMPE-Jaya is able to find the global optimum value of the function O_6 and O_{14}. Similarly, the values of standard deviation (SD) obtained by the SAMPE-Jaya algorithm for the same cases are better or equal in 10 cases out of 15. For rest of the cases (O_4, O_8, O_{12} and O_{13}), performance of the SAMPE-Jaya algorithm is competitive except for the objective O_9. These results prove the robustness of the Jaya algorithm and its variants over the other reported methods used for the optimization of these problems.

The performance of the SAMPE-Jaya algorithm and its comparison with the other optimization algorithms for maximum function evaluations of 200,000 is presented in Table 3.3. An observation can be made from this table that the mean function value obtained by the SAMPE-Jaya algorithm is better or equal in 12 cases out of 15 in comparison to the other approaches. The present method is able to strike the global optimum value of the function O_3, O_6 and O_{14}. Similarly, the values of SD obtained by SAMPE-Jaya algorithm for the same cases are better or equal in 11 cases out of 15. For rest of the cases, performance of the proposed SAMPE-Jaya algorithm is competitive except for the objective O_9. These results prove the robustness of the SAMPE-Jaya algorithm over the other reported methods for optimization of these problems.

3.2 Applications on CEC 2015's Computationally Expensive Benchmark Problems

These problems are taken from CEC 2015. Table 3.4 presents the fifteen CEC 2015 benchmark functions. All the functions are to be minimized. Problems 1–9 are shifted and rotated bench mark problems; Problems 10–12 are hybrid function and Problems 13–15 are composite functions. Detailed information and guidelines for CEC 2015 problems can be found from the literature (Ngo et al. 2016). As per the guidelines of CEC 2015, computational experiments are carried out. Maximum function evaluations (MFE) of 500 and 1500 are considered as one of the stopping criteria for 10 dimension and 30-dimension problems respectively. Second stopping criterion is, while the error value (current optimum value—global optimum value) is less than 1.00E-03. Average of the minimum error value is recorded over 20 independent runs. These values are used for the performance comparison of the optimization algorithms.

The computational results achieved by Jaya algorithm and its variants are compared with EPSO, DE, ($\mu + \lambda$)-evolutionary strategy (ES), specialized and generalized parameters experiments of covariance matrix adaption evolution

Table 3.4 Summary of the CEC2015 expensive optimization test problems (Ngo et al. 2016; Reprinted with permission from Elsevier)

Categories	No.	Functions	Related basic functions	Fi*
Unimodal functions	1	Rotated Bent Cigar function	Bent Cigar function	100
	2	Rotated discus function	Discus function	200
Simple multimodal functions	3	Shifted and rotated Weierstrass function	Weierstrass function	300
	4	Shifted and rotated Schwefel's function	Schwefel's function	400
	5	Shifted and rotated Katsuura function	Katsuura function	500
	6	Shifted and rotated HappyCat function	HappyCat function	600
	7	Shifted and rotated HGBat function	HGBat function	700
	8	Shifted and rotated expanded Griewank's plus Rosenbrock's function	Griewank's function Rosenbrock's function	800
	9	Shifted and rotated expanded Scaffer's F6 function	Expanded Scaffer's F6 Function	900
Hybrid functions	10	Hybrid function 1 ($N = 3$)	Schwefel's Function Rastrigin's function High conditioned elliptic function	1000
	11	Hybrid function 2 ($N = 4$)	Griewank's function Weierstrass function Rosenbrock's function Scaffer's F6 function	1100
	12	Hybrid function 3 ($N = 5$)	Katsuura function HappyCat function Griewank's function Rosenbrock's function Schwefel's function Ackley's function	1200
Composition functions	13	Composition function 1 (N = 5)	Rosenbrock's function High conditioned elliptic function Bent Cigar function Discus function High conditioned elliptic function	1300
	14	Composition function 2 ($N = 3$)	Schwefel's function Rastrigin's function High conditioned elliptic function	1400
	15	Composition function 3 ($N = 5$)	HGBat function Rastrigin's function Schwefel's function Weierstrass function High conditioned elliptic function	1500

strategy (CMAES-S and CMAES-G) (Andersson et al. 2015). The comparison of the computational results is presented in Table 3.5. An observation can be made from this table that the results achieved by the SAMPE-Jaya algorithm are better in 12 cases for 10-dimension and 9 cases for 30-dimension problems. It can also be seen from Table 3.5 that the results obtained by the SAMPE-Jaya algorithm are better or competitive in comparison to other algorithms used for this problem. Bold values in Table 3.5 show the minimum value of the mean error for each function.

Furthermore, computational complexity of the SAMPE-Jaya algorithm is evaluated for 30-dimensions and 10-dimensions problems as per the guidelines of the CEC2015. The computational complexity of the algorithms is defined as the complexity in the computational time when the dimensions of the problem are increased. The test function provided by CEC2015 is run on the same computer which is used for optimization of problems of CEC2015 and the time (T_0) taken by this function is found to be 0.3447 s. Next step is to record the average processing time (T_1) for each problem (both for 10d and 30d problems). The complexity of the of the algorithm (T_1/T_o) is calculated and is shown in Table 3.4. In Table 3.4, the values T_B/T_A reveal the complexity of the algorithm when problem changes from 10d to 30d. The value of T_B/T_A equal to 1 means, it is having zero complexity when problem changes from 10d to 30d.

The values greater than 1 reveal complexity of computational time of the SAMPE-Jaya algorithm. Functions FCEC3 and FCEC5 are having the higher complexity in terms of computational time because these are multi-modal functions. Similarly, hybrid functions FCEC11 and FCEC12 and composite functions FCEC13-FCEC15 have shown the higher complexity in terms of computational time. For the remaining problems the value of T_B/T_A is almost equal to 3.5 which reveal that the computational complexity of the present algorithm is increased 3.5 times when the dimension of the problem is increased from 10 to 30. The computational complexity of the SAMPE-Jaya algorithm is about 3 for the problems FCEC1, FCEC2, FCEC4, FCEC6, FCEC7, FCEC8, FCEC9, FCEC10 FCEC13 and FCEC14. It shows that computational complexity of the SAMPE-Jaya algorithm is increased about three times when the dimension of the problems changes from 10 to 30.

The computational complexity of SAMPE-Jaya algorithm is more than 4 for the problems FCEC3, FCEC5, FCEC11, FCEC12 and FCEC 15. This increment in the computational complexity is due to the complexity and multimodality of these problems. However, the computational complexity of the SAMPE-Jaya algorithm is less as compared to SAMP-Jaya and EPSO algorithms. As the computational complexity of the other algorithms for CEC 2015 problems are not available (except EPSO) in literature, therefore, the computational complexity of the SAMPE-Jaya algorithm cannot be compared with others. It can also be observed from Table 3.4 that the computational complexity (T_B/T_A) of the SAMPE-Jaya algorithm is less in comparison to EPSO for all the problems of CEC 2015.

An observation can made from Tables 3.2, 3.3, 3.4, 3.5 and 3.6 that the performance of the SAMPE-Jaya algorithm is better or competitive in comparison to the other algorithms. However, it becomes necessary to prove the significance of the proposed approach over the other algorithms by means of some statistical test.

Table 3.5 Comparative results for mean error obtained by different approaches for CEC 2015 problems (Rao and Saroj 2018; Reprinted with permission from Springer Nature)

Function	Dimension	DE (Ngo et al. 2016)	(μ + λ) ES (Ngo et al. 2016)	CMAES-S (Andersson et al. 2015)	CMAES-G (Andersson et al. 2015)	EPSO (Ngo et al. 2015)	Jaya algorithm	SAMP-Jaya	SAMPE-Jaya
FCEC1	10	3.4143E+09	2.6325E+09	3.6620E+07	6.5990E+07	1.5785E+09	1.3175E+07	6.40376E+06	**3.0377E+06**
	30	2.3911E+10	3.5775E+10	**6.8700E+07**	1.1080E+08	8.4866E+09	7.1308E+08	3.02517E+08	6.2057E+08
FCEC2	10	7.4931E+04	4.8418E+04	5.8080E+04	1.0240E+05	1.8953E+04	5.4161E+03	**1.9699E+03**	2.5711E+03
	30	1.8254E+05	1.6179E+05	2.3630E+05	2.9530E+05	6.3748E+04	1.3021E+04	**1.1894E+03**	**1.1894E+03**
FCEC3	10	3.1093E+02	3.1048E+02	6.1200E+02	6.1570E+02	3.1009E+02	3.0566E+02	3.0021E+02	**3.0005E+02**
	30	3.4190E+02	3.4353E+02	6.3390E+02	6.5270E+02	3.3800E+02	3.0231E+02	3.0114E+02	**3.0008E+02**
FCEC4	10	2.2974E+03	1.4368E+03	3.1890E+03	4.1090E+03	2.0662E+03	4.6250E+02	**4.2458E+02**	**4.2458E+02**
	30	7.9627E+03	7.0557E+03	8.6730E+03	1.2040E+04	6.6946E+03	8.2815E+02	**7.77849E+02**	**7.77849E+02**
FCEC5	10	5.0286E+02	5.0318E+02	1.0010E+03	1.0060E+03	5.0305E+02	5.0001E+02	**5.00007E+02**	**5.00007E+02**
	30	5.0431E+02	5.0499E+02	1.0010E+03	1.0080E+03	5.0430E+02	**5.0000E+02**	**5.00000E+02**	**5.00000E+02**
FCEC6	10	6.0286E+02	6.0223E+02	1.2010E+03	1.2010E+03	6.0234E+02	6.0528E+02	**6.01758E+02**	**6.01758E+02**
	30	6.0365E+02	6.0433E+02	1.2010E+03	1.2010E+03	6.0276E+02	**6.0209E+02**	6.07713E+02	6.04121E+02
FCEC7	10	7.2588E+02	7.1668E+02	1.4010E+03	1.4020E+03	7.1725E+02	734.419551	706.745007	**705.0833**
	30	7.5438E+02	7.8216E+02	1.4010E+03	1.4010E+03	**7.2189E+02**	1.2025E+03	1062.726345	1.03644E+03
FCEC8	10	4.1637E+03	1.1691E+03	**1.6130E+03**	1.6480E+03	1.6412E+03	2.109E+03	1.760E+03	1.760E+03
	30	7.9963E+05	7.3789E+06	**1.7670E+03**	2.3210E+03	1.2746E+05	1.400E+05	7.109E+05	7.109E+05
FCEC9	10	9.0415E+02	9.0414E+02	1.8080E+03	1.8080E+03	9.0407E+02	903.426016	903.105930	**903.02836**
	30	9.1394E+02	9.1408E+02	1.8270E+03	1.8280E+03	9.1372E+**02**	913.311482	**913.281842**	**912.43551**
FCEC10	10	1.3622E+06	1.4666E+06	1.7440E+05	1.7700E+06	1.3052E+06	43885.7623	22782.9766	**11669.4733**
	30	3.8759E+07	9.5323E+07	3.6310E+06	1.4730E+07	2.6363E+07	31356.0475	*85,248.5215*	**29088.6247**
FCEC11	10	1.1229E+03	1.1150E+03	2.2120E+03	2.2190E+03	**1.1140E+03**	1119.28310	1118.847733	1118.847733
	30	1.2870E+03	1.4378E+03	2.2460E+03	2.2580E+03	1.2288E+03	1165.52400	1164.500690	**1163.230763**
FCEC12	10	1.5980E+03	1.5797E+03	2.7390E+03	2.9810E+03	1.5291E+03	1423.57853	1352.880566	**1213.143563**
	30	3.0110E+03	3.8087E+03	3.4540E+03	4.0940E+03	2.4432E+03	1490.04466	1284.020952	**1211.077290**

(continued)

Table 3.5 (continued)

Function	Dimension	DE (Ngo et al. 2016)	(μ + λ) ES (Ngo et al. 2016)	CMAES-S (Andersson et al. 2015)	CMAES-G (Andersson et al. 2015)	EPSO (Ngo et al. 2015)	Jaya algorithm	SAMP-Jaya	SAMPE-Jaya
FCEC13	10	1.7969E+03	**1.6663E+03**	3.2580E+03	3.3000E+03	1.6932E+03	1663.94774	1663.187051	**1.662289769**
	30	1.9613E+03	2.2208E+03	3.3840E+03	3.4260E+03	1.8839E+03	1716.39830	1700.033267	**1688.291320**
FCEC14	10	1.6135E+03	1.6148E+03	3.2090E+03	3.2170E+03	1.6064E+03	*1446.11153*	***1426.164181***	***1426.164181***
	30	1.7479E+03	1.8406E+03	3.2660E+03	3.3000E+03	1.7016E+03	***1675.18244***	1699.661232	1699.661232
FCEC15	10	1.9452E+03	1.9740E+03	3.7770E+03	3.9020E+03	**1.8662E+03**	1863.52977	**1862.935212**	**1862.1706944**
	30	2.9304E+03	2.9154E+03	4.4270E+03	4.8360E+03	2.7488E+03	1932.59722	1936.961342	1919.523203

Note Bold value shows better solution

Table 3.6 Computational complexity of the SAMP-Jaya and SAMPE-Jaya algorithms (Rao and Saroj 2018; Reprinted with permission from Springer Nature)

Function	d = 10		d = 30		Computational complexity		
	$T_1(s)$	$T_A = T_1/T_0$	$T_1(s)$	$T_B = T_1/T_0$	SAMPE-Jaya (T_B/T_A)	SAMP-Jaya	EPSO (Ngo et al. 2016)
1	0.039641	0.115003771	0.11094	0.321844116	2.798553	3.588393	3.605
2	0.030814	0.089393772	0.102472	0.297278503	3.325495	3.413454	3.432
3	0.152056	0.441127647	1.053516	3.056328595	6.928445	7.429281	7.968
4	0.058882	0.170823325	0.14269	0.413952713	2.42328	3.547105	3.591
5	0.248303	0.720346872	1.373376	3.984265835	5.531038	7.271586	7.511
6.	0.038857	0.112728653	0.120896	0.350727686	3.111256	3.351844	3.385
7	0.031087	0.090187603	0.112622	0.326725172	3.622728	3.400896	3.444
8	0.033716	0.097813654	0.129546	0.375821971	3.842224	3.443678	3.592
9	0.029994	0.087015182	0.108133	0.313702253	3.605144	3.52979	3.692
10	0.034821	0.101020404	0.128561	0.37296509	3.691978	3.500343	3.679
11	0.060534	0.175613577	0.250034	0.725366212	4.130468	4.04166	6.158
12	0.055663	0.161483706	0.254652	0.738763659	4.57485	4.676693	4.785
13	0.071664	0.207903684	0.208266	0.604194179	2.906125	3.327989	5.702
14	0.072664	0.210806112	0.263034	0.763081714	3.619827	3.406342	5.270
15	0.188662	0.547323276	1.225011	3.55384663	6.49314	6.325926	8.036

Table 3.7 Friedman rank test for uni-modal and multimodal problems with 50,000 function evaluations (Rao and Saroj 2018; Reprinted with permission from Springer Nature)

Problem	50,000 function evaluations					
Algorithm	RGA	GSA	EPSO	Jaya	SAMP-Jaya	SAMPE-Jaya
Friedman ranks	4.8333	4.9	3.0333	3.7	2.6667	1.8667
p-value	4.16E−06					
χ^2	32.7822					

Therefore, a well-known statistical method known as 'Friedman test' (Joaquin et al. 2016) is used to compare the performance of the SAMPE-Jaya algorithm with the other algorithms. Mean value of the fitness function obtained by different methods is considered for the test. This method first finds the rank of algorithms for the individual problems and then calculates the average rank to get the final rank of the each algorithm for the considered problems. This test is performed with assuming χ^2 distribution and with $k - 1$ degree of freedom.

Table 3.7 presents the mean rank of the algorithms for the uni-modal and multimodal problems with maximum function evaluations of 50,000. An observation can be made from Table 3.7 that the performance of the SAMPE-Jaya algorithm is better than the other methods. It has obtained higher rank in comparison to the other algorithms. Figure 3.1 presents the comparison of the same on the bar chart.

Table 3.8 presents the mean rank of the algorithms for the uni-modal and multimodal problems with maximum function evaluations of 200,000. An observation can be made from the Table 3.8 that the performance of the SAMPE-Jaya algorithm is better than the other methods. In this case also, the SAMPE-Jaya algorithm is having higher rank in comparison to the rest of the algorithms. The comparison of the ranks is presented on the bar chart, as shown in Fig. 3.2.

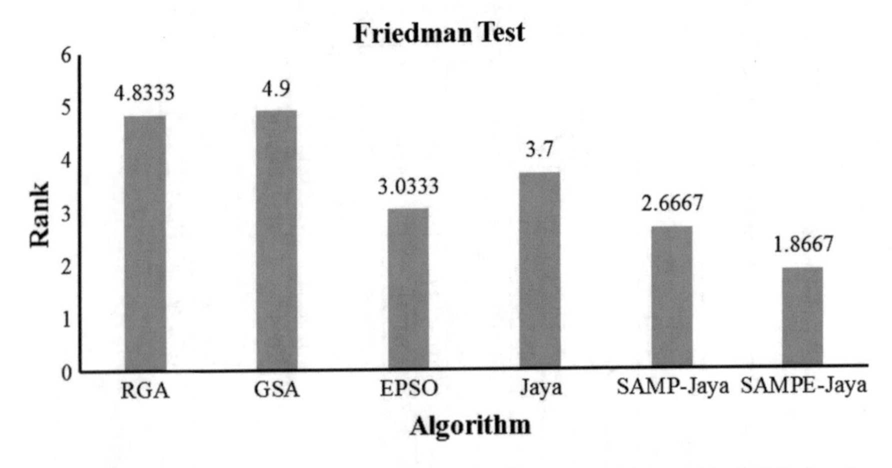

Fig. 3.1 Friedman rank test for uni-modal and multimodal problems with 50,000 function evaluations (Rao and Saroj 2018; Reprinted with permission from Springer Nature)

Table 3.8 Friedman rank test for uni-modal and multimodal problems with 200,000 function evaluations (Rao and Saroj 2018; Reprinted with permission from Springer Nature)

Problem	200,000 function evaluations									
Algorithm	PSO	CPSO	CLPSO	FFIPS	F−PSO	AIWOP	EPSO	Jaya	SAMP−Jaya	SAMPE−Jaya
Friedman ranks	7.3	7.6333	5.8	7.3	8.1	5.2	4.4667	3.866	2.9	2.4333
p−value	3.13E−12									
χ^2	73.4921									

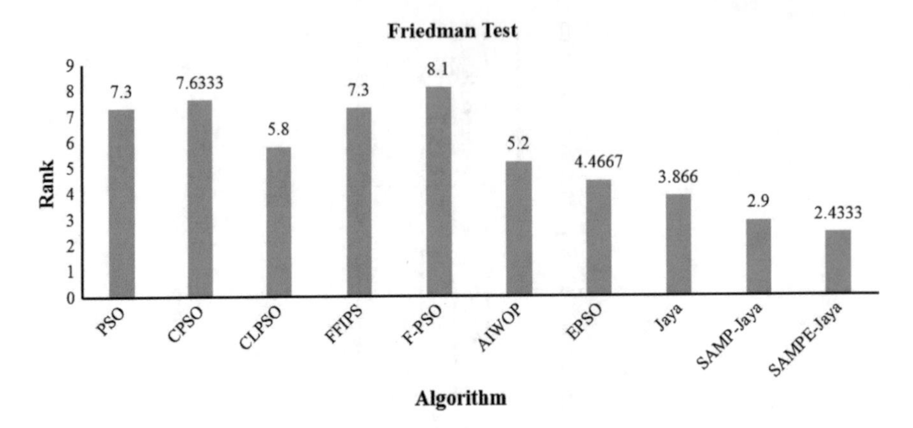

Fig. 3.2 Friedman rank test for uni-modal and multimodal problems with 200,000 function evaluations (Rao and Saroj 2018; Reprinted with permission from Springer Nature)

Table 3.9 Friedman rank test for CEC 2015 problems with 10-dimensions (Rao and Saroj 2018; Reprinted with permission from Springer Nature)

Algorithm	DE	(μ + λ) −ES	CMAES-S	CAMES-G	EPSO	Jaya	SAMP-Jaya	SAMPE-Jaya
Friedman ranks	5.9333	4.6	6.2667	7.4667	4.1333	3.8	2.1333	1.6667
p−value	9.06E−13							
χ^2	71.0511							

The performance comparison of the present algorithm on CEC 2015 problems are shown in Tables 3.9 and 3.10 for 10-dimension and 30-dimension problems respectively. Average rank of the SAMPE-Jaya algorithm for the 10 dimension problem is 1.6667, which is better than the rest of algorithms. Similarly, average rank of the SAMPE-Jaya algorithm for the 30-dimension problem is 2.2, which is better than the rest of algorithms. Thus, the Friedman rank test confirms that the performance of the SAMPE-Jaya algorithm is better for the considered unconstrained benchmark problems and it has obtained the highest rank in comparison to the other approaches for the all the cases considered for the comparison.

Table 3.10 Friedman rank test for CEC 2015 problems with 30-dimensions (Rao and Saroj 2018; Reprinted with permission from Springer Nature)

Algorithm	DE	(μ + λ) −ES	CMAES-S	CAMES-G	EPSO	Jaya	SAMP-Jaya	SAMPE-Jaya
Friedman ranks	5.3333	6.0667	6	6.9333	3.9333	2.9333	2.6	2.2
p−value	3.87E−10							
χ^2	57.9488							

Table 3.11 Performance of SAMPE-Jaya on large scale problems (Rao and Saroj 2018; Reprinted with permission from Springer Nature)

Function		Dimension		
		100	500	1000
Rosenbrock	SAMPE-Jaya	**4.68E−27**	**2.73E−15**	**1.60E−19**
	CCPSO	7.73E−14	7.73E−14	5.18E−13
	MLCC	9.02E−15	4.30E−13	8.46E−13
	Sep-CMA−ES	9.02E−15	2.25E−14	7.81E−13
Rastrigin	SAMPE-Jaya	3.25E+01	7.80E+01	1.01E+02
	CCPSO	**6.08E+00**	**5.79E+01**	**7.82E+01**
	MLCC	2.31E+01	6.67E+01	1.09E+02
	Sep-CMA-ES	2.31E+01	2.12E+02	3.65E+02
Griewank	SAMPE-Jaya	**9.15E−04**	**1.85E+02**	**8.01E+02**
	CCPSO	4.23E+02	7.24E+02	1.33E+03
	MLCC	1.50E+02	9.25E+02	1.80E+03
	Sep-CM-ES	4.31E+00	2.93E+02	9.10E+02

CCPSO Cooperatively coevolving particle swarm optimization; *MLCC* multilevel cooperative coevolution; *Sep-CMA-ES* separable covariance matrix adaptation evolution strategy

3.3 Application of SAMPE-Jaya Algorithm on Large Scale Problems

The performance of the SAMPE-Jaya algorithm is tested on three large scale problems taken from the literature (Cheng and Jin 2015a). The dimensions of the considered problems are: 100, 500 and 1000. Table 3.11 presents comparison of the SAMPE-Jaya algorithm with other algorithms. It can be observed from Table 3.11 that the performance of SAMPE-Jaya algorithm is better for Rosenbrock and Griewank function and competitive for Rastrigin function as compared to the other algorithms for the considered large scale problems. Hence, it can be concluded based on these results that the proposed SAMPE-Jaya algorithm is performing satisfactorily for the large scale problems also.

3.4 Comparison of Performance of SAMP-Jaya Algorithm with Four Recent Algorithms

In this section, comparison of the performance of the SAMP-Jaya algorithm is carried out with four recently proposed algorithms namely across neighborhood search (ANS) optimization algorithm, multi-population ensemble of mutation differential evolution (MEMDE), social learning particle swarm optimization algorithm (SL-PSO), competitive swarm optimizer and other algorithms (Rao and Saroj 2017). The comparison of performance of these algorithms with the SAMP-Jaya algorithm is presented in Tables 3.12, 3.13, 3.14, 3.15 and 3.16. The symbols "+",

Table 3.12 Comparative results of SAMP Jaya algorithm with ANS and other algorithms (Rao and Saroj 2017; Reprinted with permission from Elsevier)

Function	30D	SAMP−Jaya	ANS	jDE	JADE	CLPSO	FIPS	FDR	CPSO	ABC
f_1	**M**	1.91E−182	**2.11E−245**	6.32E−125	3.88E−131	6.68E−38	3.41E−27	1.38E−136	4.20E−12	6.63E−16
	SD	1.87E−181	3.13E−244	1.43E−124	1.22E−131	5.41E−38	2.78E−27	1.38E−137	6.15E−12	8.12E−17
f_2	**M**	**2.52E−25**	8.34E+00	4.86E−05	3.98E−01	1.95E+01	2.21E+01	2.25E+01	3.52E+01	8.14E+00
	SD	6.193E−25	9.22E+00	8.01E−05	1.26E+00	4.31E+00	4.31E−01	3.26E+00	7.29E+00	3.22E+00
f_3	**M**	**1.24E−22**	5.36E−20	4.15E−09	9.02E−15	8.83E−03	3.58E−06	9.54E−06	4.10E−05	3.59E+00
	SD	1.32E−25	6.44E−21	8.25E−10	1.85E−14	2.70E−03	1.60E−06	1.10E−05	1.17E−04	5.12E−01
f_4	**M**	6.81E−141	**7.91E−168**	2.35E−63	6.16E−49	1.89E−23	4.91E−17	2.03E−70	1.22E−07	1.49E−15
	SD	6.811−138	8.22E−167	1.35E−63	1.95E−48	9.30E−24	1.53E−17	5.80E−70	6.56E−08	1.68E−16
f_5	**M**	**0.00E+00**	**0.00**	**0.00**	**0.00**	**0.00**	**0.00**	**0.00**	**0.00**	**0.00**
	SD	0.00E+00	0.00	0.00	0.00	0.00	0.00	0.00	0.00	0.00
f_6	**M**	**2.180−111**	1.54E−03	2.45E−03	6.49E−04	3.36E−03	2.62E−03	2.68E−03	9.26E−03	1.54E−01
	SD	1.19E−110	5.23E−04	8.12E−04	3.31E−04	6.42E−04	1.18E−03	1.13E−03	3.64E−03	3.23E−02
f_7	**M**	43.12	**0.00**	**0.00**	**0.00**	**0.00**	6.50E+01	1.85E+01	1.08E−08	1.24E−15
	SD	26.25	0.00	0.00	0.00	0.00	1.60E+02	1.19E+01	1.42E−08	1.21E−15
f_8	**M**	**0.00**	**0.00**	**0.00**	**0.00**	**0.00**	6.50E+01	1.85E+01	1.08E−04	2.51E−14
	SD	0.00	0.00	0.00	0.00	0.00	1.60E+02	1.19E+01	1.42E−08	1.38E−14
f_9	**M**	**3.55E−15**	**3.55E−15**	**3.55E−15**	4.544E−15	4.26E−15	1.74E−14	2.59E−14	3.93E−07	5.15E−14
	SD	0.00E+00	0.00	0.00	0.00	1.45E−15	5.59E−15	1.34E−14	4.76E−07	5.25E−15
f_{10}	**M**	**0.00E+00**	**0.00**	**0.00**	**0.00**	9.99E−17	4.19E−08	1.25E−02	1.88E−02	4.10E−14
	SD	0.00E+00	0.00	0.00	0.00	3.15E−16	1.24E−07	7.9E−03	2.51E−02	2.44E−14

(continued)

Table 3.12 (continued)

Function	30D	SAMP−Jaya	ANS	jDE	JADE	CLPSO	FIPS	FDR	CPSO	ABC
f_{11}	**M**	**0.00**	1.57E−32	1.57E−32	1.57E−32	1.57E−32	2.44E−27	1.57E−32	1.17E−14	5.96E−16
	SD	0.00	2.72E−48	2.88E−48	2.88E−48	2.88E−48	2.83E−27	2.88E−48	1.65E−14	5.65E−16
f_{12}	**M**	**1.54E−33**	1.35E−32	1.35E−32	1.35E−32	1.35E−32	1.25E−27	5.49E−03	4.88E−13	6.66E−16
	SD	2.25E−38	2.88E−32	2.88E−32	2.88E−32	2.88E−32	6.42E−28	5.79E−03	1.26E−14	8.45E−17
	+/=/−		5/4/3	7/4/1	8/3/1	9/2/1	11/1/0	10/1/1	10/1/1	10/1/1

ACR Across neighbourhood search; *jDE* differential evolution with self-adapting control parameters; *JADE* adaptive differential evolution with an external archive; *CLPSO* comprehensive learning particle swarm optimizer; *FIPS* fully informed particle swarm; *FDR* fitness-distance-ratio based particle swarm optimization; *CPSO* cooperative particle swarm optimization; *ABC* artificial bee colony

Table 3.13 Comparative results of SAMP-Jaya algorithm with MEMDE and other algorithms (Rao and Saroj 2017; Reprinted with permission from Elsevier)

	30D	SAMP-Jaya	MPEDE		JADE		SHADE		EPSDE		jDE	
F1	**M**	0.00E+00	0.00E+00	=	0.00E+00	=	0.00E+00	=	0.00E+00	=	0.00E+00	=
	SD	0.00E+00	0.00E+00		0.00E+00		0.00E+00		0.00E+00		0.00E+00	
F2	**M**	0.00E+00	1.01E−26	+	1.26E−28	+	4.51E−29	+	8.32E−26	+	3.45E−06	+
	SD	0.00E+00	2.05E−26		1.22E−28		7.28E−29		2.66E−26		2.76E−06	
F3	**M**	2.43E+03	1.01E+01	−	8.42E+03	+	6.20E+03	+	6.34E+05	+	2.44E+05	+
	SD	1.53E+01	8.32E+00		6.58E+03		5.14E+03		3.44E+06		3.22E+05	
F4	**M**	2.051E−16	6.61E−16	+	4.13E−16	+	7.03E−16	+	3.88E+02	+	4.78E−02	+
	SD	1.15E−16	5.68E−16		3.45E−16		1.01E−15		3.13E+03		2.12E−02	
F5	**M**	4.14E−06	7.21E−06	+	7.59E−08	+	3.15E−10	−	1.38E+03	+	5.56E+02	+
	SD	3.12E−07	5.12E−06		5.65E−07		6.91E−10		7.43E+02		5.62E+02	
F6	**M**	5.41E−20	9.65E+00	+	1.16E+01	+	2.64E−27	−	6.44E−01	+	2.65E+01	+
	SD	3.25E−21	4.65E+00		3.16E+01		1.32E−26		1.24E+00		2.32E+01	
F7	**M**	283.20	2.36E−03	−	8.27E−03	−	2.17E−03	−	1.58E−02	−	1.14E−02	−
	SD	128.50	1.15E−03		8.22E−03		4.29E−03		2.54E−02		7.28E−03	
F8	**M**	2.01E+01	2.09E+01	+	2.09E+01	+	2.05E+01	+	2.09E+01	+	2.09E+01	+
	SD	2.00E−02	5.87E−01		1.68E−01		3.39E−01		2.84E−01		4.54E−01	
F9	**M**	14.32	0.00E+00	−	0.00E+00	−	0.00E+00	−	0.00E+00	−	0.00E+00	−
	SD	1.02E−02	0.00E+00		0.00E+00		0.00E+00		0.00E+00		0.00E+00	
F10	**M**	3.70E+01	1.52E+01	−	2.42E+01	−	1.62E+01	−	5.24E+01	+	5.46E+01	+
	SD	2.15E+00	2.98E+00		5.44E+00		3.35E+00		4.64E+01		8.85E+00	
F11	**M**	7.50E+00	2.58E+01	+	2.57E+01	+	2.71E+01	+	3.77E+01	+	2.88E+01	+
	SD	2.01E+00	3.11E+00		2.21E+00		1.57E+00		6.22E+00		2.61E+00	
F12	**M**	5.062E+00	1.17E+03	+	6.45E+03	+	2.90E+03	+	3.67E+04	+	8.23E+03	+
	SD	2.009E+00	8.66E+02		2.89E+03		3.11E+03		5.66E+03		8.54E+03	

(continued)

Table 3.13 (continued)

	30D	SAMP-Jaya	MPEDE		JADE		SHADE		EPSDE		jDE	
F13	**M**	1.950E+00	2.92E+00	+	1.47E+00	–	1.15E+00	+	2.04E+00	+	1.67E+00	+
	SD	2.50E−01	6.33E−01		1.15E−01		9.33E−02		2.12E−01		1.56E−01	
F14	**M**	0.00E+00	1.23E+01	+	1.123E+01	+	1.25E+01	+	1.35E+01	+	1.30E+01	+
	SD	0.00E+00	4.22E−01		3.21E−01		3.67E−01		2.35E−01		2.23E−01	
	+/=/–	**–**	**9/1/4**		**9/1/4**		**8/1/5**		**11/1/2**		**11/1/2**	

Note *MPEDE* multi-population ensemble DE; *JADE* adaptive differential evolution with optimal external achieve; *SHADE* success-history based Adaptive DE; *EPSDE* DE algorithm with ensemble of parameters and mutation and cross over

Table 3.14 Comparative results of SAMP-Jaya algorithm with SL-PSO and other algorithms (Rao and Saroj 2017; Reprinted with permission from Elsevier)

	30-D	*SAMP-Jaya*	*SL-PSO*		*GPSO*		*LPSO*		*FIPS*		*DMS-PSO*		*CLPSO*	
f_1	**M**	7.057E−90	**4.24E−90**	=	1.25E−61	+	8.48E−35	+	6.20E−70	+	3.30E−14	+	4.76E−19	+
	SD	3.445E−89	**5.26E−90**		2.82E−61		2.85E−34		1.44E−69		1.27E−13		1.92E−19	
f_2	**M**	**9.500E−96**	1.50E−46	+	7.33E+00	+	6.67E−01	+	1.13E−38	+	8.48E−11	+	7.54E−12	+
	SD	**4.371E−95**	5.34E−47		1.39E+01		2.58E+00		5.70E−39		1.84E−10		2.50E−12	
f_3	**M**	**0.00E+00**	4.66E−07	+	4.22E+03	+	3.65E−01	+	1.21E+00	+	9.79E+01	+	1.13E+03	+
	SD	**0.00E+00**	2.48E−07		5.08E+03		3.83E−01		6.59E−01		7.31E+01		2.89E+02	
f_4	**M**	6.91E−08	**1.17E−24**	–	8.49E−07	+	4.42E−05	+	2.37E+00	+	1.90E+00	+	4.31E+00	+
	SD	6.91E−08	**8.37E−25**		1.01E−06		2.32E−05		1.17E+00		7.85E−01		6.84E−01	
f_5	**M**	**2.52E−25**	2.15E+01	+	6.05E+03	+	5.18E+01	+	3.53E+01	+	5.60E+01	+	9.28E+00	+
	SD	**6.19E−25**	3.41E+00		2.32E+04		3.68E+01		2.71E+01		3.28E+01		1.03E+01	
f_6	**M**	**0.00E+00**	**0.00E+00**	=	**0.00E+00**	=	**0.00E+00**	=	**0.00E+00**	=	**5.33E−01**	+	**0.00E+00**	=
	SD	**0.00E+00**	**0.00E+00**		**0.00E+00**		**0.00E+00**		**0.00E+00**		**9.15E−01**		**0.00E+00**	
f_7	**M**	59.2456	1.50E+03	–	5.62E+03	+	3.07E+03	+	2.98E+03	–	5.74E−08	–	**6.06E−13**	–
	SD	230,896	9.10E+01		2.19E+03		7.80E+02		7.87E+02		6.02E−10		**8.88E−13**	
f_8	**M**	**4.44E−15**	1.55E+01	+	4.65E+01	+	5.02E+01	+	3.86E+01	+	2.70E−13	+	5.83E−09	+
	SD	**0.00E+00**	3.19E+00		2.55E+01		2.25E+01		1.04E+01		8.41E−13		5.02E−09	
f_9	**M**	**0.00E+00**	5.51E−15	+	1.36E14	+	7.67E+00	+	6.69E−15	+	6.11E−09	+	2.99E−10	+
	SD	**0.00E+00**	1.59E−15		4.34E15		9.79E+00		1.83E−15		1.89E−08		9.47E−11	
f_{10}	**M**	1.570E−18	**0.00E+00**	–	1.21E−02	+	2.46E−03	+	2.07E−13	+	1.76E−02	+	8.40E−12	+
	SD	6.494E−17	**0.00E+00**		1.58E−02		6.64E−03		5.03E−13		2.56E−02		1.45E−11	
f_{11}	**M**	**1.54E−33**	1.57E−32	+	6.91E−03	+	1.57E−32	+	1.57E−32	+	9.32E−15	+	3.61E−20	+
	SD	2.250E−34	0.00E+00		2.68E−02		2.83E−48		2.83E−48		3.61E−14		1.87E−20	
f_{12}	**M**	**0.00E+00**	1.35E−32	+	7.32E−04	+	7.32E−04	+	1.35E−32	+	1.47E−03	+	3.31E−19	+
	SD	**0.00E+00**	0.00E+00		2.84E−03		2.84E−03		2.83E−48		3.87E−03		8.67E−20	
+/=/–			8/1/3		11/1/0		11/1/0		10/1/1		13/0/1		10/1/1	

Note *SL-PSO* social learning PSO; *GPSO* global version PSO; *LPSO* local version PSO; *FIPS* fully informed PSO; *DMS-PSO* dynamic multi-swarm PSO;

Table 3.15 Comparative results of SAMP-Jaya algorithm with CSO and other algorithms for 100-dimension problems (Rao and Saroj 2017; Reprinted with permission from Elsevier)

100-D	SAMP Jaya	CSO	CCPSO2	MLCC	Sep-CMA-ES	EPUS-PSO	DMS-PSO
f1	4.68E−27(6.82E −27)	9.11E−29(1.10E −28)	7.73E−14 (3.23E−14)	9.02E−15 (5.53E−15)	9.02E−15 (5.53E−15)	7.47E−01 (1.70E−01)	**0.00E+00 (0.00E+00)**
f2	32.5651 (0.005)	3.35E+01(5.38E +00)	**6.08E+00 (7.83E+00)**	2.31E+01(1.39E +01)	2.31E+01(1.39E +01)	1.86E+01(2.26E +0)	**3.65+00(7.30E −01)**
f3	**9.15E−04 (3.27E−03)**	3.90E+02(5.53E +02)	4.23E+02(8.65E +02)	1.50E+02(5.72E +01)	4.31E+00(1.26E +01)	4.99E+03(5.35E +03)	**2.83E+02 (9.40E+02)**
f4	**0.00E+00 (0.00E +00)**	5.60E+01(7.48E +00)	3.98E−02 (1.99E−01)	4.39E−13 (9.21E−14)	2.78E+02(3.43E +01)	4.71E+02(5.94E +01)	**1.83E+02 (2.16E+01)**
f5	**0.00E+00 (0.00E +00)**	**0.00E+00(0.00E +00)**	3.45E−13 (4.88E−03)	3.41E−14 (1.16E−14)	2.96E−04 (1.48E−03)	3.72E−01 (5.60E−02)	0.00E+00(0.00E +00)
f6	**0.00E+00 (0.00E +00)**	1.20E−014 (1.52E−15)	1.44E+03(1.04E +01)	1.11E−13 (7.87E−15)	2.12E+01(4.02E −01)	2.06E+0(4.40E −01)	0.00E+00(0.00E +00)
+/=/−	**–**	**4/1/1**	**5/0/1**	**5/0/1**	**5/0/1**	**5/0/1**	**2/2/2**

Note *CSO* competitive swarm optimizer; *CCPSO2* cooperatively coevolving particle swarm optimization 2; *MLCC* multilevel cooperative coevolution; *Sep-CMA-ES* separable covariance matrix adaptation evolution strategy; *EPUS-PSO* efficient population utilization strategy for particle swarm optimizer; *DMS-PSO* dynamic multi swarm PSO

Table 3.16 Comparative results of SAMP-Jaya algorithm with CSO and other algorithms for 500-dimension problems (Rao and Saroj 2017; Reprinted with permission from Elsevier)

100-D	SAMP Jaya	CSO	CCPSO2	MLCC	Sep-CMA-ES	EPUS-PSO	DMS-PSO
f1	4.68E−27(6.82E −27)	9.11E−29(1.10E −28)	7.73E−14 (3.23E−14)	9.02E−15 (5.53E−15)	9.02E−15 (5.53E−15)	7.47E−01 (1.70E−01)	**0.00E+00 (0.00E+00)**
f2	32.5651 (0.005)	3.35E+01(5.38E +00)	**6.08E+00 (7.83E+00)**	2.31E+01(1.39E +01)	2.31E+01(1.39E +01)	1.86E+01(2.26E +0)	**3.65+00(7.30E −01)**
f3	**9.15E−04 (3.27E−03)**	3.90E+02(5.53E +02)	4.23E+02(8.65E +02)	1.50E+02(5.72E +01)	4.31E+00(1.26E +01)	4.99E+03(5.35E +03)	**2.83E+02 (9.40E+02)**
f4	**0.00E+00 (0.00E +00)**	5.60E+01(7.48E +00)	3.98E−02 (1.99E−01)	4.39E−13 (9.21E−14)	2.78E+02(3.43E +01)	4.71E+02(5.94E +01)	**1.83E+02 (2.16E+01)**
f5	**0.00E+00 (0.00E +00)**	**0.00E+00(0.00E +00)**	3.45E−13 (4.88E−03)	3.41E−14 (1.16E−14)	2.96E−04 (1.48E−03)	3.72E−01 (5.60E−02)	0.00E+00(0.00E +00)
f6	**0.00E+00 (0.00E +00)**	1.20E−014 (1.52E−15)	1.44E+03(1.04E +01)	1.11E−13 (7.87E−15)	2.12E+01(4.02E −01)	2.06E+0(4.40E −01)	0.00E+00(0.00E +00)
+/=/−	**–**	**4/1/1**	**5/0/1**	**5/0/1**	**5/0/1**	**5/0/1**	**2/2/2**

Note *CSO* competitive swarm optimizer; *CCPSO2* cooperatively coevolving particle swarm optimization 2; *MLCC* multilevel cooperative coevolution; *Sep-CMA-ES* separable covariance matrix adaptation evolution strategy; *EPUS-PSO* efficient population utilization strategy for particle swarm optimizer; *DMS-PSO* dynamic multi swarm PSO

"−" and "=" indicate that the performance of the SAMP-Jaya algorithm is "better than", "worse than" and "similar to" respectively as compared to the corresponding peer algorithm.

Table 3.12 presents the comparison of the performance of the SAMP-Jaya algorithm with ANS and other algorithms on 12 unconstrained benchmark problems considered from the work of Wu (2016). The termination criterion used is same as that considered in the work of Wu (2016). In order to make the fair comparison of the performance of the algorithms 25 independent runs are carried out and the mean function value (M) and standard deviation (SD) values are recorded. It can be observed from Table 3.12 that the SAMP-Jaya algorithm has performed better in comparison to rest of the algorithms. It has performed better in 5 cases, competitive in 4 cases and inferior in 3 cases, in comparison to the performance of ANS algorithm. Similarly, the performance of SAMP-Jaya algorithm is much better in comparison to the performance of jDE, JADE, CLPSO, FIPS, FDR, CPSO and ABC algorithms. It has obtained global optimum value of the objective function for the functions f5, f8, f10 and f11.

Table 3.13 presents the comparison of the performance of the SAMP-Jaya algorithm with multi-population ensemble of mutation differential evolution (MEMDE) and other algorithms on 14 unconstrained benchmark problems considered from the work of Wu et al. (2017). The termination criterion used is same as that considered in the work of Wu et al. (2017). In order to make fair comparison of the performance of the algorithms 25 independent runs are carried out and the mean function value (M) and standard deviation (SD) values are recorded. It can be observed from Table 3.13 that the SAMP-Jaya algorithm has performed better in comparison to the rest of the algorithms. It has performed better in 9 cases, competitive in 1 case and inferior in 4 cases, in comparison to the performance of MEMDE algorithm. Similarly, the performance of SAMP-Jaya algorithm is better or competitive in comparison to the performance of JADE, SHADE, EPSDE and jDE. The proposed SAMP-Jaya algorithm has obtained global optimum value of the objective function for the functions F1, F2 and F14.

Table 3.14 presents the comparison of the performance of the SAMP-Jaya algorithm with social learning particle swarm optimization algorithm (SL-PSO) and other algorithms on 12 unconstrained benchmark problems considered from the work of Cheng and Jin (2015a). The termination criterion used is same as that considered in the work of Cheng and Jin (2015a). In order to make fair comparison of the performance of the algorithms 30 independent runs are carried out and the mean function value (M) and standard deviation (SD) values are recorded. It can be observed from the Table 3.14 that the SAMP-Jaya algorithm has performed better in comparison to rest of the algorithms. It has performed better in 8 cases, competitive in 1 case and inferior in 3 cases, in comparison to the performance of SL-PSO algorithm. Similarly, the performance of SAMP-Jaya algorithm is better or competitive in comparison to the performance of JADE, SHADE, EPSDE and jDE. It has obtained global optimum value of the objective function for functions f3, f6, f8 and f12.

Tables 3.15 and 3.16 present the comparison of the performance of the SAMP-Jaya algorithm with competitive swarm optimizer (CSO) and other algorithms on 6 unconstrained benchmark problems with 100-dimension 500-dimension respectively. The test problems are considered from the work of Cheng and Jin [29]. The termination criterion used is same as that considered in the work of Cheng and Jin (2015b).

It can be observed from Table 3.15 that for the 100-dimension problems, the SAMP-Jaya algorithm has performed better in comparison to the rest of the algorithms. For these problems the SAMP-Jaya algorithm has performed better in 4 cases, competitive in 1 case and inferior in 1 case, in comparison to the performance of CPSO algorithm. Similarly, the performance of the SAMP-Jaya algorithm is better or competitive in comparison to the performance of CPSO2, MLCC, Sep-CMA-ES, EPUS-PSO and DMS-PSO algorithms. It has obtained global optimum value of the objective function for the functions f5 and f6.

Similarly, it can be observed from Table 3.16 that for 500-dimension problems also, the SAMP-Jaya algorithm has performed better in comparison to rest of the algorithms. For these problems the SAMP-Jaya algorithm has performed better in 4 cases and inferior in 2 cases, in comparison to the performance of CPSO algorithm. Similarly, the performance of the SAMP-Jaya algorithm is better or competitive in comparison to the performance of CPSO2, MLCC, Sep-CMA-ES, EPUS-PSO and DMS-PSO algorithms. It has obtained global optimum value of the objective function for function f4, f5 and f6.

It can be concluded based upon the comparison of these results that the proposed SAMP-Jaya algorithm performs significantly better as compared to the performance of across neighborhood search (ANS) optimization algorithm, multi-population ensemble of mutation differential evolution (MEMDE), social learning particle swarm optimization algorithm (SL-PSO) and competitive swarm optimizer and other algorithms for the considered problems. Similarly, it has been shown that SAMPE-Jaya, another variant of Jaya algorithm, can perform well as compared to the ANS, MEMDE and SL-PSO algorithms. Rao and Saroj (2018) proved the effectiveness of SAMPE-Jaya algorithm.

3.5 Application of Jaya Algorithm on Constrained Design Benchmark Functions

The main difference between constrained and unconstrained optimization problems is that a penalty function is used in the constrained optimization problem to take care of the violation of each constraint and the penalty is operated upon the objective function. Rao and Waghmare (2017) presented the results of application of Jaya algorithm on 21 constrained design benchmark optimization problems and 4 mechanical design problems. Out of these, 5 constrained design benchmark optimization problems are presented in this section for demonstration.

The problem 3.1 is a minimization problem and the objective function involves thirteen variables and nine linear inequality constraints. The Jaya algorithm is applied to solve this problem and its performance is compared with eight other optimization algorithms. Just like any other algorithm, Jaya algorithm requires proper tuning of common control parameters such as population size and number of generations to execute the algorithm effectively. After making a few trials, population size of 50 and function evaluations of 1500 are considered. The results of Elitist TLBO, DETPS, TLBO, M-ES, PESO, CDE, CoDE and ABC are referred from Rao and Waghmare (2014), Zhang et al. (2013), Rao et al. (2011), Mezura-Montes and Coello (2006), Zavala et al. (2005), Becerra and Coello (2006), Huang et al. (2007) and Karaboga and Basturk (2007) respectively.

Problem 2 is a maximization problem involving ten variables and a nonlinear constraint. After making a few trials, population size of 50 and function evaluations of 25,000 are considered. Problem 3 is a minimization problem which involves seven variables and four nonlinear inequality constraints. After making a few trials, population size of 10 and function evaluations of 30,000 are considered. Problem 4 is a linear minimization problem which involves eight variables and three nonlinear inequality and three linear inequality constraints. After making a few trials, population size of 10 and function evaluations of 99,000 are considered. Problem 5 is a maximization problem involving three design variables and 729 nonlinear inequality constraints. After making a few trials, population size of 50 and function evaluations of 5000 are considered.

Problem 1

$$\min f(x) = 5\sum_{i=1}^{4} x_i - 5\sum_{i=1}^{4} x_i^2 - \sum_{i=5}^{13} x_i \tag{3.1}$$

$$\text{s.t.}\, g_1(x) = 2x_1 + 2x_2 + x_{10} + x_{11} - 10 \le 0 \tag{3.2}$$

$$g_2(x) = 2x_1 + 2x_3 + x_{10} + x_{12} - 10 \le 0 \tag{3.3}$$

$$g_3(x) = 2x_2 + 2x_3 + x_{11} + x_{12} - 10 \le 0 \tag{3.4}$$

$$g_4(x) = -8x_1 + x_{10} \le 0 \tag{3.5}$$

$$g_5(x) = -8x_2 + x_{11} \le 0 \tag{3.6}$$

$$g_6(x) = -8x_3 + x_{12} \le 0 \tag{3.7}$$

$$g_7(x) = -2x_4 - x_5 + x_{10} \le 0 \tag{3.8}$$

$$g_8(x) = -2x_6 - x_7 + x_{11} \le 0 \tag{3.9}$$

$$g_9(x) = -2x_8 - x_9 + x_{12} \leq 0 \tag{3.10}$$

where $0 \leq xi \leq 1$, i = 1, 2, 3, …, 9; $0 \leq xi \leq 100$, i = 10, 11, 12; $0 \leq x13 \leq 1$.

The optimal solution is f(x*) = − 15 at x* = (1, 1, 1, 1, 1, 1, 1, 1, 1, 3, 3, 3, 1).

Problem 2

$$\max f(x) = \left(\sqrt{n}\right)^n \prod_{i=1}^{n} x_i \tag{3.11}$$

$$s.t. h(x) = \sum_{i=1}^{n} x_i^2 - 1 = 0 \tag{3.12}$$

where n = 10 and $0 \leq xi \leq 10$, i = 1, 2, 3, …, n. The global maximum f(x*) = 1 at x* = (1/n0.5, 1/n0.5, …)

Problem 3

$$\begin{aligned} \min f(x) = {} & (x_1 - 10)^2 + 5(x_2 - 12)^2 + x_3^4 + 3(x_4 - 11)^2 \\ & + 10x_5^6 + 7x_6^2 + x_7^4 - 4x_6x_7 - 10x_6 - 8x_7 \end{aligned} \tag{3.13}$$

$$s.t. g_{1(x)} = -127 + 2x_1^2 + 3x_2^4 + x_3 + 4x_4^2 + 5x_5 \leq 0 \tag{3.14}$$

$$g_{2(x)} = -282 + 7x_1 + 3x_2 + 10x_3^2 + x_4 - x_5 \leq 0 \tag{3.15}$$

$$g_{3(x)} = -196 + 23x_1 + x_2^2 + 6x_6^2 - 8x_7 \leq 0 \tag{3.16}$$

$$g_{4(x)} = 4x_1^2 + x_2^2 - 3x_1x_2 + 2x_3^2 + 5x_6 - 11x_7 \leq 0 \tag{3.17}$$

Where $-10 \leq xi \leq 10$, i = 1, 2, 3, …, 7. The optimal solution is f(x*) = 680.6300573 at x* = (2.330499, 1.951372, −0.4775414, 4.365726, −0.6244870, 1.1038131, 1.594227).

Problem 4

$$\min f(x) = x_1 + x_2 + x_3 \tag{3.18}$$

$$s.t.\ g_{1(x)} = -1 + 0.0025(x_4 + x_6) \leq 0 \tag{3.19}$$

$$g_{2(x)} = -1 + 0.0025(x_5 + x_7 - x_4) \leq 0 \tag{3.20}$$

$$g_{3(x)} = -1 + 0.01(x_8 - x_5) \leq 0 \quad (3.21)$$

$$g_{4(x)} = -x_1x_6 + 833.3325x_4 + 100x_1 - 83333.333 \leq 0 \quad (3.22)$$

$$g_{5(x)} = -x_2x_7 + 1250x_5 + x_2x_4 - 1250x_4 \leq 0 \quad (3.23)$$

$$g_{6(x)} = -x_3x_8 + 1250000 + x_3x_5 - 2500x_5 \leq 0 \quad (3.24)$$

Where 100 ≤ x1 ≤ 10,000, 1000 ≤ xi ≤ 10,000, i = 2, 3, 100 ≤ xi 10,000, i = 4, 5, …, 8. The optimal solution is f(x*) = 7049.248021 at x* = (579.3066, 1359.9709, 5109.9707, 182.0177, 295.601, 217.982, 286.165, 395.6012).

Problem 5

$$\max f(x) = \frac{100 - (x_1 - 5)^2 - (x_2 - 5)^2 - (x_3 - 5)^2}{100} \quad (3.25)$$

$$\text{s.t.}\, g(x) = (x_1 - p)^2 - (x_2 - q)^2 - (x_3 - r)^2 \leq 0 \quad (3.26)$$

Where 0 ≤ xi ≤ 10, i = 1, 2, 3, p, q, r = 1, 2, 3, …, 9. The optimal solution is f(x*) = 1 at x* = (5, 5, 5).

The comparison of statistical results of the considered nine algorithms for the test problems 1–5 are presented in Table 3.17 where “Best”, “Mean” and “Worst” represent best solution, mean best solution and worst solution respectively over 30 independent runs. For problem 1 the Jaya algorithm finds the same global optimum solution as that given by elitist TLBO, DETPS, TLBO, M-ES, PESO, CDE, CoDE and ABC but the Jaya algorithm uses 31.25, 21.42, 74.92, 30.24 and 31.25% function evaluations to obtain competitive results as compared to M-ES, PESO, CDE, CoDE and ABC respectively. While the elitist TLBO requires approximately 94, 94, 86, 96, 44 and 33% function evaluations than ABC, CoDE, CDE, PESO, M-ES, TLBO, DETPS respectively. Jaya algorithm is superior in terms of standard deviation and robustness for problem 1 than elitist TLBO algorithm.

For problem 2, the Jaya algorithm finds the better quality of solution than DETPS and CDE. The results of Jaya algorithm are same as the results of elitist TLBO, TLBO, M-ES, and ABC algorithms but it uses 35.71, 27.53, 25, 10.41, 7.41, 24.97 and 10.41% function evaluations to obtain competitive results as compared to elitist TLBO, DETPS, TLBO, M-ES, PESO, CDE, CoDE and ABC respectively. The Jaya algorithm also computes function value in less time as compared to the other optimization algorithms considered.

For problem 3 the results of Jaya algorithm are same as the results of elitist TLBO, DETPS, TLBO, PESO and CDE algorithms in terms of the best solution but Jaya algorithm uses 99.23, 92.06, 30, 12.5, 8.57, 29.97, 12.5 and 30% function evaluations to obtain competitive results as compared to elitist TLBO, DETPS, TLBO, M-ES, PESO, CDE, CoDE and ABC respectively. For problem 3, solutions

Table 3.17 Comparison of statistical results of nine algorithms for test problems 1-5 (Rao and Waghmare 2017; Reprinted with permission from Taylor & Francis)

Problem		Jaya algorithm	Elitist TLBO (Rao and waghmare 2014)	DETPS (Zhang et al. 2013)	TLBO (Rao et al. 2011)	M-ES (Mezura-Montes and Coello 2005)	PESO (Zavala et al. 2005)	CDE (Becerra and Coello 2006)	CoDE (Huang et al. 2007)	ABC (Karaboga and Basturk 2007)
1	Best	**−15.0**	**−15.0**	**−15.0**	**−15.0**	**−15.0**	**−15.0**	**−15.0**	**−15.0**	**−15.0**
	Mean	**−15.0**	**−15.0**	**−15.0**	**−15.0**	**−15.0**	**−15.0**	**−15.0**	**−15.0**	**−15.0**
	Worst	**−15.0**	**−15.0**	**−15.0**	**−15.0**	**−15.0**	**−15.0**	**−15.0**	**−15.0**	**−15.0**
	SD	**0.000**	1.9e−6	–	–	–	–	–	–	–
	FE	75,000	**14,009**	20,875	25,000	240,000	350,000	100,100	2,48,000	240,000
2	Best	**1.000**	**1.000**	**1.001**	**1.000**	**1.000**	**1.005**	0.995	–	**1.000**
	Mean	**1.000**	**1.000**	0.992	**1.000**	**1.000**	**1.005**	0.789	–	**1.000**
	Worst	**1.000**	**1.000**	0.995	**1.000**	**1.000**	**1.005**	0.640	–	**1.000**
	SD	**0.000**	**0.000**	–	–	–	–	–	–	–
	FE	**25,000**	69,996	90,790	100,000	240,000	350,000	100,100	–	240.000
3	Best	**680.630**	**680.630**	**680.630**	**680.630**	680.632	**680.630**	**680.630**	680.771	680.634
	Mean	680.639	680.631	**680.630**	680.633	680.643	**680.630**	**680.630**	681.503	680.640
	Worst	680.651	680.633	**680.630**	680.638	680.719	**680.630**	**680.630**	685.144	680.653
	SD	**2.9e−2**	3.4e−3	–	–	–	–	–	–	–
	FE	**30,000**	30,019	32,586	100,000	240,000	350,000	100,100	240,000	100,000
4	Best	**7049.248**	**7049.248**	7049.257	**7049.248**	7051.903	7049.459	**7049.248**	–	7053.904
	Mean	7056.632	7050.261	7050.834	7083.673	7253.047	7099.101	**7049.248**	–	7224.407
	Worst	7087.620	7055.481	7063.406	7224.497	7099.101	7251.396	**7049.248**	–	7604.132
	SD	**2.6e−2**	2.8e−2	–	–	–	–	–	–	–
	FE	**99,000**	99,987	100,000	100,000	240,000	350,000	100,100	–	240,000
5	Best	**1.000**	**1.000**	**1.000**	**1.000**	**1.000**	**1.000**	**1.000**	**1.000**	**1.000**

(continued)

Table 3.17 (continued)

Problem		Jaya algorithm	Elitist TLBO (Rao and waghmare 2014)	DETPS (Zhang et al. 2013)	TLBO (Rao et al. 2011)	M-ES (Mezura-Montes and Coello 2005)	PESO (Zavala et al. 2005)	CDE (Becerra and Coello 2006)	CoDE (Huang et al. 2007)	ABC (Karaboga and Basturk 2007)
	Mean	**1.000**	**1.000**	**1.000**	**1.000**	**1.000**	**1.000**	**1.000**	**1.000**	**1.000**
	Worst	**1.000**	**1.000**	**1.000**	**1.000**	**1.000**	**1.000**	**1.000**	**1.000**	**1.000**
	SD	**0.000**	**0.000**	–	–	–	–	–	–	–
	FE	**5000**	5011	6,540	50,000	240,000	350,000	100,100	240,000	100,000

Elitist TLBO Elitist teaching-learning-based optimization; *DETPS* differential evolution algorithms and tissue P systems; *TLBO* teaching-learning-based optimization; *M-ES* multimembered evolution strategy; *PESO* particle evolutionary swarm optimization; *CDE* cultural differential evolution; *CoDE* co-evolutionary differential evolution; *ABC* artificial bee colony; *FE* function evaluations; – result is not available given by Jaya algorithm, elitist TLBO, M-ES, CoDE and ABC algorithms are inferior to that given by DETPS, PESO and CDE in terms of mean and worst solutions. But to obtain the global optimum solution DETPS, PESO and CDE needed more function evaluations than Jaya algorithm.

For problem 4, the results of Jaya algorithm are same as the results of elitist TLBO, TLBO and CDE algorithms in terms of the best solution but Jaya algorithm uses 99.01, 99 and 98.90% function evaluations to obtain competitive results compared to elitist TLBO, TLBO and CDE respectively. However, the solution given by CDE is better than that given by Jaya algorithm, elitist TLBO, DETPS, TLBO, M-ES, PESO, CoDE and ABC algorithms in terms of mean and worst solutions. But CDE requires more function evaluations than Jaya algorithm to obtain the global optimum solution. Jaya algorithm uses 99.01, 99, 99, 41.25, 28.28, 98.90 and 41.25% function evaluations to obtain competitive results compared to elitist TLBO, DETPS, TLBO, M-ES, PESO, CDE and ABC respectively.

For problem 5, Jaya algorithm finds the same global optimum solution as elitist TLBO, DETPS, TLBO, M-ES, PESO, CDE, CoDE and ABC algorithms but Jaya algorithm uses 99.78, 76.45, 10, 2.08, 1.42, 4.99, 2.08 and 5% function evaluations to obtain competitive results compared to elitist TLBO, DETPS, TLBO, M-ES, PESO, CDE, CoDE and ABC respectively.

Rao (2016) conducted computational experiments on 24 constrained benchmark functions given in CEC 2006 (Liang et al. 2006; Karaboga and Basturk 2007; Karaboga and Akay 2011) using Jaya algorithm. The objectives and constraints of the considered 24 benchmark functions of CEC 2006 have different characteristics such as linear, nonlinear, quadratic, cubic and polynomial. The number of design variables and their ranges are different for each problem. The results obtained by using the Jaya algorithm are compared with the results obtained by the other optimization algorithms such as homomorphous mapping (HM), adaptive segregational constraint handling evolutionary algorithm (ASCHEA), simple multi-membered evolution strategy (SMES), genetic algorithm (GA), particle swarm optimization (PSO), differential evolution (DE), artificial bee colony (ABC), biogeography based optimization (BBO) and elitist heat transfer search (HTS) algorithm available in the literature. A common platform is provided by maintaining the identical function evolutions for different algorithms considered for the comparison. Thus, the consistency in the comparison is maintained while comparing the performance of Jaya algorithm with other optimization algorithms. However, in general, the algorithm which requires less number of function evaluations to get the same best solution can be considered as better as compared to the other algorithms. However, to maintain the consistency in the comparison of competitive algorithms, a common experimental platform is provided by setting the maximum number of function evaluations as 240,000 for each benchmark function and the static penalty method is used as a constraint handling technique. Just like other algorithms, the proposed Jaya algorithm is executed 100 times for each benchmark function and the mean results obtained are compared with the other algorithms for the same number of runs.

In general, there is a need to conduct a number of trials of each algorithm considering different strategies with different combinations of control parameters (like population size and number of generations) even for the same number of function evaluations to arrive at a proper combination of the control parameters to effectively run the algorithm. The results of the Jaya algorithm for each benchmark function by employing appropriate population sizes while maintaining the function evaluations of 240,000 revealed that the Jaya algorithm has obtained global optimum values for all the benchmark functions except for G10, G14, G21, G22 and G23. However, in these cases also, the respective global optimum values obtained by Jaya algorithm is superior to the remaining algorithms. Furthermore, the 'mean' values of all the 24 benchmark functions obtained by Jaya algorithm are better than all other algorithms.

It may be concluded from the results of application of Jaya algorithm and its variants such as SAMP-Jaya algorithm and SAMPE-Jaya algorithm on the considered unconstrained and constrained optimization problems including the large scale problemsthat these algorithms are more powerful and more effective.

References

Andersson, M., Bandaru, S., Ng, A. H. C., & Syberfeldt, A. (2015). Parameter tuned CMA-ES on the CEC'15 expensive problems. In *IEEE Congress on Evolutionary Computation*. Japan: Sendai.

Becerra, R., & Coello, C. A. C. (2006). Cultured differential evolution for constrained optimization. *Computer Methods in Applied Mechanics and Engineering, 195,* 4303–4322.

Bergh, F. V., & Engelbrecht, A. P. (2004). A cooperative approach to particle swarm optimization. *IEEE Transactions on Evolutionary Computation, 8*(3), 225–239.

Cheng, R., & Jin, Y. (2015a). A Competitive swarm optimizer for large scale optimization. *IEEE Transactions on Cybernetics, 45*(2), 191–204.

Cheng, R., & Jin, Y. (2015b). A social learning particle swarm optimization algorithm for scalable optimization. *Information Sciences, 291,* 43–60.

Eberhart, R. C., & Kennedy, J. (1995). A new optimizer using particle swarm theory. *Sixth International Symposium on Micro machine and Human Science* (pp. 39–43). Japan: Nagoya.

Haupt, R. L., & Haupt, S. E. (2004). *Practical genetic algorithms*. Hoboken, New Jersey: Wiley.

Huang, F. Z., Wang, L., & He, Q. (2007). An effective co-evolutionary differential evolution for constrained optimization. *Applied Mathematical Computations, 186,* 340–356.

Joaquin, D., Salvador, G., Daniel, M., & Francisco, H. (2016). A practical tutorial on the use of nonparametric statistical tests as a methodology for comparing evolutionary and swarm intelligence algorithms. *Swarm and Evolutionary Computation, 1*(1), 3–18.

Karaboga, D., & Akay, B. (2011). A modified Artificial Bee Colony (ABC) algorithm for constrained optimization problems. *Applied Soft Computing, 11,* 3021–3031.

Karaboga, D., & Basturk, B. (2007). Artificial bee colony (ABC) optimization algorithm for solving constrained optimization problems. *LNAI 4529* (pp. 789–798). Berlin: Springer.

Liang, J. J., Runarsson, T. P., Mezura-Montes, E., Clerc, M., Suganthan, P. N., Coello, C. A. C., & Deb, K. (2006) *Problem definitions and evaluation criteria for the CEC 2006 special session on constrained real-parameter optimization*, Technical Report, Nanyang Technological University, Singapore. http://www.ntu.edu.sg/home/EPNSugan.

Liang, J. J., & Qin, A. K. (2006). Comprehensive learning particle swarm optimizer for global optimization of multimodal functions. *IEEE Transactions on Evolutionary Computation, 10* (3), 281–295.

Mendes, R., Kennedy, J., & Neves, J. (2004). The fully informed particle swarm: Simpler, may be better. *IEEE Transactions on Evolutionary Computation, 8*(3), 204–210.

Mezura-Montes, E., & Coello, C. A. C. (2006). A simple multi membered evolution strategy to solve constrained optimization problems. *IEEE Transactions on Evolutionary Computation, 9,* 1–17.

Ngo, T. T., Sadollah, A. J., & Kim, H. (2016). A cooperative particle swarm optimizer with stochastic movements for computationally expensive numerical optimization problems. *Journal of Computational Science, 13,* 68–82.

Nickabadi, A., Ebadzadeh, M. M., & Safabakhsh, R. (2011). A novel particle swarm optimization algorithm with adaptive inertia weight. *Applied Soft Computing, 11*(4), 3658–3670.

Oca, M. A., & Stutzle, T. (2009). Frankenstein's PSO: A composite particle swarm optimization algorithm. *IEEE Transactions on Evolutionary Computation, 13*(5), 1120–1132.

Rao, R. V. (2016). Jaya: A simple and new optimization algorithm for solving constrained and unconstrained optimization problems. *International Journal of Industrial Engineering Computations, 7*(1), 19–34.

Rao, R. V., & Saroj, A. (2017). A self-adaptive multi-population based Jaya algorithm for engineering optimization. *Swarm and Evolutionary Computation*. https://doi.org/10.1016/j.swevo.2017.04.008.

Rao, R. V., & Saroj, A. (2018). An elitism based self-adaptive multi-population Jaya algorithm and its applications. *Soft Computing*, 1–24. https://doi.org/10.1007/s00500-018-3095-z.

Rao, R. V., Savsani, V. J., & Vakharia, D. P. (2011). Teaching-learning-based optimization: A novel method for constrained mechanical design optimization problems. *Computer-Aided Design, 43*(3), 303–315.

Rao, R. V., & Waghmare, G. G. (2014). Complex constrained design optimization using an elitist teaching-learning-based optimization algorithm. *International Journal of Metaheuristics, 3*(1), 81–102.

Rao, R. V., & Waghmare, G. G. (2017). A new optimization algorithm for solving complex constrained design optimization problems. *Engineering Optimization, 49*(1), 60–83.

Rashedi, E., Nezamabadi-pour, H., & Saryazdi, S. (2009). GSA: A gravitational search algorithm. *Information Sciences, 179*(13), 2232–2248.

Wu, G. (2016). Across neighbourhood search for numerical optimization. *Information Sciences, 329,* 597–618.

Wu, G., Mallipeddi, R., Suganthan, P. N., Wang, R., & Chen, H. (2017). Differential evolution with multi-population based ensemble of mutation strategies. *Information Sciences, 329,* 329–345.

Zavala, A. E. M., Aguirre, A. H., & Diharce, E. R. V. (2005). Constrained optimization via evolutionary particle swarm optimization algorithm (PESO). *Proc* (pp. 209–216). Washington D.C.: GECCO.

Zhang, G., Cheng, J., Gheorghe, M., & Meng, Q. (2013). A hybrid approach based on differential evolution and tissue membrane systems for solving constrained manufacturing parameter optimization problems. *Applied Soft Computing, 13*(3), 1528–1542.

Chapter 4
Single- and Multi-objective Design Optimization of Heat Exchangers Using Jaya Algorithm and Its Variants

Abstract This chapter presents design optimization case studies of shell-and-tube and plate-fin heat exchangers. The single objective and multi-objective design optimization case studies are solved by the Jaya algorithm and its variants such as self-adaptive Jaya, SAMP-Jaya and SAMPE-Jaya. The results of application of Jaya algorithm and its variants are compared with those of the other state-of-the-art optimization algorithms and the performance supremacy of the Jaya algorithm and its variants is established.

4.1 Heat Exchangers

A heat exchanger is a heat-transfer devise that is used for transfer of internal thermal energy between two or more fluids available at different temperatures. Heat exchangers are used as process equipment in the process, power, petroleum, transportation, air conditioning, refrigeration, cryogenic, heat recovery, alternate fuels, and other industries. Common examples of heat exchangers familiar to us in day-to-day use are automobile radiators, condensers, evaporators, air re-heaters, and oil coolers (Shah and Sekulic 2003). The design of a heat exchanger involves a consideration of both the heat transfer rates between the fluids and the mechanical pumping power expended to overcome fluid friction and move the fluids through the heat exchanger. The heat transfer rate per unit of surface area can be increased by increasing fluid-flow velocity and this rate varies as something less than the first power of the velocity. The friction power expenditure is also increases with flow velocity, but in this case the power varies by as much as the cube of the velocity and never less than the square. It is this behavior that allows the designer to match both heat transfer rate and friction (pressure-drop) specifications, and it is this behavior that dictates many of the characteristics of different classes of heat exchanger.

R. Venkata Rao, *Jaya: An Advanced Optimization Algorithm and its Engineering Applications*, https://doi.org/10.1007/978-3-319-78922-4_4

4.1.1 Shell-and-Tube Heat Exchanger

Shell-and-tube heat exchangers consist of a series of tubes. One set of these tubes contains the fluid that must be either heated or cooled. The second fluid runs over the tubes that are being heated or cooled so that it can either provide the heat or absorb the heat required. A set of tubes is called the tube bundle and can be made up of several types of tubes: plain, longitudinally finned, etc. Shell-and-tube heat exchangers are typically used for high pressure applications (with pressures greater than 30 bar and temperatures greater than 260 °C). This is because the shell-and-tube heat exchangers are robust due to their shape. Figure 4.1 gives the idea about the construction of the STHE.

4.1.2 Plate-Fin Heat Exchanger

Plate-fin heat exchanger (PFHE) is a compact heat exchanger which is widely used for the thermal systems. PFHE is used to recover the thermal energy between two fluids available at different temperatures. The design of heat exchanger includes many geometric and operating parameters for a heat exchanger geometry which fulfils the thermal energy demand within the given constraints. These heat exchangers are used as process equipment in the industries of petroleum, petrochemical, chemical and power generation. Extended surfaces or fin elements are introduced for increasing the heat transfer. Fin height, fin pitch, fin offset length, hot and cold stream flow length and non-flow length are the major parameters of the PFHE. The selection of proper combination of these parameters depends upon the pressure drops, temperatures, thermal stresses, and dynamic properties of fluids. A designer applies his/her cognition, experience, and judgment for assigning the parameters and designing an effective heat exchanger. However, it is difficult even

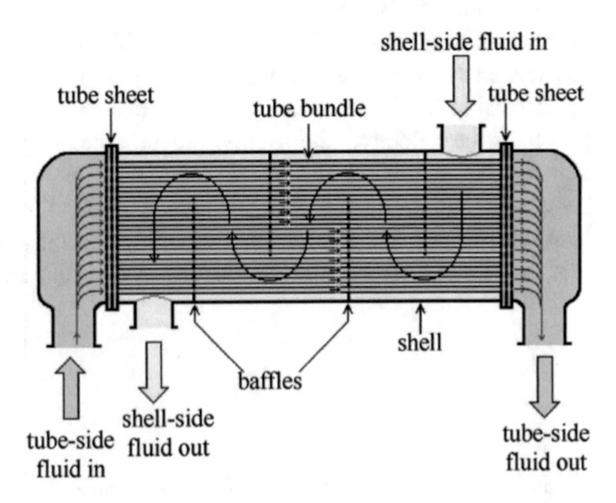

Fig. 4.1 A shell-and-tube heat exchanger

for the experienced designer to consider all these parameters together due to the size and complexity of the designing task.

Various efforts are made by the researchers for developing the systematic design approaches for seeking the best possible heat exchanger that provides the optimum heat duty while meeting a set of specified constraints. In the pursuit of improved designs, much research has been carried out with the objective functions aimed at decreasing total cost and heat transfer area. In recent years the application of advanced optimization algorithms for design problems of PFHE has gained much momentum.

A PFHE is said to be a compact if it incorporates a heat transfer surface having a surface area density greater that about 700 m^2/m^3 or a hydraulic diameter is 5 mm for operating in a gas or air stream and 400 m^2/m^3 or higher for operating in a liquid or phase changer stream. A plate-fin heat exchanger is a type of compact exchanger that consists of a stack of alternate flat plates called parting sheets and corrugated fins brazed together as a block (Shah and Sekulic 2003). Streams exchange heat by flowing along the passages made by the fins between the parting sheets. The fins serve both as a secondary heat transfer surface and mechanical support for the internal pressure between layers. Figure 4.2 gives the idea about the construction of the PFHE.

The core of matrix, parting sheets, and side bars is laid up by hand, clamped firmly in a jig, and then brazed into a rigid structure, either in a molten salt bath or in a vacuum furnace. Due to such type of processes, a strong rigid structure of extremely high volumetric heat transfer surface density is achieved. However, the units are limited in overall size and materials of construction and cannot be cleaned mechanically. The most common area of application is in cryogenic processing such as liquefied natural gas production, hydrogen purification and helium separation and liquefaction. Plate-fin exchangers are manufactured in virtually all shapes and sizes and are made from a variety of materials.

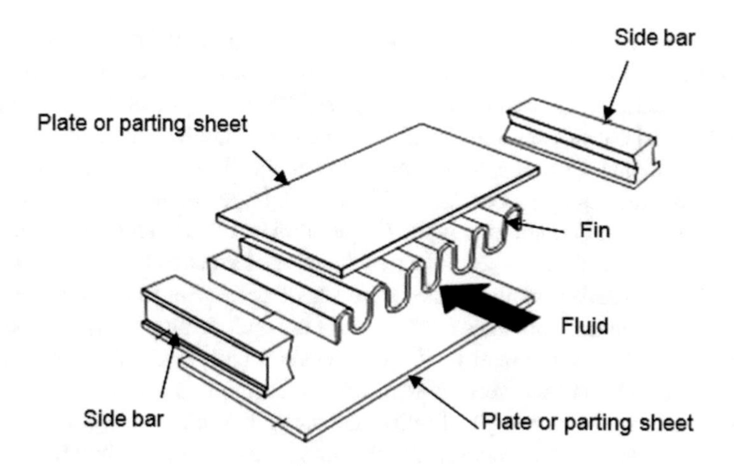

Fig. 4.2 Basic components of a plate-fin heat exchanger

In recent years, much literature has appeared on optimization of heat exchangers. Reneaume and Niclout (2003) had used simulated annealing (SA) and branch and bound algorithms to solve corresponding optimization problem of PFHE. Jorge et al. (2004) presented an optimization algorithm for the configuration design of plate heat exchangers (PHEs). Firstly, the configuration was represented by a set of six distinct parameters and a methodology to detect equivalent configurations was presented. The problem of optimizing the PFHE configuration was formulated as the minimization of the heat transfer area. Peng and Ling (2008) presented the application of genetic algorithm combined with back propagation neural networks for the optimal design of PFHE. The major objectives in the PFHE design considered were the minimum total weight and total annual cost for a given constrained conditions.

Yu et al. (2008) presented the multi-objective optimal design of PFHE by considering fuzzy optimization technology by particle swarm optimization. The minimum weight, higher heat efficiency and reasonable pressure loss were considered as multi-objectives of PFHE to be optimized. Xie et al. (2008) had presented application of a genetic algorithm for optimization of a plate–fin compact heat exchanger. A generalized procedure had developed to carry out the optimization to find the minimum volume and/or annual cost and pressure drop of the heat exchangers, respectively, based on the ε-NTU and the genetic algorithm technique. Caputo et al. (2008) used GA for cost optimization of STHS. Han et al. (2008) used particle swarm optimization (PSO) algorithm for rolling fin-tube heat exchanger optimization. Miyazaki and Akisawa (2009) applied PSO to obtain the optimum cycle time of single stage adsorption chiller. The authors had investigated the influences of heat exchanger parameters, such as heat capacity and NTU, on the optimum performance of an adsorption chiller. Hosseini and Ceylan (2009) obtained the heat transfer coefficient and pressure drop on the shell side of a STHE experimentally for three different types of copper tubes (smooth, corrugated and with micro-fins). Also, experimental data has been compared with theoretical data available.

Mishra and Das (2009) used GA for optimal design of PFHE. The authors had considered minimization of total annual cost and total thermo economic cost as an objective function. Ravagnani et al. (2009) used PSO for optimal design of STHEs. The author had considered minimization of area or minimization of cost as per the availability of data. Mishra et al. (2009) used genetic algorithm to carry out second law based optimization of cross flow PFHE having large number of design variables of both discrete and continuous type. The authors had considered minimization of entropy generation units as an objective function. Lemouedda et al. (2010) presented an optimization of the angle of attack of delta-winglets in an elementary model of plate-fin-and-tube heat exchangers. This optimization study was based on the strategy of Pareto optimality. The optimization process consisted of a combination of a CFD analysis, genetic algorithm and appropriate approximation techniques. In this study, investigated inline and staggered tube arrangements.

Patel and Rao (2010) proposed the use of a non-traditional optimization technique; called particle swarm optimization (PSO), for design optimization of STHE s from economic view point. Minimization of total annual cost was considered as an objective function. Rao and Patel (2010) demonstrated the use of particle swarm optimization (PSO) algorithm for thermodynamic optimization of a cross flow plate-fin heat exchanger. Minimization of total volume, and minimization of total annual cost were considered as an objective functions and were treated individually. Sanaye and Hajabdollahi (2010) presented thermal modeling and optimal design of compact heat exchangers. Non-dominated sorting genetic-algorithm (NSGA-II) was applied to obtain the maximum effectiveness and the minimum total annual cost. Najafi et al. (2011) presented a multi-objective genetic algorithm applied to plate-fin heat exchanger. Air is considered as its working fluid on both sides. Geometric parameters of PFHE like cold and hot stream flow length, fin height, fin frequency, fin thickness, fin lance length and number of fin layers on hot side are considered as design variables. Given geometric parameters were optimized for minimum total annual cost and maximum heat transfer rate. Sahin et al. (2011) used artificial bee colony (ABC) algorithm for design optimization of STHS in economic point of view.

Rao and Patel (2013) used modified TLBO for multi-objective optimization of heat exchangers. Plate-fin heat exchanger and STHE are considered for the optimization. Maximization of heat exchanger effectiveness and minimization of total cost of the exchanger are considered as the objective functions. Hadidi and Nazari (2013a, b) used biogeography-based optimization (BBO) and imperialist competitive algorithm (ICA), which yielded improved quality solutions and with more choices for the design optimization of STHE with respect to traditional methods.

Turgut et al. (2014) achieved the economic optimization of the STHE design using intelligent tuned harmony search (ITHS) algorithm and improved ITHS (IITHS) algorithm. Kang et al. (2015) used a multi-objective optimization method of heat exchanger network (HEN) on the basis of partition of minimum approach temperature intervals to minimize the total annual cost and accumulated CO_2 emissions in construction and operation phases. Cavazzuti et al. (2015) presented in which a finned concentric pipes heat exchanger simulated by using CFD, and optimized by the Nelder and Mead simplex downhill optimization algorithm. Mohanty (2016) had used Firefly algorithm for economic optimization of shell-and-tube heat exchanger.

Lee and Kim (2015) optimized to enhance heat transfer, and also to reduce the pressure drop based on three-dimensional Reynolds averaged Naviere-Stokes analysis and a multi-objective genetic algorithm with surrogate modeling. Two objective functions related to heat transfer and pressure drop in the cold channels were optimized simultaneously. Du et al. (2015) applied Taguchi method to investigate the influence of five geometric parameters on the comprehensive performance of overlapped helical baffled heat exchangers. Kang et al. (2015) proposed a multi-objective optimization model on a HEN retrofit with a heat pump. The objectives of optimization were minimizing the total annual cost for the retrofit and maximizing the total annual CO_2 emission reduction simultaneously. Wang and Li (2015) reviewed the layer pattern thermal design and optimization of the multi-stream PFHE.

Ayala et al. (2016) proposed multi-objective fee search approach combined with differential evaluation (MOFSDE) method had to solve the PFHE and the STHE multi-objective problems. Improvement in the results is observed using the MOFSDE algorithm as compared to the NSGA-II approach showing the improvement potential of the proposed algorithm for such thermodynamic optimization. Gharbi et al. (2015) conducted a two-dimensional numerical CFD model using finite volume discretization to evaluate the performance of heat exchangers with circular and non-circular shapes.

Turgut and Çoban (2016) investigated the thermal design of PFHE by using Hybrid Chaotic Quantum behaved Particle Swarm Optimization (HCQPSO) algorithm. Wong et al. (2016) used Excel-based multi-objective optimization (EMOO) program, based on the elitist non dominated sorting genetic algorithm with different termination criterion for design and optimization of STHEs. Rao and Saroj (2017a, b, c, d) presented the results of application of Jaya algorithm and its variants for the design optimization of shell-and-tube and plate-fin heat exchangers.

4.2 Modeling of Heat Exchangers and the Case Studies

4.2.1 Modeling of Shell-and-Tube Heat Exchanger

According to flow regime, the tube side heat transfer coefficient (h_t) is computed from following correlation (Caputo et al. 2008).

$$h_t = \frac{k_t}{d_i}\left[3.657 + \frac{0.067\left(Re_t Pr_t\left(\frac{d_i}{L}\right)\right)^{1.33\frac{1}{3}}}{1 + 0.1Pr_t\left(Re_t\left(\frac{d_i}{L}\right)\right)^{0.3}}\right] \tag{4.1}$$

(If $Re_t < 2{,}300$),

$$h_t = \frac{k_t}{d_i}\left[\frac{\left(\frac{f_t}{8}\right)(Re_t - 1{,}000)Pr_t}{1 + 12.7(f_t/8)^{1/2}\left(Pr_t^{\frac{2}{3}} - 1\right)}\left(1 + \frac{d_i}{L}\right)^{0.67}\right] \tag{4.2}$$

(If $2{,}300 < Re_t < 10{,}000$),

$$h_t = 0.027\frac{k_t}{d_0}Re_t^{0.8}Pr_t^{\frac{1}{3}}\left(\frac{\mu_t}{\mu_{wt}}\right)^{0.14} \tag{4.3}$$

(if $Re_t > 10{,}000$),

Where, *ft* is the Darcy friction factor and is given as,

$$f_t = (1.82 \log 10^{Re_t} - 1.64)^{-2} \tag{4.4}$$

Re_t is the tube side Reynolds number and given by,

$$Re_t = \frac{\rho_t v_t d_i}{\mu_t} \tag{4.5}$$

Flow velocity for tube side is found by,

$$v_t = \left(\frac{m_t}{\left(\frac{\pi}{4}\right) d_t^2 \rho_t}\right)\left(\frac{n}{N_t}\right) \tag{4.6}$$

N_t is the number of tubes and n is the number of tube passes which can be found approximately from the following equation (Incropera and DeWitt 1996; Shah and Bell 2009)

$$N_t = C\left(\frac{D_s}{d_0}\right)^{n_1} \tag{4.7}$$

where, C and n_1 are coefficients that are taking values according to low arrangement and number of passes. These coefficients are shown in table for different flow arrangements. Pr_t is the tube side Prandtl number and given by,

$$Pr_t = \frac{\mu_t C_{pt}}{k_t} \tag{4.8}$$

Also, $d_i = 0.8 d_o$

Kern's formulation for segmental baffle shell-and-tube exchanger is used for computing shell side heat transfer coefficient h_s is given by (Kern 1950)

$$h_s = 0.36 \frac{k_t}{d_e} Re_s^{0.55} Pr_s^{\frac{1}{3}} \left(\frac{\mu_s}{\mu_{wts}}\right)^{0.14} \tag{4.9}$$

where, d_e is the shell hydraulic diameter and computed as,

$$d_e = \frac{4\left(S_t^2 - \left(\frac{\pi d_o^2}{4}\right)\right)}{\pi d_o} \tag{4.10}$$

(For square pitch)

$$d_e = \frac{4\left(0.43S_t^2 - \left(\frac{0.5\pi d_o^2}{4}\right)\right)}{0.5\pi d_o} \tag{4.11}$$

(For triangular pitch)
Cross section area normal to flow direction is determined by (Selbas et al. 2006)

$$A_s = D_s B\left(1 - \frac{d_o}{S_t}\right) \tag{4.12}$$

The standard values for C and n_1 are selected from Table 4.9 (Shah and Bell 2009) (Table 4.1).

Flow velocity for the shell side can be obtained from,

$$v_s = \frac{m_s}{\rho_s A_s} \tag{4.13}$$

Reynolds number for shell side is,

$$Re_s = \frac{m_s d_e}{A_s \mu_s} \tag{4.14}$$

Prandtl number for shell side follows,

$$Pr_s = \frac{\mu_s C_{ps}}{k_s} \tag{4.15}$$

The overall heat transfer coefficient (U) depends on both the tube side and shell side heat transfer coefficients and fouling resistances are given by (Selbas et al. 2006)

$$U = \frac{1}{\left(\frac{1}{h_s}\right) + R_{fs} + (d_o/d_i)\left(R_{ft} + \left(\frac{1}{h_t}\right)\right)} \tag{4.16}$$

Table 4.1 Standard pitch designation

No. of passes	Triangle tube pitch $S_t = 1.25d_o$		Square tube pitch $S_t = 1.25d_o$	
	C	n_1	C	n_1
1	0.319	2.142	0.125	2.207
2	0.249	2.207	0.156	2.291
4	0.175	2.285	0.158	2.263
6	0.0743	2.499	0.0402	2.617
8	0.0365	2.675	0.0331	2.643

Considering the cross flow between adjacent baffle, the logarithmic mean temperature difference (LMTD) is determined by,

$$\mathrm{LMTD} = \frac{(T_{hi} - T_{co}) - (T_{ho} - T_{ci})}{\ln((T_{hi} - T_{co}) - (T_{ho} - T_{ci}))} \tag{4.17}$$

The correction factor F for the flow configuration involved is found as a function of dimensionless temperature ratio for most flow configuration of interest (Edwards 2008).

$$F = \sqrt{\frac{R^2+1}{R-1}} \frac{\ln\left(\frac{1-P}{1-PR}\right)}{\ln\left(\left(2 - PR + 1 - \sqrt{R^2+1}\right)/\left(2 - PR + 1 + \sqrt{R^2+1}\right)\right)} \tag{4.18}$$

where R is the correction coefficient given by,

$$R = \frac{(T_{hi} - T_{ho})}{(T_{co} - T_{ci})} \tag{4.19}$$

P is the efficiency given by,

$$P = \frac{(T_{co} - T_{ci})}{(T_{hi} - T_{ci})} \tag{4.20}$$

Considering overall heat transfer coefficient, the heat exchanger surface area (A) is computed by,

$$A = \frac{Q}{\mathrm{UFLMTD}} \tag{4.21}$$

For sensible heat transfer, the heat transfer rate is given by,

$$Q = m_h C_{ph}(T_{hi} - T_{ho}) = m_c C_{pc}(T_{co} - T_{ci}) \tag{4.22}$$

Based on total heat exchanger surface area (*A*), the necessary tube length (*L*) is,

$$L = \frac{A}{\pi d_o N_t} \tag{4.23}$$

Tube side pressure drop includes distributed pressure drop along the tube length and concentrated pressure losses in elbows and in the inlet and outlet nozzle.

$$\Delta P_t = \Delta P_{tubelength} + \Delta P_{tubeelbow} \tag{4.24}$$

$$\Delta P_t = \frac{\rho_s v_t^2}{2}\left(\frac{L}{d_i} f_t + p\right) n \tag{4.25}$$

Different values of constant p are considered by different authors (Kern 1950) assumed p = 4, while assumed p = 2.5.

The shell side pressure drop is,

$$\Delta P_s = f_s \left(\frac{\rho_s v_s^2}{2}\right)\left(\frac{L}{B}\right)\left(\frac{D_s}{d_e}\right) \tag{4.26}$$

where,

$$f_s = 2 b_o Re_s^{-0.15} \tag{4.27}$$

And b_o = 0.72 valid for Re_s < 40,000

$$P = \frac{1}{\eta}\left(\frac{m_t}{\sigma_t}\Delta P_t + \frac{m_s}{\sigma_s}\Delta P_s\right) \tag{4.28}$$

The objective function total cost C_{tot}, includes capital investment (C_i), energy cost (C_e), annual operating cost (C_o) and total discounted operating cost (C_{od}) is given by (Caputo et al. 2008),

$$C_{tot} = C_i + C_{od} \tag{4.29}$$

Adopting Hall's correlation (Taal et al. 2003) the capital investment C_i is computed as a function of the exchanger surface area.

$$C_i = a_1 + a_2 * A^{a_3} \tag{4.30}$$

where a_1 = 8,000, a_2 = 259.2 and a_3 = 0.93 for exchanger made with stainless steel for shell-and-tubes (Taal et al. 2003).

The total discounted operating cost related to pumping power to overcome friction losses is computed from the following equation,

$$C_o = P C_e H \tag{4.31}$$

$$C_{od} = \sum_{x=1}^{ny} \frac{C_o}{(1+i)^x} \tag{4.32}$$

Based on all above calculations, total cost is computed from Eq. (4.29). The procedure is repeated computing new value of exchanger area (A), exchanger length (L), total cost (C_{tot}) and a corresponding exchanger architecture meeting the specifications. Each time the optimization algorithm changes the values of the design variables d_o, D_s and B in an attempt to minimize the objective function.

4.2.1.1 Case Study 1: 4.34 (MW) Duty Methanol–Brackish Water Exchanger

Table 4.2 shows input parameters and physical properties for 4.34 (MW) duty Methanol-Brackish water heat exchanger (Sinnot et al. 1996). Sea water is allocated to tube side as the mass flow rate of sea water is much higher compared to methanol. Also it is easy to clean tubes from sludge by chemical wash. The procedure explained in the mathematical model is used for calculating other geometric parameters, pressure drops on both shell-and-tube side and overall heat transfer coefficient of heat exchanger. The shell side inside diameter is not more than 1.5 m, the tubes outer diameter ranges from 0.015 to 0.15 m and the baffle spacing should not be more than 0.5 m. The pressure drops and calculated heat transfer is used to find total annual operating and overhead costs. *do*, *Ds* and *B* are considered as decision variables. The process input and physical properties are as follows.

Tube side and shell side velocities and the ratio of baffle spacing to shell side inner diameter (R_{bs}) are considered as constraints. Tube side velocities for water and similar fluids ranges from 0.5 to 2.5 m/s, shell side velocities generally ranges from 0.2 to 1.5 m/s, and R_{bs} ranges from 0.2 to 1. Other design parameters like *Nt*, *L* and *di* are outcome of mathematical model-1 based of *do*, *Ds and B*. Given specifications are as follows:

- All the values of discounted operating costs are computed with $ny = 10$ year.
- Annual discount rate $(i) = 10\%$.
- Energy cost $(Ce) = 0.12$ €/kW h.
- An annual amount of work hours $H = 7{,}000$ h/year.

Minimization of total annual cost is considered as objective. The same model were previously attempted by different researchers using GA (Caputo et al. 2008), PSO (Rao and Patel 2010), ABC (Sahin et al. 2011), ICA (Hadidi et al. 2013a), BBO (Hadidi et al. 2013b), CSA (Asadi et al. 2014), FFA (Mohanty 2016), ITHS (Turgut et al. 2014), I-ITHS (Turgut et al. 2014), CI (Dhavale et al. 2016). The same configuration is retained in the present approach. Table 4.3 shows the results of optimal heat exchanger geometry using different optimization methods. Result

Table 4.2 Process parameters and physical properties for case study 4.2.1.1

Fluid allocation	Shell side	Tube side
Fluid	Methanol	Sea water
Mass flow rate (kg/s)	27.80	68.90
Inlet temperature (°C)	95.00	25.00
Outlet temperature (°C)	40.00	40.00
Density (kg/m^3)	750	995
Specific heat (kJ/(kg K)	2.84	4.2
Viscosity (Pa s)	0.00034	0.0008
Thermal conductivity (W/m K)	0.19	0.59
Fouling factor (m^2 K/W)	0.00033	0.0002

Table 4.3 Optimal heat exchanger geometry using different optimization methods for case study 4.2.1.1

Parameter	Original design	GA	PSO	ABC	ICA	BBO	CSA	FFA	ITHS	I-ITHS	CI	TLBO	Jaya	Self-adaptive Jaya
D_s (m)	*0.894*	0.830	0.810	1.3905	0.879	0.801	0.826	0.858	0.7620	0.7635	0.7800	0.858	0.7820	0.7820
L (m)	*4.830*	3.379	3.115	3.963	3.107	2.040	2.332	2.416	2.0791	2.0391	1.9363	2.416	2.4487	2.4487
B (m)	*0.356*	0.500	0.424	0.4669	0.500	0.500	0.414	0.402	0.4988	0.4955	0.5	0.402	0.4912	0.4912
d_o (m)	0.020	0.016	0.015	0.0104	0.015	0.010	0.0151	0.01575	0.0101	0.0100	0.10	0.01575	0.015	0.015
P_t (m)	0.025	0.020	0.0187	–	0.01875	0.0125	0.0187	0.01968	0.1264	0.0125	0.0125	0.01968	0.019	0.019
N_t	918	1567	1,658	1,528	1,752	3,587	1,754	1,692	3,454	3,558	3734.1233	1,692	1,532	1,532
v_t (m/s)	0.75	0.69	0.67	0.36	0.699	0.77	0.65	0.656	0.7820	0.7744	0.7381	0.656	0.7994	0.7994
Re_t	14,925	10,936	10,503	–	10,429	7,643	10,031	10,286	7,842.52	7,701.29	7,342.7474	10,286	11,931	11,931
Pr_t	5.7	5.7	5.7	–	5.7	5.7	5.7	5.7	5.700	5.700	5.6949	5.7	5.7	5.7
h_t (W/m^2 K)	3,812	3,762	3,721	3,818	3,864	4,314	6,104	6,228	4,415.918	4,388.79	4,584.7085	6,228	3,668	3,668
f_t	0.028	0.031	0.0311	–	0.031	0.034	0.032	0.03119	0.03540	0.03555	0.0343	0.03119	0.021	0.021
ΔP_t (Pa)	6,251	4,298	4,171	3,043	5,122	6,156	4,186	4,246	6,998.70	6,887.63	5,862.7287	4,246	4,755	4,755
d_e (m)	0.014	0.011	0.0107	–	0.011	0.007	0.0107	0.0105	0.00719	0.00711	0.0071	0.0105	0.0106	0.0106
v_s (m/s)	0.58	0.44	0.53	0.118	0.42	0.46	0.56	0.54	0.48755	0.48979	0.4752	0.54	0.481	0.481
Re_s	18,381	11,075	12,678	–	9,917	7,254	13,716	12,625	7,736.89	7,684.054	7,451.39	12,625	11,351	11,351
Pr_s	5.1	5.1	5.1	–	5.1	5.1	5.1	5.1	5.08215	5.08215	5.0821	5.1	5.1	5.1
h_s (W/m^2 K)	1,573	1,740	1,950.8	3,396	1,740	2,197	2,083	1,991	2,213.89	2,230.913	2,195.9461	1,991	1,936	1,936
f_s	0.330	0.357	0.349	–	0.362	0.379	0.351	0.349	0.3759	0.37621	0.3780	0.349	0.2978	0.2978
ΔP_s (Pa)	35,789	13,267	20,551	8,390	12,367	13,799	22,534	18,788	14,794.94	14,953.91	13,608.44	18,788	19,178	19,178
U (W/m^2 K)	615	660	713.9	832	677	755	848.2	876.4	760.594	761.578	764.50	876.4	726	726
A (m^2)	278.6	262.8	243.2	–	256.6	229.95	209.1	202.3	228.32	228.03	227.16	202.3	202	202
C_i (V)	51,507	49,259	46,453	44,559	48,370	44,536	40,343.7	39,336	44,301.66	44,259.01	44,132.51	39,336	39,741	39,741
C_o (V/year)	2,111	947	1,038.7	1,014.5	975	984	1,174.4	1,040	964.164	962.4858	955.91	1,040	661.67	661.67
C_{od} (V)	12,973	5,818	6,778.2	6,233.8	5,995	6,046	7,281.4	6,446	5,924.373	5,914.058	5,873.66	6,446	4,928.5	4,928.5
C_{tot} (V)	64,480	55,077	53,231	50,793	543,66	50,582	47,625.1	45,782	50,226	50,173	50,006.17	45,782	**44,669.5**	**44,669.5**

GA (Caputo et al. 2008); PSO (Rao and Patel 2010); ABC (Sahin et al. 2011); ICA (Hadidi et al. 2013a); BBO (Hadidi et al. 2013b); CSA (Asadi et al. 2014); FFA (Mohanty 2016); ITHS & I–ITHS (Turgut et al. 2014) and CI (Dhavale et al. 2016). Bold value indicates the best results show that a significant reduction in the number of tubes and increase in the tube side flow velocity which consecutively reduces the tube side heat transfer coefficient. The combined effect of capital investment and operating costs led to a reduction of the total cost of about 18.8, 16.0, 12.05, 17, 11.68, 6.20, 2.42, 11.06, 10.96, 10.6, and 2.42% as compared with those given by GA, PSO, ABC, ICA, BBO, CSA, FFA, ITHS, I-ITHS, CI and TLBO approach, respectively.

For the same case study, the SAMPE-Jaya algorithm is also applied and it has obtained a reduced total cost of $42,015.98in comparison to the other algorithms. Total cost of the STHE is reduced by 34.83, 23.71, 21.06, 17.28, 16.93, 11.77, 8.22, 16.34, 16.25, 15.97, 8.22, 5.93% as compared to the results of original design, GA, PSO, ABC, ICA, BBO, CSA, FFA, ITHS, I-ITHS, CI, TLBO and Jaya algorithms. The surface area required for the heat transfer is reduced by 39.58, 36.24, 31.10, 34.69, 27.13, 19.86, 17.17, 26.61, 26.51, 26.23, 17.17 and 17.04% as compared to the results of original design, GA, PSO, ICA, BBO, CSA, FFA, ITHS, I-ITHS, CI, TLBO and Jaya algorithms. This reduces the value of C_i of the STHE layout. The operational cost of the design using SAMPE-Jaya algorithm is decreased by 65.34, 22.72, 33.66, 27.87, 25, 25.63, 38.25, 30.2, 24.10, 23.97, 23.45, 30.25 and 8.77% as compared to the results of original design, GA, PSO, ABC, ICA, BBO, CSA, FFA, ITHS, I-ITHS, CI, TLBO and Jaya algorithms.

4.2.1.2 Case Study 2: 1.44 (MW) Duty Kerosene-Crude Oil Heat Exchanger

The case study is based on the design problem discussed by Sinnot et al. (1996) and described in Table 4.4. The procedure explained in the mathematical model is used for calculating other geometric parameters, pressure drops on both shell-and-tube side and overall heat transfer coefficient of heat exchanger. The shell side inside diameter is not more than 1.5 m, the tubes outer diameter ranges from 0.051 to 0.081 m and the baffle spacing should not be more than 0.5 m. Tube side and shell side velocities and the ratio of baffle spacing to shell side inner diameter (R_{bs}) are considered as constraints. Tube side velocities for water and similar fluids ranges from 0.5 to 2.5 m/s, shell side velocities generally ranges from 0.2 to 1.5 m/s, and R_{bs} ranges from 0.2 to 1. The process input and physical properties are given in Table 4.4. Given specifications are as follows:

- All the values of discounted operating costs are computed with ny = 10 year.
- Annual discount rate (i) = 10%.
- Energy cost (Ce) = 0.12 €/kW h.
- An annual amount of work hours H = 7,000 h/year.

The best results obtained by Jaya algorithm for this case study are compared with the literature results using GA (Caputo et al. 2008), PSO (Rao and Patel 2010), ABC (Sahin et al. 2011), ICA (Hadidi et al. 2013a), BBO (Hadidi et al. 2013b), CSA (Asadi et al. 2014), FFA (Mohanty 2016), ITHS (Turgut et al. 2014), I-ITHS (Turgut et al. 2014), CI (Dhavale et al. 2016) and the results are presented in

Table 4.4 Process parameters and physical properties for case study 4.2.1.2

Fluid location	Shell side	Tube side
Fluid	Kerosene	Crude oil
Mass flow rate (kg/s)	5.52	18.80
Inlet temperature ($^\circ_C$)	199.00	37.80
Outlet temperature ($^\circ_C$)	93.30	76.70
Density (kg/m^3)	850	995
Specific heat (kJ/(kg K)	2.47	2.05
Viscosity (Pa s)	0.0004	0.00358
Thermal conductivity (W/m K)	0.13	0.13
Fouling factor (m^2 K/W)	0.00061	0.00061

Table 4.5. Two tube side passages (triangle pitch pattern) and one shell side passage are assumed for this exchanger. The shell side and tube side pressure losses are increased and as a result the operating expenses are increased. The reduction of the total cost obtained by Jaya and self-adaptive Jaya algorithm is about 13.86, 12.2, 13.5, 13.5, 11.7,11.5, 10.01, 13.4, 13.3, 9.38, and 10.01% compared to that given by literature, GA, PSO, ABC, ICA, BBO, CSA, FFA, ITHS, I-ITHS, CI and TLBO approaches, respectively because of reduction in heat exchanger area.

4.2.1.3 Case Study 3: 0.46 (MW) Duty Distilled Water-Raw Water Exchanger

The case study is based on the design problem discussed in Kern (1950), described in Table 4.6. The procedure explained in the mathematical model is used for calculating other geometric parameters, pressure drops on both shell-and-tube side and overall heat transfer coefficient of heat exchanger. The shell side inside diameter is not more than 1.5 m, the tubes outer diameter ranges from 0.015 to 0.61 m and the baffle spacing should not be more than 0.5 m. Tube side and shell side velocities and the ratio of baffle spacing to shell side inner diameter (R_{bs}) are considered as constraints. Tube side velocities for water and similar fluids ranges from 0.5 to 2.5 m/s, shell side velocities generally ranges from 0.2 to 1.5 m/s, and R_{bs} ranges from 0.2 to 1.

Given specifications are as follows:

- All the values of discounted operating costs are computed with ny = 10 year.
- Annual discount rate (i) = 10%.
- Energy cost (Ce) = 0.12 €/kW h.
- An annual amount of work hours H = 7,000 h/year.

The process input and physical properties are given in Table 4.6.

The original design proposed by Caputo et al. (2008) using GA approach assumed a heat exchanger with four tube side passages (with square pitch pattern) and one shell side passage. The same configuration is retained in the present approach. Table 4.7 shows the results of optimal heat exchanger geometry using

Table 4.5 Optimal heat exchanger geometry using different optimization methods for case study 4.2.1.2

Parameter	Original design	GA	PSO	ABC	ICA	BBO	CSA	FFA	ITHS	I-ITHS	CI	TLBO	Jaya	Self-adaptive Jaya
D_s (m)	0.539	0.63	0.59	0.3293	0.3293	0.74	0.63	0.7276	0.32079	0.31619	0.4580	0.7276	0.4003	0.4003
L (m)	4.88	2.153	1.56	3.6468	3.6468	1.199	1.412	1.64	5.15184	5.06235	1.3833	1.64	2.44	2.44
B (m)	0.127	0.12	0.1112	0.0924	0.0924	0.1066	0.1118	0.1054	0.24725	0.24147	0.125	0.1054	0.2171	0.2171
d_o (m)	0.025	0.02	0.015	0.0105	0.0105	0.015	0.015	0.01575	0.01204	0.01171	0.010	0.01575	0.015	0.015
P_t (m)	0.031	0.025	0.0187	–	–	0.0188	0.0187	0.01968	0.01505	0.01464	0.0125	0.01968	0.0245	0.0245
N_t	158	391	646	511	511	1,061	876	924	301	309	1,152.88	924	350	350
v_t (m/s)	1.44	0.87	0.93	0.43	0.43	0.69	0.69	0.677	0.8615	0.8871	0.6522	0.677	0.9547	0.9547
Re_t	8,227	4,068	3,283	–	–	2,298	2,528	2,408	2,306.77	2,303.46	1,450.017	2,408	3,284	3,284
Pr_t	55.2	55.2	55.2	–	–	55.2	55.2	55.2	56.4538	56.4538	56.4538	55.2	55.2	55.2
h_t (W/m^2 K)	619	1,168	1,205	2,186	2,186	1,251	1,305	1,262	1,398.85	1,435.68	1,639.221	1,262	1,228.57	1,228.57
f_t	0.033	0.041	0.044	–	–	0.05	0.046	0.049	0.04848	0.04854	0.0591	0.049	0.0445	0.0445
ΔP_t (Pa)	49,245	14,009	16,926	1,696	1,696	5,109	9,643	9,374	10,502.45	11,165.45	5,382.93	9,374	16,520	16,520
d_e (m)	0.025	0.019	0.0149	–	–	0.0149	0.0149	0.0156	0.01585	0.01527	0.0114	0.0156	0.01068	0.01068
v_s (m/s)	0.47	0.43	0.495	0.37	0.37	0.432	0.435	0.4	0.40948	0.42526	0.5672	0.4	0.3734	0.3734
Re_s	25,281	18,327	15,844	–	–	13,689	13,942	14,448	10,345.294	10,456.39	8,568.03	1,4448	8,474	8,474
Pr_s	7.5	7.5	7.5	–	–	7.5	7.5	7.5	7.6	7.6	7.6	7.5	7.5	7.5
h_s (W/m^2 K)	920	1,034	1,288	868	868	1,278	1,221	1,156	1,248.86	1,290.789	2,062.19	1,156	1,181.56	1,181.56
f_s	0.315	0.331	0.337	–	–	0.345	0.348	0.3422	0.35987	0.35929	0.3702	0.3422	0.370	0.370
ΔP_s (Pa)	24,909	15,717	21,745	10,667	10,667	15,275	14,973	12,768	14,414.26	15,820.74	36,090.09	12,768	9,294	9,294
U (W/m^{2}K)	317	376	409.3	323	323	317.7	367.5	347.6	326.071	331.358	381.68	347.6	454.16	454.16
A (m^2)	61.5	52.9	47.5	61.566	61.566	60.35	53.6	56.6	58.641	57.705	50.09	56.6	42.82	42.82
C_i (V)	19,007	17,599	16,707	19,014	19,014	18,799	18,052.5	18,202	18,536.55	18,383.46	17,129.85	18,202	15,956	15,956
C_o (V/year)	1,304	440	523.3	197.319	197.319	164.4	293.6	210.2	272.576	292.7937	352.885	210.2	256.08	256.08
C_{od} (V)	8,012	2,704	3,215.6	1,211.3	1,211.3	1,010.2	1,721.1	1,231	1,674.86	1,799.09	2,163.32	1,231	1,531.55	1,531.55
C_{tot} (V)	27,020	20,303	19,922.6	20,225	20,225	19,810	19,773.6	19,433	20,211	20,182	19,298.18	19,433	**17,487.55**	**17,487.55**

GA (Caputo et al. 2008); PSO (Rao and patel 2010); ABC (Sahin et al. 2011); ICA (Hadidi et al. 2013a); BBO (Hadidi et al. 2013b); CSA (Asadi et al. 2014); FFA (Mohanty 2016); ITHS & I–ITHS (Turgut et al. 2014) and CI (Dhavale et al. 2016). Bold value indicates the best results

Table 4.6 Process parameters and physical properties for case study 4.2.1.3

Fluid Location	Shell side	Tube side
Fluid	Distilled water	Raw water
Mass flow rate (kg/s)	22.07	35.31
Inlet temperature (°C)	33.90	23.90
Outlet temperature (°C)	29.40	26.70
Density (kg/m^3)	995	999
Specific heat (kJ/(kg K)	4.18	4.18
Viscosity (Pa s)	0.0008	0.00092
Thermal conductivity (W/m K)	0.62	0.62
Fouling factor (m^2 K/W)	0.00017	0.00017

different optimization methods. Results show that a significant reduction in the number of tubes and increase in the tube side flow velocity which consecutively reduces the tube side heat transfer coefficient. The reduction in tube side flow velocity reduces the tube side pressure drop while the high shell side flow velocity increases the shell side pressure drop. The combined effect of capital investment and operating costs led to a reduction of the total cost of about 17.3, 15.17, 11.5, 10.7, 12.5, 12.4, and 17.19%, as compared with those given by literature, GA, PSO, ABC, BBO, FFA, ITHS, I-ITHS and CI approach, respectively.

It may be noted that the results for Jaya and self-adaptive Jaya algorithms are exactly equal. However, in Jaya algorithm we need to perform experiments by changing the population size to achieve the optimal solution but in self-adaptive Jaya algorithm it is not necessary to change the population size manually, thus reducing the number of experiments and computational time.

4.3 Plate-Fin Heat Exchangers

4.3.1 Case Study 1 of PFHE

The present application example is taken from the work of Xie et al. (2008). A cross flow plate-fin heat exchanger as shown in Fig. 4.3, having hot gas and air on both the sides of heat exchanger needs to be designed and optimized for minimum volume and minimum total annual cost. The input parameters values considered for the two fluids and economic data are shown in Table 4.8. Plain triangular fins and offset-strip fins are employed on gas side and air side respectively. The plate thickness is 0.4 mm, and both fins and plates are made from aluminum. The maximum allowable pressure drops on the hot and cold sides are 0.3 and 2 k Pa, respectively. So, the objective is to find out the hot stream flow length (L_a), cold stream flow length (L_b) and no flow length (L_c) of heat exchanger which result in minimum volume and also minimum total annual cost.

Table 4.7 Optimal heat exchanger geometry using different optimization methods for case study 4.2.1.3

Parameter	Original design	GA	ABC	PSO	BBO	ITHS	I-ITHS	CI	TLBO	Jaya	Self-adaptive Jaya
D_s (m)	0.387	0.62	0.0181	1.0024	0.55798	0.5726	0.5671	0.5235	0.5733	0.5733	0.5733
L (m)	4.88	1.548	1.45	2.4	1.133	0.9737	0.9761	1.1943	0.604	0.604	0.604
B (m)	0.305	0.44	0.423	0.354	0.5	0.4974	0.4989	0.5000	0.5	0.5	0.5
d_o (m)	0.019	0.016	0.0145	0.103	0.01	0.0101	0.01	0.0100	0.015	0.015	0.015
P_t (m)	0.023	0.02	0.0187	–	0.0125	0.0126	0.0125	0.0125	0.0191	0.0191	0.0191
N_t	160	803	894	704	1,565	1,845	1,846	1,548.6665	773	773	773
v_t (m/s)	1.76	0.68	0.74	0.36	0.898	0.747	0.761	0.9083	0.808	0.808	0.808
Re_t	36,409	9,487	9,424	–	7,804	6,552	6,614	7,889.7151	10,537	10,537	10,537
Pr_t	6.2	6.2	6.2	–	6.2	6.2	6.2	6.2026	6.2	6.2	6.2
h_t (W/m^2 K)	6,558	6,043	5,618	4,438	9,180	5,441	5,536	4,901.72	5,611	5,611	5,611
f_t	0.023	0.031	0.0314	–	0.0337	0.0369	0.0368	0.0336	0.0309	0.0309	0.0309
ΔP_t (Pa)	62,812	3,673	4,474	2,046	4,176	3,869	4,049	6,200.047	3,653	3,653	3,653
d_e (m)	0.013	0.015	0.01	0.0071	0.0071	0.0071	0.0071	0.0071	0.0102	0.0102	0.0102
v_s (m/s)	0.94	0.41	0.375	0.12	0.398	0.3893	0.3919	0.4237	0.3870	0.3870	0.3870
Re_s	16,200	8,039	4,814	–	3,515	3,473	3,461	3,746.028	5,140	5,140	5,140
h_s (W/m^2 K)	5,735	3,476	4,088.3	5,608	4,911	4,832	4,871	5,078.10	4,212	4,212	4,212
f_s	0.337	0.374	0.403	–	0.423	0.4238	0.4241	0.4191	0.399	0.399	0.399
ΔP_s (Pa)	67,684	4,365	4,271	27,166	5,917	4,995	5,062	6,585.24	1,932.6	1,932.6	1,932.6
U (W/m^2 K)	1,471	1,121	1,177	1,187	1,384	1,220	1,229	1,198.4141	1,535	1,535	1,535
A (m^2)	46.6	62.5	59.2	54.72	55.73	57.3	56.64	58.097	45.65	45.65	45.65
C_i (V)	16,549	19,163	18,614	17,893	18,059	18,273	18,209	18,447.63	15,828	15,828	15,828
C_o (V/year)	4,466	272	276	257.82	203.68	231	231	383.469	135	135	135
C_{od} (V)	27,440	1,671	1,696	1,584.2	1,251.5	1,419	1,464	2,356.256	1,399.3	1,399.3	1,399.3
C_{tot} (V)	43,989	20,834	20,310	19,478	19,310	19,693	19,674	20,803.89	**17,227.3**	**17,227.3**	**17,227.3**

GA (Caputo et al. 2008); ABC (Hadidi et al. 2013a); PSO (Rao and Patel 2010); BBO (Hadidi et al. 2013b); FFA (Mohanty 2016); ITHS & I−ITHS (Turgut et al. 2014) and CI (Dhavale et al. 2016). Bold value indicates the best results

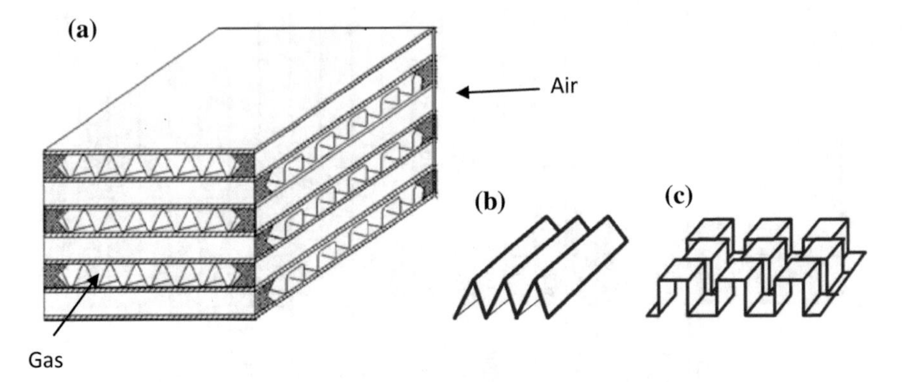

Fig. 4.3 PFHE geometry **a** core of two-stream plate–fin heat exchanger; **b** plain triangular fin; **c** offset-strip fin (Xie et al. 2008; reprinted with permission from Elsevier)

Total volume and total annual cost are as follows, i.e.

$$V = L_a \times L_b \times L_c, \tag{4.33}$$

$$TAC = C_{in} + C_{op}, \tag{4.34}$$

where, C_{in} and C_{op} are:

$$C_{in} = C_A \times A^n \tag{4.35}$$

$$C_{op} = \left\{ k_{el}\tau \frac{\Delta PV_t}{\eta} \right\}_h + \left\{ k_{el}\tau \frac{\Delta PV_t}{\eta} \right\}_c \tag{4.36}$$

The two objectives are treated as single objectives and solved for the minimum values of respective objective functions. Here the two single objectives are solved by TLBO, Jaya algorithm and self-adaptive Jaya algorithm with given constraints using a priori approach. The two objectives are combined together to form a single objective function with the help of preference-based multi-objective method with the use of weights as shown in Eq. (4.37). The normalized combined objective function (Z) is formulated considering different weightages to all objectives and is given by the following equation:

Minimize,

$$Z = w_1 \frac{V}{V_{min}} + w_2 \frac{TAC}{TAC_{min}} \tag{4.37}$$

where, V_{min} = minimum total volume obtained by treating minimizing the total volume as the single objective function, TAC_{min} = minimum total cost obtained by treating minimizing the total cost as the single objective function, w_1 and w_2 are equal weightages assigned to the objectives function TAC and V.

Table 4.8 Operating parameters and economic data for PFHE

Operating parameters	Hot fluid	Cold fluid
Volumetric flow rate, V_t (m^3/s)	1.2	0.6
Inlet temperature, T_1 (K)	513	277
Outlet temperature, T_2 (K)	324	–
Inlet pressure, P_1 (k Pa)	110	110
Economic data		
Cost per unit area, C_A (\$/m^2)	100	
Hours of operation per year, τ (h/year)	650	
Price of electrical energy, k_{el} (\$/MW h)	30	
Pump efficiency, η	0.5	
Exponent of nonlinear increase with area increase, n	0.6	

The weights are assigned depending on the importance given to the individual objective function but in the present case study the multi-objective function is solved by assigning equal weights i.e. $w_1 = w_2 = 0.5$.

The normalized combined objective function (Z) is written as:

Minimize,

$$Z = 0.5\left(\frac{V}{0.1025}\right) + 0.5\left(\frac{TAC}{2661.29}\right) \tag{4.38}$$

Bound and constraints:

- The upper and lower bounds of design variables imposed for PFHE are:
- Hot-stream flow length (L_a), ranging from 0.2 to 0.4 m.
- Cold-stream flow length (L_b), ranging from 0.5 to 0.7 m.
- No flow length (L_c), ranging from 0.75 to 1 m.

The constraints are, $\Delta P_{\mathrm{h}} \leq 0.3$ kPa, $\Delta P_c \leq 2$ kPa.

In the present work, the design optimization of plate-fin heat exchanger is attempted using the Jaya, self-adaptive Jaya and TLBO algorithms. Table 4.9 shows the optimized values of the design variables of the considered example using Jaya, self-adaptive Jaya and TLBO algorithm for various objective functions and the comparison with the results obtained by Xie et al. (2008) using GA and by Rao and Patel (2010) using PSO approach.

Results show that for minimum volume consideration hot stream flow length and no flow length are reduced (about 1.42 and 11.6% respectively) while the cold stream flow length is increased (about 0.06%) in the present approach as compared to GA approach considered by Xie et al. (2008).

Results show that for minimum volume consideration hot stream flow length is increased (about 3.05%) while the cold stream flow length and no flow length are reduced (about 2.49 and 4.26% respectively) in the present approach as compared to PSO approach considered by Rao and Patel (2010). The effect of these changes in

Table 4.9 Comparison of the heat exchanger design geometries with individual objectives PFHE

Parameter	Minimum volume consideration			Minimum cost consideration		
	GA (Xie et al. 2008)	PSO (Rao and Patel 2010)	TLBO, Jaya and self-adaptive Jaya	GA (Xie et al. 2008)	PSO (Rao and Patel 2010)	TLBO, Jaya and self-adaptive Jaya
L_a (m)	0.2390	0.2284	0.2356	0.2350	0.2190	0.2156
L_b (m)	0.5000	0.5131	0.5003	0.5000	0.5000	0.5000
L_c (m)	0.9840	0.9083	0.8696	1.000	1.0000	0.9601
ΔP_h(kPa)	0.2888	0.2999	0.1266	0.2979	0.2640	0.1002
ΔP_c(kPa)	1.9918	2.000	1.9975	1.8995	2.0000	1.9977
V (m^3)	0.1131	0.1064	**0.102510**	0.1175	0.1095	0.1055
Total annual cost ($/year)	3,078.68	3,072.2	2,722.61	3,047.67	3,017.9	**2,661.29**

heat exchanger length results in reduction in volume of the heat exchanger by about 9.67 and 3.66% in the present approach as compared to the GA approach by Xie et al. (2008) and PSO approach by Rao and Patel (2010).

For minimum total cost consideration, about 12.67 and 11.81% reduction in total cost is observed due to reduction in hot stream flow length in the present approach as compared to the GA approach (Xie et al. 2008) and PSO approach (Rao and Patel 2010). Xie et al. (2008) used a population size of 50 and number of generations of 1,000 and therefore the total no. of function evaluations was 50,000 to obtain the results using GA approach and Patel and Rao (2010) used a population size 50 and number of generations of 100 and therefore the total no. of function evaluation was 5,000 to obtain the results using PSO approach.

In the present approach, self-adaptive Jaya algorithm requires less number of function evaluations to obtain optimal solution. With small population size less number of function evaluations reduces the computational effort. It is observed that various objective functions considered in the present example converge within about 20 generations while Xie et al. (2008) reported about 300 generations using GA and Rao and Patel (2010) reported about 40 generations using PSO approach.

The best results obtained by Jaya, self-adaptive Jaya and TLBO algorithms for the combined objective function are compared with the results obtained by genetic algorithm, and are presented in Table 4.10. The best solution obtained by Jaya, self-adaptive Jaya and TLBO algorithm shows 11.84% decrease in total volume and 12.41% decrease in the total annual cost when compared with the results given by Xie et al. (2008) using GA. The comparisons with PSO are not possible as details are not available at Rao and Patel (2010).

It may be noted that the results for TLBO, Jaya and self-adaptive Jaya algorithms are exactly equal probably because of the reason that the achieved solution is global optimum. However, in TLBO and Jaya algorithm we need to perform experiments

Table 4.10 Comparison of combined objective function of the PFHE design geometries

Parameter	GA (Xie et al. 2008)	Jaya, self-adaptive Jaya and TLBO algorithms
Hot stream flow length, L_a (m)	0.238	0.2286
Cold stream flow length, L_b (m)	0.5	0.5008
No-flow length, L_c (m)	0.979	0.8971
Hot side pressure drop, ΔP_h (k Pa)	0.2885	0.1163
Cold side pressure drop, ΔP_c (k Pa)	1.9951	1.9999
Total volume, V (m^3)	0.1165	**0.1027**
Total annual cost, *TAC* ($)	3,051.25	**2,672.41**

by changing the population size to achieve the optimal solution but in self-adaptive Jaya algorithm it is not necessary to change the population size manually, thus reducing the number of experiments and computational time.

4.3.2 Case Study 2 of PFHE

This case study is considered from the literature (Mishra et al. 2009). The entropy generation minimization is considered as an objective function for a gas-to-air cross flow PFHE. The specific heat duty of the heat exchanger is 160 kW. The maximum dimension of the heat exchanger is 1 × 1 m^2 and maximum number of hot layers is limited to 10. The specification and input data of the PFHE are shown in Table 4.11.

Various efforts are made by the researchers for developing optimal design of the heat exchanger considering different objectives (Rao and Saroj 2017a). PSO algorithm was used for the optimization entropy generation unit (N_s) for given heat

Table 4.11 Input parameters/specification of the PFHE (Mishra et al. 2009)

Parameters	Hot fluid	Cold fluid
Temperature at inlet, T_i (K)	513	277
Rate of Mass flow, m (kg/s)	0.8962	0.8296
Specific heat, Cp (J/kg K)	1,017.7	1,011.8
Inlet pressure, P_i (Pa)	10^5	10^5
Dynamic viscosity, m (Ns/m^2)	241	218.2
Density, (kg/m^3)	0.8196	0.9385
Prandtl number, Pr	0.6878	0.6954
Heat duty of the exchanger, Q (kW)	160	

duty requirement (Rao and Patel 2010). GA was used for optimization of PFHE design by minimization the N_s for a given heat duty and satisfying the constraints imposed on it (Mishra et al. 2009). Cuckoo search algorithm (CSA) was used for the minimization of rate of entropy generation for a PFHE (Wang and Li 2015).

Entropy generation (S_g) minimization follows the 2nd law of the thermodynamics. It is used for the thermodynamic optimization of a thermal system which is having thermodynamic inadequacy to the irreversibility because of fluid flow and transfer of heat and mass. The thermodynamic irreversibility or entropy generation units (N_s) show the quantity of loss of useful work, which is unavailable because of the system irreversibility. In a PFHE, irreversibility is caused because of the finite temperature difference in heat transfer of the fluid streams and the pressure drops along them. Optimizing a PFHE or any other thermal system on the basis of entropy generation minimization corresponds to the minimization of the quantity of unavailable (loss) work by considering the finite size bounds of definite devices and finite time bounds of definite process (Mishra et al. 2009). Figure 4.4 presents the detailed layout of plate-fin heat exchanger (Rao and Saroj 2017a).

The entropy generation rate for an ideal gas with finite temperature difference ($S_{\Delta T}$) and finite pressure drop ($S_{\Delta P}$) is calculated as (Wang and Li 2015):

$$S_{\Delta T} = \int_i^o \left(\frac{mc_P dT}{T}\right)_{h,c} = m_h c_{P,h} \ln\left(\frac{T_{h,o}}{T_{h,i}}\right) + m_b c_{P,c} \ln\left(\frac{T_{c,o}}{T_{c,i}}\right) \quad (4.39)$$

$$S_{\Delta P} = \int_i^o \left(-\frac{mRdP}{P}\right)_{h,c} = m_h R_h \ln\left(\frac{P_{h,o}}{P_{h,i}}\right) - m_c R_c \ln\left(\frac{P_{c,o}}{P_{c,i}}\right) \quad (4.40)$$

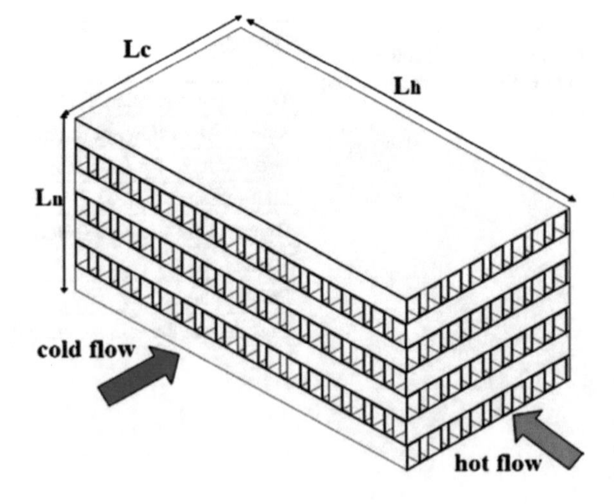

Fig. 4.4 Layout of plate-fin heat exchanger (Rao and Saroj 2017d; reprinted with permission from Taylor & Francis)

Here, m, C_p, T and P are the mass flow rate, specific heat, temperature and pressure respectively. Subscripts h and c denotes the hot and cold side of the fluid.

The total entropy generation rate (S) in PFHEs is calculated as:

$$S = S_{\Delta T} + S_{\Delta P} \quad (4.41)$$

The N_s is the dimensionless value for entropy generation rate and it calculated as follows:

$$N_S = \frac{S}{\max(C_a, C_b)} = N_{S,\Delta T} + N_{S,\Delta P} \quad (4.42)$$

The aim of this case study is to minimize the value of N_s which given by Eq. (4.42). The irreversibility due the temperature difference is much larger than irreversibility due to the pressure drop. A lower value of $S_{\Delta T}$ is related to heat transfer, results in higher thermal efficiency while the entropy generation due to the fluid friction increases (Wang and Li 2015). Therefore, in this study two types of entropy generations are taken as two distinct conflicting objectives. A priori approach is used for optimization of total entropy generation and it is defined as:

$$O_{min} = w \times N_{S,\Delta T} + (1 - w)N_{S,\Delta P} + \Delta g(x)^2 \cdot P \quad (4.43)$$

Here, w is the weight for the objective, here same weights for both the objectives are considered in order to make the fare comparison of results with the previous results; $\Delta g(x)$ is the violation of the constraint and P is known as penalty factor.

The design of the PFHE is governed by seven design variables and the ranges of the design variables are as follows (Mishra et al. 2009):

- Stream flow length of hot side, L_h (m): $0.1 \le L_h \le 1$.
- Stream flow length of cold side, L_c (m): $0.1 \le L_c \le 1$.
- Fin height, b (mm): $2 \le b \le 10$.
- Fin thickness, t_f (mm): $0.1 \le t_f \le 0.2$.
- Frequency of fin (n): $100 \le n \le 1{,}000$.
- Offset length, x (mm): $1 \le x \le 10$.
- Fin layers number (N_P–): $1 \le N_p \le 200$.

The design of PFHE is also imposed by an inequality constraint:

$$Q(\text{kW}) - 160\,(\text{kW}) \ge 0 \quad (4.6)$$

Now, the SAMPE-Jaya algorithm is used for the optimization of PFHE design. The termination criterion used in this section is maximum function evaluation of 5,000, which is same as used by the previous researches. The design suggested by present approach is compared with the design suggested by GA, PSO and CSA and comparison is shown in Table 4.12. It can be observed that the values of $N_{S,\Delta T}$ and $N_{S,\Delta P}$ are reduced by 1.075 and 69% in comparison to the results of CSA. It can also be observed from Table 4.12 that the value of N_S is decreased by 8.11, 27.49 and 39.17% in comparison to the results of the CSA, PSO and GA. The value of L_c is increased by 0.603, 8.10 and 10.86% in comparison to GA, PSO and GA. Similarly, value of L_h is increased by 12.73, 3.5 and 9.6%. The fin frequency of the present design is also increased to 980. These increments in the length of flow and fin frequency increase the convective heat transfer of the heat exchanger. It can also be observed from Table 4.12 that value of pressure drops of hot fluid and cold fluid are increased in comparison to the previous designs. However, difference between outlet pressure of hot fluid and cold is minimum in comparison to the other designs. This results in the reduction of the entropy generation rate due to finite pressure difference. Figure 4.5 presents the conversion of the SAMP-Jaya algorithm for PFHE design problem. It can be observed from the figure that the convergence of algorithm takes place in a few iterations.

Furthermore, for checking the diversity of the search process, standard deviation of the objective function values corresponding population individuals is calculated after each iteration is recorded and it is shown in Table 4.13. It can be observed that the value of standard deviation is not zero and is different for each iteration, which shows that the algorithm is continuously exploring the search process and it does not fall in the trap of local minima. The early convergence of Fig. 4.5 shows that the algorithm has reached to global optimum value or near global optimum value of the objective function in few iterations.

Table 4.12 Comparison of the results for entropy generation minimization of PFHE

Parameters	GA (Mishra et al. 2009)		PSO (Rao and Patel 2010)		CSA (Wang and Li 2015)		SAMPE-Jaya algorithm	
	Cold fluid	Hot fluid	Cold fluid	Hot fluid	Cold fluid	Hot fluid	Cold fluid	Hot fluid
t_f (mm)	0.146	0.146	0.10	0.10	0.09	0.09	0.1997	0.1997
L (m)	0.994	0.887	0.925	0.966	0.902	0.912	1	1
x (mm)	6.30	6.30	9.80	9.80	9.9	10	10	9.9998
n (fin/m)	534	534	442	442	419	501	980	980
b (mm)	9.53	9.53	9.80	9.80	10	9.9	10	10
ΔP (k Pa)	7.100	4.120	1.750	11.10	3.723	6.062	8.189	7.551
$N_{S,\Delta P}$	–		–		0.0043		0.0013	
$N_{S,\Delta T}$	–		–		0.0376		0.0372	
N_S	0.0633		0.0531		0.0419		0.0385	
E	0.9161		0.9161		0.9161		0.9161	

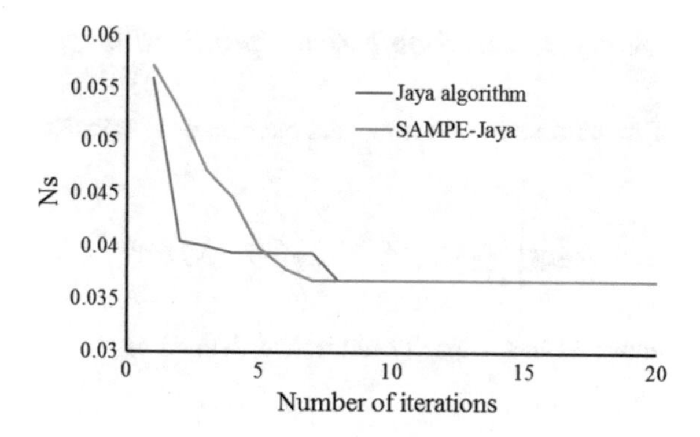

Fig. 4.5 Convergence of Jaya and SAMPE-Jaya algorithms for PFHE design

Table 4.13 Standard deviation of the solutions in each iteration using Jaya and SAMPE-Jaya algorithms

Iteration No.	Jaya algorithm	SAMPE-Jaya	Iteration No.	Jaya algorithm	SAMPE-Jaya
1	0.013282746	0.00881	21	0.01075	0.009908
2	0.015870823	0.004585	22	0.01075	0.012269
3	0.016974039	0.005458	23	0.010741	0.01441
4	0.017538914	0.005755	24	0.010744	0.014195
5	0.019188605	0.005618	25	0.01019	0.013251
6	0.02219307	0.003984	26	0.006347	0.014204
7	0.0231661	0.003225	27	0.001762	0.015378
8	0.023694038	0.006591	28	0.001499	0.015716
9	0.023991376	0.004209	29	0.000142	0.018139
10	0.022965606	0.010176	30	6.61E-05	0.015063
11	0.020751619	0.009398	31	3.34E-05	0.01633
12	0.015849161	0.009426	32	1.57E-05	0.01541
13	0.014387688	0.011408	33	1.01E-05	0.013725
14	0.014624281	0.011391	34	1.09E-05	0.011788
15	0.014662846	0.009828	35	1.09E-05	0.01331
16	0.014673993	0.009521	36	1.80E-05	0.007235
17	0.014674564	0.015726	37	2.47E-05	0.010926
18	0.014123168	0.011112	38	2.88E-05	0.01239
19	0.010769027	0.011303	39	1.97E-05	0.008539
20	0.010749522	0.008125	40	1.32E-05	0.012941

4.3.3 Case Study 3: Multi-objective Optimization of PFHE

Effectiveness of an unmixed cross-flow heat exchanger is calculated as (Incropera et al. 2010),

$$\varepsilon = 1 - \exp\left[\left(\frac{1}{C}\right) NTU^{0.22}\left\{\exp\left(-C \cdot NTU^{0.78}\right) - 1\right\}\right] \tag{4.44}$$

Here, C is known as heat capacity ration and defined as:

$$C = \frac{C_{\min}}{C_{\max}} = \frac{\min(C_h, C_c)}{\max(C_h, C_c)} \tag{4.45}$$

Here, suffix h and c denotes the hot and cold side respectively. Figure 4.4 presents the layout of a PFHE.

Outlet temperatures of hot fluid ($T_{h,o}$) and cold fluid ($T_{c,o}$) are calculated as:

$$T_{h,o} = T_{h,i} - \varepsilon C_{min}/C_h\left(T_{h,i} - T_{c,i}\right) \tag{4.46}$$

$$T_{c,o} = T_{c,i} + \varepsilon\, C_{min}/C_c\left(T_{h,i} - T_{c,i}\right) \tag{4.47}$$

Now, the number of transfer units (NTU) can be calculated as:

$$\frac{1}{NTU} = \frac{C_{\min}}{UA_t} \tag{4.48}$$

A_t is the total heat transfer area of plate-fin heat exchanger and U is known as overall heat transfer co-efficient. It is defined as:

$$\frac{1}{UA} = \frac{1}{(hA)_h} + \frac{1}{(hA)_c} \tag{4.49}$$

Convective-heat transfer coefficient is calculated as:

$$h = j.G.C_p.P_r^{-2/3} \tag{4.50}$$

Here, j is known as Colburn factor; G is mass flux and defined as:

$$\begin{aligned} j = {} & 0.6522(\mathrm{Re})^{-0.5403}(\alpha)^{-0.1541}(\delta)^{0.1499}(\gamma)^{-0.0678} \\ & \times \left[1 + 5.269 \times 10^{-5}(\mathrm{Re})^{1.34}(\alpha)^{0.504}(\delta)^{0.456}(\gamma)^{-1.055}\right]^{0.1} \end{aligned} \tag{4.51}$$

$G = m/A_f$ (10). Here, α, δ and γ are the geometrical parameters of PFHE. *Re* is known as Reynolds number and defined as:

$$R_e = \frac{G \cdot D_h}{\mu} \tag{4.52}$$

Here μ is dynamic viscosity and D_h is known as hydraulic diameter and can be evaluated as:

$$D_h = \frac{4(c - t_f)(b - t_f)x}{2((c - t_f)x + (b - t_f)x + t_f(b - t_f)) + t_f(c - t_f) - t_f^2} \tag{4.53}$$

Here, t_f, b, c and x are the thickness, height, pitch and length of the fin, respectively.

Here, A_f is known as free flow area and evaluated as:

$$A_{f,h} = L_h N_p (b_h - t_{f,h})(1 - n_h . t_{f,h}) \tag{4.54}$$

$$A_{f,c} = L_{c.} (N_p + 1)(b_h - t_{f,h})(1 - n_h . t_{f,h}) \tag{4.55}$$

Here, L_c and L_h are the hot and cold flow length. Heat transfer area of hot side and cold side is calculated as:

$$A_h = L_h L_c N_p [1 + 2n_h (b_h - t_{f,h})] \tag{4.56}$$

$$A_c = L_h L_c (Np + 1)[1 + 2n_c (b_c - t_{f,c})] \tag{4.57}$$

Now, the total heat transfer area is calculated as:

$$A_t = A_h + A_c \tag{4.58}$$

And the rate of heat transfer is evaluated as:

$$Q = \varepsilon C_{min} (T_{h,i} - T_{c,i}) \tag{4.59}$$

Due to friction, pressure drop is caused. Hot and cold side pressure drop is evaluated as:

$$\Delta P_h = \frac{2 f_h L_h G_h^2}{\rho_h D_{h.h}} \tag{4.60}$$

$$\Delta P_c = \frac{2 f_c L_c G_c^2}{\rho_c D_{h,c}} \tag{4.61}$$

Here, for a off-strip fins fanning factor f is evaluated as (Incropera et al. 2010):

$$f = 9.6243(\mathrm{Re})^{-0.7422}(\alpha)^{-0.1856}(\delta)^{0.3053}(\gamma)^{-0.2659} \times \left[1 + 7.669 \times 10^{-8}(\mathrm{Re})^{4.429}(\alpha)^{0.920}(\delta)^{3.767}(\gamma)^{0.236}\right]^{0.1} \quad (4.62)$$

The considered design optimization problem of PFHE is governed by seven restricted design variables and nine constraints.

The ranges of the design variables are as follows (Yousefi et al. 2012):

(a) Stream flow length of hot side, L_h (m): $0.1 \leq L_h \leq 1$.
(b) Stream flow length of cold side, L_c (m): $0.1 \leq L_h \leq 1$.
(c) Fin height, b (mm): $2 \leq b \leq 10$.
(d) Fin thickness, t_f (mm): $0.1 \leq t_f \leq 0.2$.
(e) Frequency of fin (n): $100 \leq n \leq 1{,}000$.
(f) Offset length, x (mm): $1 \leq x \leq 10$.
(g) Fin layers number (N_P−): $1 \leq N_p \leq 200$.

Out of these constraints N_P is a discrete variable and rest of the variables are continuous in nature.

Nine constraints are imposed on the PFHE design in order get the specific duty of heat exchanger with limitations on mathematical model and geometries are defined as follows:

The value of *Re* for hot and cold steam flow must be in the following range:

$$120 \leq \mathrm{Re}_h \leq 10^4$$
$$120 \leq \mathrm{Re}_c \leq 10^4$$

The equations used for the calculation of Colburn factor and fanning factor are to be used only when the values *Re* of the suggested design falls in the above given ranges.

The geometrical parameters of the PFHE must be in the following ranges:

$$0.134 \leq \alpha \leq 0.997$$
$$0.041 \leq \gamma \leq 0.121$$
$$0.012 \leq \delta \leq 0.048$$

No-flow length (L_n) of PFHE is also restricted:

$$L_n = 1.5$$

The value of L_n is evaluated with the help of following equation:

$$L_n = b - 2t_p + N_p(2b + 2t_p) \quad (4.63)$$

Heat duty required for the PFHE is also taken as constraint in order to meet the minimum heat duty:

$$Q \geq 1069.8\ \text{kW}$$

Design of PFHE is also restricted by the maximum allowable pressure drop of hot side and cold side. It is defined as follows:

$$\Delta P_h \leq 9.5\ \text{kPa}$$
$$\Delta P_c \leq 8\ \text{kPa}$$

The following four objectives are considered for the optimization of PFHE.

4.3.3.1 Minimization of Total Annual Cost

First test case is the minimization of total annual cost which is the sum of initial investment cost C_i and operational cost of C_{op}. Detailed mathematical model for the calculation of these costs is described as follows (Hadidi 2016):

$$C_{in} = a \cdot C_a \cdot A_t^n \tag{4.64}$$

$$C_{op} = \left[k_{el}\tau\frac{\Delta P_h}{\eta}\frac{m_h}{\rho_h}\right] + \left[k_{el}\tau\frac{\Delta P_c}{\eta}\frac{m_c}{\rho_c}\right] \tag{4.65}$$

$$C_{tot} = C_{in} + C_{op} \tag{4.66}$$

In the above equations, C_a is cost per unit of A_t; n is exponent value; k_{el} is electricity price; τ is hours of operation and η is known as compressor efficiency. In Eq. (4.67), *a* is known as annual cost coefficient and described as follows:

$$a = \frac{i}{1-(1+i)^{-ny}} \tag{4.67}$$

where, *i* is rate of interest and *ny* is time of depreciation.

Values of constants used in the above equations are given in Table 4.14.

Table 4.14 Coefficients of cost for PFHE (Yousefi et al. 2012)

Economic parameters	Value
C_a (\$/m^2)	90
τ (h)	5,000
k_{el} (\$/MW h)	20
H	0.60
N	0.6
Ny	10
i (%)	10

4.3.3.2 Minimization of Total Heat Transfer Area

Minimization of heat transfer area (A_t) required for proper heat transfer is the second objective of this study. The design equation is linked with investment cost of the considered PFHE. In this case constraint on no-flow length is not required.

4.3.3.3 Minimization of Total Pressure Drop

Third objective is also to be minimized which is a combined function pressure drops of cold side and hot side fluid. The objective of this case is linked with the operating cost of the PFHE system. A combined normalized function of pressure drops is in the optimization study and it is defined by the following equation:

$$O(x) = \frac{\Delta P_h}{P_{h,\max}} + \frac{\Delta P_c}{P_{c,\max}} \tag{4.68}$$

4.3.3.4 Maximization of Effectiveness of the Heat Exchanger

Effectiveness maximization is considered as the fourth objective. Calculation of the effectiveness of the heat exchanger is based upon Eq. (4.44).

Application of Jaya algorithm to the individual objectives lead to most minimum C_{tot} of 915.66 ($/year), most minimum A_t of 107.92 (m^2), most minimum total pressure drop in terms of $O(x)$ of 0.054315 and most maximum effectiveness ε of 0.94509 (Rao and Saroj 2017d). Now, multi-objective design optimization considering all these four objectives *simultaneously* is carried out using the Jaya algorithm.

Multi-objective optimization is defined as the searching the best compromise solution among the several objectives (i.e. maximization or minimization) while satisfying a number of constraints. In this work four conflicting objectives namely minimization of the total cost (annual investment cost and operational cost), total surface area, total pressure drop and maximization of heat exchanger effectiveness are optimized simultaneously. A priori approach is used in this work for multi-objective design optimization of PFHE. The normalized combined objective function is formulated by considering different weights for the objectives. The formulated combined objective is to be minimized and it is defined by Eq. (4.69).

$$\begin{aligned} Minimize,\ Z(x) = {} & w_1 * \left(C_{tot}/C_{tot,min}\right) + w_2 * \left(At/A_{t,min}\right) + w_3 * \left(O(x)/O(x)_{,min}\right) \\ & - w_{14} * \left(\varepsilon/\varepsilon_{,max}\right) + \sum_{j=1}^{m} P\left(C_j(Y)\right)^2 \end{aligned} \tag{4.69}$$

where w_1, w_2, w_3 and w_4 are the weight factors for C_{tot}, A_t, O(x) and ε respectively. In the present study equal weightages are assigned to each objective (i.e. $w_1 = w_2 = w_3 = w_4 = 0.25$). $C_{tot,min}$, $A_{t,min}$, $O(x)_{,min}$ are the most minimum values of the three objective functions and $\varepsilon_{,max}$ is the most maximum value of the objective function when these are optimized independently and these values are 915.66 ($/year), 107.92 ($m^2$), 0.054315 and 0.94509 respectively. *P* is the penalty parameters with a large value. A static penalty for constraint handling is used with the Jaya algorithm for solving the constrained optimization problems. The multi-objective

Table 4.15 Multi-objective optimization results

Parameters	Value
L_h (m)	0.69824
L_c (m)	0.74735
b (m)	0.01
t_f (m)	0.00015387
n (m^{-1})	255.93
x (m)	0.003382
Np	71
D_h (mm)	5.2518
A_{ffh} (m^2)	0.50188
A_{ffc} (m^2)	0.4755
G_h (kg/m^2 s)	3.3076
G_c (kg/m^2 s)	4.2061
Re_h	433.18
Re_c	657.42
f_h	0.1164
f_c	0.087775
ΔP_h (k Pa)	0.53782
ΔP_c (k Pa)	0.45855
j_h	0.023993
h_h (W/m^2 K)	109.73
j_c	0.019836
h_c (W/m^2 K)	114.21
NTU	6.7697
ε	0.82055
A_h (m^2)	223.78
A_c (m^2)	226.93
A_t (m^2)	450.71
C_{in} ($/year)	572.91
C_{op} ($/year)	394.93
C_{tot} ($/year)	967.84
O(*x*)	0.61393
Z(*x*)	1.6157

optimization results obtained by the Jaya algorithm are shown in Table 4.15 (Rao and Saroj 2017d).

It can be observed that the values obtained for C_{tot}, A_t, total pressure drop in terms of $O(x)$ and effectiveness ε are the optimum compromise values satisfying all the objectives, constraints and ranges of the variables *simultaneously* and represent the realistic design values.

In all the case studies considered for the single and multi-objective design optimization of shell-and-tube and plate-fin heat exchangers it is observed that the Jaya algorithm and its variants have provided better results as compared to the other well known advanced optimization algorithms.

References

Asadi, A., Songb, Y., Sundenc, B., & Xie, G. (2014). Economic optimization design of shell-and-tube heat exchangers by a cuckoo-search-algorithm. *Applied Thermal Engineering, 73*(1), 1032–1040.

Ayala, H. V. H., Keller, P., Morais, M. D. F., Mariani, V. C., Coelho, L. D. S., & Rao, R. V. (2016). Design of heat exchangers using a novel multiobjective free search differential evaluation paradigm. *Applied Thermal Engineering, 94,* 170–177.

Caputo, A. C., Pelagagge, P. M., & Salini, P. (2008). Heat exchanger design based on economic optimization. *Applied Thermal Engineering, 10,* 1151–1159.

Cavazzuti, M., Agnani, E., & Corticelli, M. A. (2015). Optimization of a finned concentric pipes heat exchanger for industrial recuperative burners. *Applied Thermal Engineering, 84,* 110–117.

Dhavale, S. V., Kulkarni, A. J., Shastri, A., & Kale, I. R. (2016). Design and economic optimization of shell-and-tube heat exchanger using cohort intelligence algorithm. *Natural Computing Applications*. https://doi.org/10.1007/s00521-016-2683-z.

Du, T., Du, W., Che, K., & Cheng, L. (2015). Parametric optimization of overlapped helical baffled heat exchangers by Taguchi method. *Applied Thermal Engineering, 85,* 334–339.

Edwards, J. E. (2008). *Design and rating of shell-and-tube heat exchangers*. Teesside, UK: P and I Design Ltd.

Gharbi, N. E., Kheiri, A., Ganaoui, M. E., & Blanchard, R. (2015). Numerical optimization of heat exchangers with circular and non-circular shapes. *Case Studies in Thermal Engineering, 6,* 194–203.

Hadidi, A. (2016). A robust approach for optimal design of plat fin heat exchanger using biogeography based optimization (BBO) algorithm. *Applied Energy, 150*, 196–210.

Hadidi, A., & Nazari, A. (2013a). A new design approach for shell-and-tube heat exchangers using imperialist competitive algorithm (ICA) from economic point of view. *Energy Conversion and Management, 67,* 66–74.

Hadidi, A., & Nazari, A. (2013b). Design and economic optimization of shell-and-tube heat exchangers using biogeography-based (BBO) algorithm. *Applied Thermal Engineering, 51*(1–2), 1263–1272.

Han, W. T., Tang, L. H., & Xie, G. N. (2008). Performance comparison of particle swarm optimization and genetic algorithm in rolling fin-tube heat exchanger optimization design. In *Proceedings of the ASME Summer Heat Transfer Conference*, Jacksonville, FL (pp. 5–10).

Hosseini, R., & Ceylan, H. (2009). A new solution algorithm for improving performance of ant colony optimization. *Applied Mathematics and Computation, 211,* 75–84.

Incropera, F. P., & DeWitt, D. P. (1996). *Fundamentals of heat and mass transfer*. New York: Wiley.

Incropera, F. P., Dewitt, D. P., Bergman, T. L., & Lavine, A. S. (2010). *Fundamentals of heat and mass transfer*. New York: Wiley.

Jorge, A. W., Gut, M., & Pinto, J. M. (2004). Optimal configuration design for plate heat exchangers. *International Journal of Heat and Mass Transfer, 47,* 4833–4848.

Kang, L., Liu, Y., & Liang, X. (2015). Multi-objective optimization of heat exchanger networks based on analysis of minimum temperature difference and accumulated CO_2 emissions. *Applied Thermal Engineering, 87,* 736–748.

Kern, D. Q. (1950). *Process heat transfer*. Tokyo: McGraw-Hill Book Company, Inc.

Lee, S. M., & Kim, K. Y. (2015). Multi-objective optimization of arc-shaped ribs in the channels of a printed circuit heat exchanger. *International Journal of Thermal Sciences, 94,* 1–8.

Lemouedda, A., Breuer, M., Franz, E., Botsch, T., & Delgado, A. (2010). Optimization of the angle of attack of delta-winglet vortex generators in a plate-fin-and-tube heat exchanger. *International Journal of Heat and Mass Transfer, 53,* 5386–5399.

Mishra, M., & Das, P. K. (2009). Thermo economic design-optimization of cross flow plate-fin heat exchanger using genetic algorithm. *International Journal of Exergy, 6*(6), 237–252.

Mishra, M., Das, P. K., & Sarangi, S. (2009). Second law based optimization of crossflow plate-fin heat exchanger using genetic algorithm. *Applied Thermal Engineering, 29,* 2983–2989.

Miyazaki, T., & Akisawa, A. (2009). The influence of heat exchanger parameters on the optimum cycle time of adsorption chillers. *Applied Thermal Engineering, 29*(13), 2708–2717.

Mohanty, A. K. (2016). Application of firefly algorithm for design optimization of a shell and tube heat exchanger from economic point of view. *International Journal of Thermal Sciences, 102,* 228–238.

Najafi, H., Najafi, B., & Hoseinpoori, P. (2011). Energy and cost optimization of a plate and fin heat exchanger using genetic algorithm. *Applied Thermal Engineering, 31,* 1839–1847.

Patel, V. K., & Rao, R. V. (2010). Design optimization of shell-and-tub heat exchanger using particle swarm optimization technique. *Applied Thermal Engineering, 30*(11–12), 1417–1425.

Peng, H., & Ling, X. (2008). Optimal design approach for the plate-fin heat exchangers using neural networks cooperated with genetic algorithms. *Applied Thermal Engineering, 28,* 642–650.

Rao, R. V., & Patel, V. K. (2013). Multi-objective optimization of heat exchangers using a modified teaching-learning-based optimization algorithm. *Applied Mathematical Modelling, 37* (3), 1147–1162.

Rao, R. V., & Patel, V. K. (2011). Optimization of mechanical draft counter flow wet cooling tower using artificial bee colony algorithm. *Energy Conversion and Management, 52,* 2611–2622.

Rao, R. V., & Patel, V. K. (2010). Thermodynamic optimization of cross flow plate-fin heat exchanger using a particle swarm optimization algorithm. *International Journal of Thermal Sciences, 49,* 1712–1721.

Rao, R. V., & Saroj, A. (2017a). Constrained economic optimization of shell-and-tube heat exchangers using elitist-Jaya algorithm. *Energy, 128,* 785–800.

Rao, R. V., & Saroj, A. (2017b). Economic optimization of shell-and-tube heat exchangers using Jaya algorithm with maintenance consideration. *Applied Thermal Engineering, 116,* 473–487.

Rao, R. V., & Saroj, A. (2017c). A self-adaptive multi-population based Jaya algorithm for engineering optimization. *Swarm and Evolutionary Computation*. https://doi.org/10.1016/j.swevo.2017.04.008.

Rao, R. V., & Saroj, A. (2017d). Single objective and multi-objective design optimization of plate-fin heat exchangers using Jaya algorithm. *Heat Transfer Engineering* (in press).

Ravagnani, M. A. S. S., Silva, A. P., Biscaia, E. C., & Caballero, J. A. (2009). Optimal design of shell-and-tube heat exchangers using particle swarm optimization. *Industrial and Engineering Chemistry Research, 48*(6), 2927–2935.

Reneaume, J. M., & Niclout, N. (2003). MINLP optimization of plate-fin heat exchangers. *Chemical and Biochemical Engineering Quarterly, 17,* 65–76.

Sahin, A. S., Kilic, B., & Kilic, U. (2011). Design and economic optimization of shell-and-tube heat exchangers using artificial bee colony (ABC) algorithm. *Energy Conversion and Management, 52*(11), 1417–1425.

Sanaye, S., & Hajabdollahi, H. (2010). Thermal-economic multi-objective optimization of plate-fin heat exchanger using genetic algorithm. *Applied Energy, 87,* 1893–1902.

Selbas, R., Kizilkan, O., & Reppich, M. (2006). A new design approach for shell-and-tube heat exchangers using genetic algorithms from economic point of view. *Chemical Engineering and Processing, 45,* 268–275.

Shah, R. K., & Bell, K. J. (2009). *CRC handbook of thermal engineering*. Florida: CRC Press.

Shah, R. K., & Sekulic, D. P. (2003). *Fundamentals of heat exchanger design*. New York: Wiley.

Sinnot, R. K., Coulson, J. M., & Richardson, J. F. (1996). *Chemical engineering design* (Vol. 6). Boston MA: Butterworth-Heinemann.

Taal, M., Bulatov, I., Klemes, J., & Stehlik, P. (2003). Cost estimation and energy price forecast for economic evaluation of retrofit projects. *Applied Thermal Engineering, 23,* 1819–1835.

Turgut, O. E., & Çoban, M. T. (2016). Thermal design of spiral heat exchangers and heat pipes through global best algorithm. *Heat Mass Transfer*, 1–18. https://doi.org/10.1007/s00231-016-1861-y.

Turgut, O. E., Turgut, M. S., & Coban, M. T. (2014). Design and economic investigation of shell and tube heat exchangers using improved intelligent tuned harmony search algorithm. *Ain Shams Engineering Journal, 5*(4), 1215–1231.

Wang, Z., & Li, Y. (2015). Irreversibility analysis for optimization design of plate fin heat exchangers using a multi-objective cuckoo search algorithm. *Energy Conversion and Management, 101,* 126–135.

Wong, J. Y. Q., Sharma, S., & Rangaiah, G. P. (2016). Design of shell-and-tube heat exchangers for multiple objectives using elitist non-dominated sorting genetic algorithm with termination criteria. *Applied Thermal Engineering, 93,* 888–899.

Xie, G. N., Sunden, B., & Wang, Q. W. (2008). Optimization of compact heat exchangers by a genetic algorithm. *Applied Thermal Engineering, 28,* 895–906.

Yousefi, M., Darus, A. N., & Mohammadi, H. (2012). An imperialist competitive algorithm for optimal design of plate-fin heat exchangers. *International Journal of Heat and Mass Transfer, 55,* 3178–3185.

Yu, X. C., Cui, Z. Q., & Yu, Y. (2008). Fuzzy optimal design of the plate-fin heat exchangers by particle swarm optimization, In *Proceedings of the Fifth International Conference on Fuzzy Systems and Knowledge Discovery*, Jinan, China (pp. 574–578).

Chapter 5
Single- and Multi-objective Design Optimization of Heat Pipes and Heat Sinks Using Jaya Algorithm and Its Variants

Abstract This chapter presents the application of Jaya algorithm and its variants for the single objective as well as multi-objective design optimization of heat pipes and heat sinks. Design of heat pipes and heat sinks involves a number of geometric and physical parameters with high complexity and the design processes are mostly based on trial and error. General design approaches become tedious and time consuming and these processes do not guarantee the achievement of an optimal design. Therefore, meta-heuristic based computational methods are preferred. This chapter presents the results of application of Jaya algorithm and its variants such as self-adaptive Jaya algorithm, SAMP-Jaya algorithm and SAMPE-Jaya algorithm to the design optimization problems of heat pipes and heat sinks. The results are found better than those obtained by other optimization techniques such as TLBO, Grenade Explosion Method (GEM), Niched Pareto Genetic Algorithm (NPGA), Generalized External optimization (GEO) and a hybrid multi-objective evolutionary algorithm.

5.1 Heat Pipe

Heat pipe (HP) is one of the effective thermal devices for transferring the thermal energy from one point to another. The working principle of the heat pipe is based the evaporation and condensation process. The problem of heat pipe design becomes complex because it works based on the combination of conduction and convective heat transfer process. The heat pipes can be used where space is limited and high heatflux is required. This is widely used in heat transfer systems like, space applications, cell phones, cooling of solar collectors and cooling of computers.

Figure 5.1 presents the schematic diagram of a heat pipe (Faghri 2014). The component of a heat pipe are a closed container with seal, a wick structure and a working fluid. Based on the requirements of the operational temperature various types of working fluids i.e. water, acetone, methanol, ammonia, sodium or ethanol can be used. The length of a heat pipe is divided into three parts: the evaporator section (heat addition), adiabatic (transport) section and condenser section (heat

R. Venkata Rao, *Jaya: An Advanced Optimization Algorithm and its Engineering Applications*, https://doi.org/10.1007/978-3-319-78922-4_5

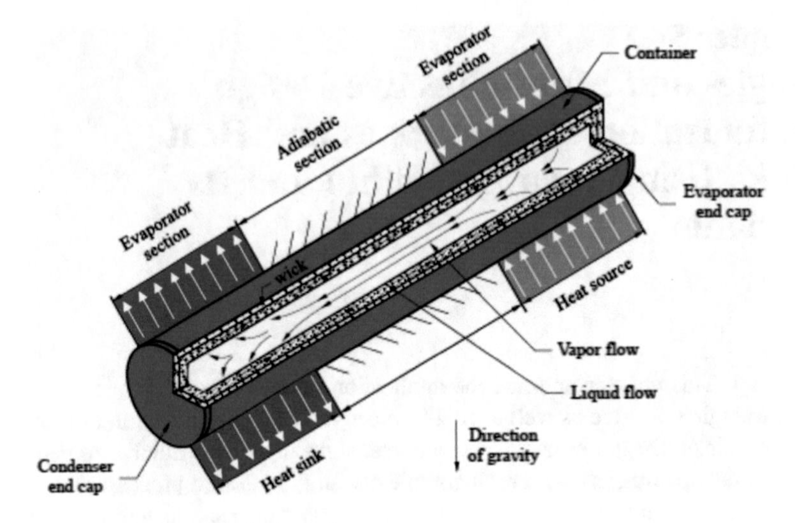

Fig. 5.1 Schematic diagram of a heat pipe

release). Depending upon the particular application a heat pipe may be having several heat sources or sinks with/without adiabatic sections. The working fluid is vaporized in the evaporator section by the externally applied heat. Due to this, heat vapour pressure is generated which drives the vapour through the adiabatic section to the condenser section.

In the condenser section, heat is provided to the sink by releasing latent heat of vaporization of the working fluid. The capillary pressure is developed due to the menisci effect in the wick. This pressure drives the condensed fluid back to the evaporator section. Hence, a heat pipe can work continuously for transferring the latent heat of vaporization from the evaporator section to the condenser section. This procedure will be continued till there is enough capillary pressure for driving the condensed fluid to back to the evaporator.

An integrated analytical tool was developed by Buksa and Hillianus (1989) for design optimization of a heat pipe radiator system. Said and Akash (1999) presented experiments on two types of heat pipes using water as a working fluid. They used a heat pipe with and without wick. Heat pipe was experimentally studied by Said and Akash (1999) with and without the consideration of wick. Heat pipe with wick structure was found better overall heat transfer coefficient as compared the heat pipe without wick. The one-dimensional mathematical model was developed by Kim et al. (2003) for heat and mass transfer in a heat pipe having grooved wick structure. The primary factors affecting the heat transfer performance of heat pipes were experimentally studied by Riegler (2003).

Generalized extremal optimization (GEO) was proposed by Sousa et al. (2004) in order to get the optimized mass of the heat pipe with three different working fluids. Shi et al. (2006) optimized the performance of miniature heat pipes with low temperature co-fired ceramics. The optimal mass characteristics for a heat pipe radiator assembly were investigated by Vlassov et al. (2006). Pareto and quasi-Pareto solutions of a multi-objective design problem were presented by Jeong et al. (2007) for the design optimization of heat pipe for a satellite. Maziuk et al. (2009) developed the software for flat miniature heat-pipe for numerical modelling of the parameters. A Niched Pareto Genetic Algorithm (NPGA) was used by Zhang et al. (2009) for design optimization of U shaped micro grooves heat pipe with objectives of total thermal resistance and transfer capability. Kiseev et al. (2010a) proposed a methodology for analysis and selection of the capillary structure pores size for loop heat pipes.

Taguchi method was used by Senthilkumar (2010) to analyse the parameters of heat pipe Kiseev et al. (2010b) presented the comparative study between the "classical" and "inverted meniscus" scheme for vaporization of loop heat pipes and heat pipes in the thermal management of electronic devices with high heat dissipation. Energy saving and dehumidification enhancement aspects of heat pipe were studied by Yau and Ahmadzadehtalpatapeh (2010). A mathematical model for minimizing the entropy generation of heat pipe was presented by Maheshkumar and Muraleedharan (2011). Taguchi method was used by Agha (2011) for analysing the consequence of the parameters of heat pipe (e.g. wick structure, heat pipe diameter, and working fluid) on the performance of system.

An analytical model was developed by Lips and Lefevre (2011) for solution of 3D temperature field and 2D pressure and velocity fields for a heat pipe. A multi-objective design optimization of multifunctional sandwich panel heat pipes was done Roper (2011). Nithiynandam and Pitchumani (2011) embedded heat pipes into latent thermal energy storage system (LTES) in order to reduce the thermal resistance of LTES by augmenting the energy transfer from the working fluid to the phase change material. A grenade explosion method (GEM) was applied to enhance the heat transfer rate and reduce the resistance of heat pipe (Rao and Rakhade 2011). A two dimensional model of heat conduction of heat pipe's wall was developed by Shabgard and Faghri (2011). A steady-state model for rotating heat pipe was presented by Bertossi et al. (2012).

Optimal design of condenser for a miniature loop heat pipes was done by Wang et al. (2012). An entrasy dissipation analysis for separated heat pipe was carried out by Dong et al. (2012). Yang et al. (2014) analysed the main characteristics of a typical pulsating heat pipe by using one of the state-of the-art mathematical models. Wang (2014) had experimentally investigated the thermal performance of heat pipes with different fan speeds and heat source areas. Cui et al. (2014) did the experiments to examine the closed-loop heat pipe with four types of working fluids, namely methanol, demonized water, ethanol and acetone. Morawietz and Hermann (2014) presented a multidisciplinary design optimization for heat pipe solar collector. Rao and More (2015) presented the application of teaching-learning-based-optimization (TLBO) algorithm for the optimal design of the heat pipes. Turgut and Çoban (2016) had presented the optimal design of heat pipe using global best algorithm (GBA). Rao and More (2017) applied Jaya and self-adaptive Jaya algorithms for the design optimization of selected thermal devices including the heat pipes.

It can be observed from the above literature review that few researchers had applied advanced optimization techniques for the design optimization of heat pipe, e.g. Generalized Extremal Optimization (GEO) algorithm (Sousa et al. 2004; Vlassov et al. 2006), Niched Pareto Genetic Algorithm (NPGA) (Zhang et al. 2009), Compromise Method (Roper 2011), Grenade Explosion Method (GEM) (Rao and Rakhade 2011), Multidisciplinary Design Optimization (MDO) (Morawietz and Hermann 2014), Multidimensional Visualization and Clustering Technique (Jeong et al. 2007), TLBO (Rao and More 2015) and GBA (Turgut and Çoban 2016). However, results of the related studies have shown that there are still chances to improve the quality of solutions. In this chapter, the Jaya algorithm and its variants are proposed for the design optimization of heat pipe.

These advanced optimization algorithms are having their individual merits but they involve the tuning of their specific parameters (except TLBO algorithm). For example, GEO algorithm needs binary string of length, mutation probability. NPGA needs proper setting of crossover probability, mutation probability, selection operator, etc.; GEM needs tuning of agent's territory radius, value of exponent, GBA needs cross over operator and mutation operator, etc. Similarly, all other optimization advanced algorithms are having their own specific parameters. These parameters are called algorithm-specific parameters and need to be controlled other than the common control parameters of number of iterations and population size. All population based algorithms need to tune the common control parameters but the algorithm-specific parameters are specific to the particular algorithm and these are also to be tuned as mentioned above.

The performance of the optimization algorithms is much affected by these parameters. The suitable setting of these parameters is very much necessary. Increase in the computational cost or tending towards the local optimal solution is caused by the improper tuning of these parameters. Hence, to overcome the problem of tuning of algorithm-specific parameters, TLBO algorithm was proposed which is an algorithm-specific parameter less algorithm (Rao 2016). Keeping in view of the good performance of the TLBO algorithm, Jaya algorithm, which is another algorithm-specific parameter-less algorithm, and its variants are used for the single and multi-objective design optimization of heat pipe.

5.1.1 Case Studies Related to Design Optimization of Heat Pipe

This section presents the mathematical model and objective functions related to the single and multi-objective optimization of heat pipe design.

5.1.1.1 Case Study 1: Single Objective Optimization of Heat Pipe

This case study was presented by Sousa et al. (2004) and Turgut and Çoban (2016). The heat pipe is a hermetically sealed tube-type container with a wick structure placed on its internal walls and filled with a working fluid. Figure 5.2 shows a conceptual drawing of a heat pipe in operation. Methanol is used as working fluid and stainless steel (SS 304) is used as the material of the container. The wick is of the mesh type and is made of SS 304.

The objective function is to minimize the total mass of the heat pipe (m_{total}), under specified boundary conditions on the condenser and desirable heat transfer rate. In this problem 18 constraints are considered which include structural, operational and dimensional constraints.

The optimization problem can then be formulated as:

Minimize

$$m_{total} = m_{cont} + m_{wd} + m_{wl} + m_{vapour} \tag{5.1}$$

where, m_{cont} is the mass of container, m_{wd} is the mass of dry wick, m_{wl} is the mass of liquid in the wick and m_{vapour} is the mass of the fluid vapour inside the heat pipe.

Operational constraints are posed to ensure that the HP will operate properly for a given heat transfer rate (Q), at a given sink temperature (T_{si}), while keeping the evaporator wall temperature (T_{so}).

The constraints are:

$$Q \leq Q_c \tag{5.2}$$

$$T_{somin} \leq T_{so} \leq T_{somax} \tag{5.3}$$

$$Q \leq Q_b \tag{5.4}$$

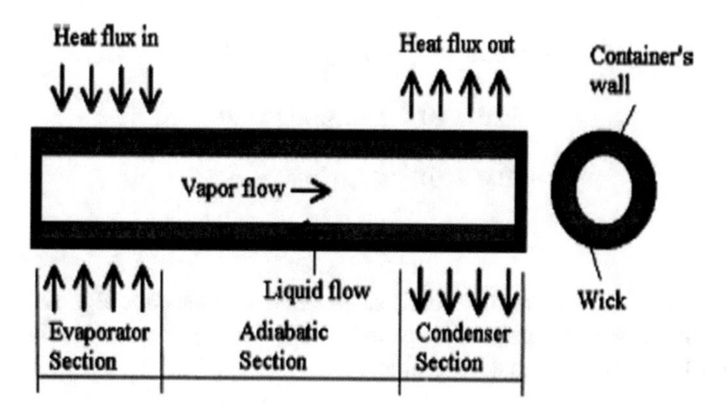

Fig. 5.2 Conceptual drawing of a conventional heat pipe in operation (Sousa et al. 2004; Reprinted with permission from Elsevier)

$$Q \leq Q_e \tag{5.5}$$

$$Q \leq Q_v \tag{5.6}$$

$$M_v \leq 0.2 \tag{5.7}$$

$$R_{ev} \leq 2300 \tag{5.8}$$

$$2d * (1 + \beta) \leq t_w \tag{5.9}$$

$$\frac{\Delta P(d_o^2 + d_i^2)}{(d_o^2 - d_i^2)} \leq \frac{uts}{4} \tag{5.10}$$

$$\frac{\Delta P(d_o^2 + 2d_i^3)}{2(d_o^3 - d_i^3)} \leq \frac{uts}{4} \tag{5.11}$$

$$0.0001 \leq \varepsilon \leq 0.9999 \tag{5.12}$$

where, Q_e is the entrainment limit, Q_v is the viscous limit, M_v is the Mach number limit, R_{ev} is the Reynolds number limit, Q_c is the capillary heat transport limit, Q_b is the boiling limit, ε is the porosity.

The design variables considered in the problem are the wick‘s mesh number (N_m), the diameter of wick (d), the diameter of the vapour core (d_v), the thickness of wick (t_w), the length of the evaporator section (L_e), the length of the condenser section (L_c) and the thickness of the container‘s wall (t_t).

$$315 \leq N_m \leq 15{,}000 \tag{5.13}$$

$$0.025 \times 10^{-3} \leq d \leq 0.1 \times 10^{-3} \tag{5.14}$$

$$5.0 \times 10^{-3} \leq d_v \leq 80 \times 10^{-3} \tag{5.15}$$

$$0.05 \times 10^{-3} \leq t_w \leq 10.0 \times 10^{-3} \tag{5.16}$$

$$50.0 \times 10^{-3} \leq L_e \leq 400 \times 10^{-3} \tag{5.17}$$

$$50.0 \times 10^{-3} \leq L_c \leq 400 \times 10^{-3} \tag{5.18}$$

$$0.3 \times 10^{-3} \leq t_t \leq 3.0 \times 10^{-3} \tag{5.19}$$

Sousa et al. (2004) used generalized external optimization algorithm for the above optimization problem. Turgut and Çoban (2016) used global based optimization for above optimization problem.

5.1.1.2 *Case Study 2: Multi-objective Optimization of Heat Pipe*

This case study was introduced by Zhang et al. (2009). In axially grooved heat pipes, the liquid flows in the capillary microgrooves and vapour flows in the vapour core in a counter-current fashion. The given geometrical structure of heat pipe with axial "Ω" shaped micro grooves as shown in Fig. 5.3. Ammonia is used as working fluid and aluminium alloy is used as the material of the container.

In this problem the two objectives are considered, one of them is maximizing the heat transfer rate and the other one is minimizing the resistance of a heat pipe. The slot width (W), wick diameter (d), slot height (δ), vapour core diameter (d_v) and groove number (N), are selected as decision variables in the optimization process. The additional assembly parameters such as the solid material, working fluid, outer diameter (d_o), length of each section and working temperature, T_{work}, have not been included in the optimization since they cannot be adjusted for a given application. The calculation conditions for heat pipe are given in Table 5.1. The heat flow paths and thermal resistances are shown in Fig. 5.4.

Using these criteria, the optimization problem can be simply formulated as a combined objective function as Eq. (5.20).

$$\text{Minimize}; Z = w_1\left(\frac{Z_1}{Z_{1min}}\right) - w_2\left(\frac{Z_2}{Z_{2max}}\right) \tag{5.20}$$

$$Z_2 = Q = f_1(W, d, \delta, d_v, N) \tag{5.21}$$

$$Z_1 = R_{tot} = f_2(W, d, \delta, d_v, N) \tag{5.22}$$

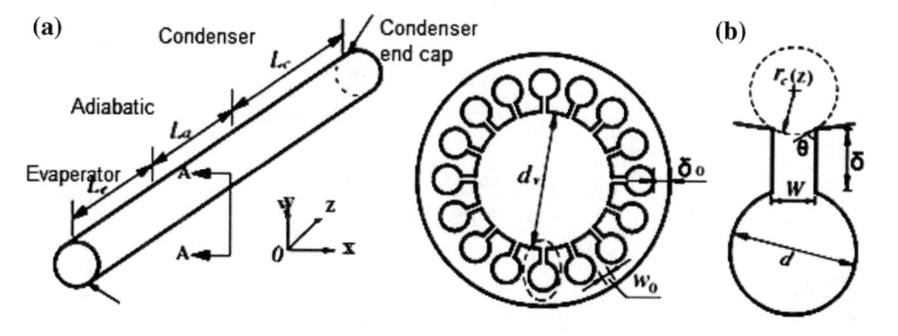

Fig. 5.3 Schematic of heat pipe with axial "Ω"-shaped micro grooves: **a** Cross section. **b** Geometric shape (Zhang et al. 2009; Reprinted with permission from Elsevier)

Table 5.1 Calculation conditions for heat pipe

L_e (m)	L_c (m)	L_a (m)	D_o (mm)	T_{work} (k)	Working fluid	Solid material
0.7	0.64	0.51	12.5	293	Ammonia	Al alloy

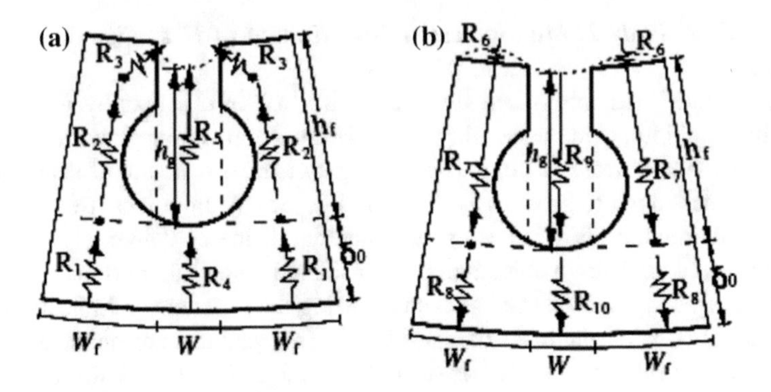

Fig. 5.4 Heat flow paths and thermal resistance **a** Evaporator section. **b** Condenser section (Zhang et al. 2009; Reprinted with permission from Elsevier)

A relationship that gives the maximum allowable heat input rate during the start-up condition is.

$$Q_{max} = 0.4\,\Pi\, r_c^2\, 0.73\, h_{fg}\, (P_v \cdot \rho_v)^{1/2} \tag{5.23}$$

where, r_c is the vapour channel radius, h_{fg} is the latent heat of vaporisation and P_v, ρ_v are the vapour pressure and vapour density in the evaporator section.

The thermal resistance of evaporator R_e, thermal resistance of condenser R_c and thermal resistance due to the vapour flow R_v can be expressed as.

$$R_e = \frac{1}{N}\frac{(R_1 + R_2 + R_3)(R_4 + R_5)}{(R_1 + R_2 + R_3) + 2(R_4 + R_5)} \tag{5.24}$$

$$R_c = \frac{1}{N}\frac{(R_6 + R_7 + R_8)(R_9 + R_{10})}{(R_6 + R_7 + R_8) + 2(R_9 + R_{10})} \tag{5.25}$$

$$R_v = \frac{T_{work}\left(P_{v,e} - P_{v,c}\right)}{\rho_v \cdot h_{fg} \cdot Q_{in}} \tag{5.26}$$

where,

$$R_1 = \frac{\delta_o}{k_s W_f L_e} \tag{5.27}$$

$$R_2 = \frac{h_g}{k_{seq} W_f L_e} \tag{528}$$

$$R_3 = \frac{1}{\propto \text{film}\left(h_f - h_g\right)L_e} \tag{5.29}$$

$$R_4 = \frac{\delta_o}{k_s W L_e} \tag{5.30}$$

$$R_5 = \frac{h_g}{k_l W L_e} \tag{5.31}$$

$$R_6 = \frac{\delta_{film}}{k_l W_f L_c} \tag{5.32}$$

$$R_7 = \frac{h_f}{k_{seq} W_f L_c} \tag{5.33}$$

$$R_8 = \frac{\delta_o}{k_s W_f L_c} \tag{5.34}$$

$$R_9 = \frac{h_g}{k_l W L_c} \tag{5.35}$$

$$R_{10} = \frac{\delta_0}{k_s W L_c} \tag{5.36}$$

The total thermal resistance R_{tot},

$$R_{tot} = R_e + R_c + R_v \tag{5.37}$$

where, T_{work} is the working temperature, $P_{v,e}$ and $P_{v,c}$ are the vapour pressure at the evaporator and condenser section, respectively. N is the groove number, ρ_v is the vapour density, Q_{in} is the heat load, h_{fg} is the latent heat, R_1–R_{10} are the thermal resistances as shown in Fig. 5.3, k_l and k_s are the conductivity of liquid and solid, $k_{s,eq}$ is the equivalent conductivity coefficient of groove fin, L_e and L_c are the length of evaporator and condenser, δ_{film} is the mean condensate liquid film thickness, the evaporating thin film heat transfer coefficient $\propto \text{film} = \frac{k_l}{0.185 W_f}$

The following constraints and boundary conditions can be introduced as,

$$d_v/2 + \delta + \left(d^2 - W^2\right)^{1/2}/4 + d/2 + \delta_o = d_o/2 \tag{5.38}$$

$$N(w_o + d) = 2\Pi\left(d_v/2 + \delta + \left(d^2 - W^2\right)^{1/2}/4\right) \tag{5.39}$$

$$W \leq d \tag{5.40}$$

N is integer

$$w_o \geq w'; \ \delta_o \geq \delta'; \ W \geq s; \ \delta \geq s; \ d_v \geq s \tag{5.41}$$

where, w', δ' and s are the minimal sizes limited by machining, and are assumed to be $w' = 0.36$ mm, $\delta' = 0.4$ mm and s = 0.22 mm, respectively. Equations (5.38) and (5.39) represent the geometrical constraints for radial and circumferential length, Eq. (5.40) is derived by the geometric shape of a "Ω" shaped groove, channel number must be an integer, and Eq. (5.41) displays the minimal size limited by machining.

Following are the upper and lower bounds of design variables imposed for heat pipe:

- Diameter of wick structure (d) ranging from 0.22 to 1.8 mm;
- Diameter of vapour core (d_v) ranging from 0.22 to 8.3 mm;
- Height of slot (δ) ranging from 0.22 to 1.1 mm;
- Width of slot (W) ranging from 0.22 to 0.65 mm;
- Number of slot (N) ranging from 15 to 29;

Zhang et al. (2009) used niched Pareto genetic algorithm (NPGA) for the above optimization problem. Rao and Rakhade (2011) used grenade explosion method for above optimization problem. Now, the Jaya algorithm and its variants of self-adaptive Jaya algorithm, SAMP-Jaya and SAMPE-Jaya algorithm are applied to these design optimization problems. The next section presents the results and discussion related to the application of the algorithms for the design optimization of heat pipe.

5.1.2 *Results and Discussion*

The computational experiments are conducted using a Notebook PC of 4 GB RAM memory, 1.9-GHz AMD A8 4500 M APU CPU.

5.1.2.1 Case Study 1: Minimization of Total Mass of Heat Pipe

For getting the performance of HP, different constant temperature (T_{si}) at condenser section are considered. These temperatures are ranging from −15 to 30 °C with the interval of 15 °C. The heat transfer rate (Q) was varied from 25.0 to 100.0 W. Three different working fluids namely Methanol, Ammonia and Ethanol are considered based on the compatibility of and application of HPs. The physical and thermal properties of these fluids are temperature dependent. Now, the SAMPE-Jaya algorithm is applied to these problems with maximum function evaluations of 25,000.

Now, the performance of the SAMPE-Jaya algorithm is tested with different combinations of population sizes (viz. 10, 20, 25, 50, 75, and 100) and different elite sizes (viz. 2, 4, 6, 8 etc.). Tables 5.2, 5.3 and 5.4 present the optimal mass of heat pipe at different values of T_{si} and Q with Methanol, Ammonia and Ethanol respectively. It can be observed from Table 5.2 that the optimal values of total mass of HP obtained by the SAMPE-Jaya algorithm is better or competitive as compared to TLBO, GBA, Jaya and SAMP-Jaya algorithms. For T_{si} = 0 °C and Q = 25.0 W, the optimal value of the objective function obtained by basic Jaya algorithm is better as compared to the other algorithms. However, SAMPE-Jaya algorithm requires less number of mean function evaluations (MFE) as compared to Jaya and SAMP-Jaya algorithms.

Table 5.3 presents different designs of HP pipe at different operational conditions while Ammonia was considered as the working fluid. It can be observed from this table that the optimum value of total mass of HP obtained by SAMPE-Jaya algorithm is less in each case as compared to the designs suggested by using Jaya and SAMP-Jaya algorithms. As the other researchers did not attempt the problem while considering Ammonia and Ethanol as the working fluids, therefore, designs obtained by the presents study cannot be compared with the previous studies. It can also be observed from Table 5.3 that the SAMPE-Jaya algorithm requires less number of mean function evaluations (MFE) as compared to the Jaya and SAMPE-Jaya algorithms.

Table 5.4 presents the different designs of HP at different operational conditions while Ethanol was considered as the working fluid. It can be observed from this table that the optimum value of total mass of HP obtained by SAMPE-Jaya algorithm is less in each case as compared to the designs suggested by using Jaya and SAMP-Jaya algorithms. It can also be observed from this table that the no. of MFEs required by the SAMPE-Jaya algorithm is less as compared to Jaya and SAMPE-Jaya algorithms.

It can also be observed from Tables 5.2, 5.3 and 5.4 that when Ammonia is considered as the working fluid then the design obtained at condenser section temperature of −15°C is cheaper (minimum total mass heat pipe) as compared the designs when the Methanol and Ethanol were considered as the working fluids. The designs obtained with Methanol and Ethanol as working fluids are almost same for the condenser section temperature −15 °C. Furthermore, for the condenser temperature from 0 to 30 °C the designs obtained while using Methanol as the working fluid are better as compared to the designs obtained while using Ethanol and Ammonia as the working fluid. The designs obtained while using Ethanol and Ammonia for the condenser section temperature 0–30 °C are almost same. Therefore, it is recommended that the Ammonia should be used as working fluid when the condenser section temperature is below 0 °C and for the higher temperature methanol should be used as the working fluid.

Figures 5.5 and 5.8 show the variation of total mass of HP for different algorithms at different temperatures of condensation section (T_{si}) with different values of heat transfer rate (Q). It can be observed that the value of total of mass HP obtained by using proposed SAMPE-Jaya algorithm is varying almost linearly with respect to variation in the heat transfer rate. It can also be observed from these figures that the

Table 5.2 Optimal mass of heat pipe with Methanol as working fluid at different values of T_{si} and Q

T_{si} (°C)	Q (W)	Method	N	D (m)	d_v (m)	t_w (m)	L_e (m)	ε	M_{total} (kg)	MFE
−15	25	TLBO	1473	3.2121E-05	0.0295	7.7965E-05	0.081	0.8948	0.5568	–
		GBA	–	–	–	–	–	–	–	–
		Jaya algorithm	7749.9	3.1275E-05	0.0297	7.0445E-05	0.0829	0.9503	0.5143	24535
		SAMP-Jaya	1174.25	0.000527	0.045526	0.001264	0.050001	0.9699	0.331081	22973.56
		SAMPE-Jaya	1457.445	0.00043	0.035747	0.001031	0.05	0.9699	**0.244033**	**19730.21**
	50	TLBO	8316	3.4416E-05	0.0346	9.6686E-05	0.0776	0.9565	0.645	–
		GBA	–	–	–	–	–	–	–	–
		Jaya algorithm	9939.8	2.3817E-05	0.0329	9.5959E-05	0.1178	0.9202	0.6045	23256.25
		SAMP-Jaya	5365.925	0.0001	0.052708	0.003483	0.055985	0.9999	0.576062	14296.88
		SAMPE-Jaya	944.8135	0.000647	0.057974	0.001554	0.05	0.9699	**0.454201**	**11084.38**
	75	TLBO	9654	3.1034E-05	0.0362	9.9797E-05	0.1353	0.935	0.7049	–
		GBA	–	–	–	–	–	–	–	–
		Jaya algorithm	8833.2	3.1475E-05	0.0357	9.0437E-05	0.1132	0.8991	0.6987	16964.58
		SAMP-Jaya	5319.474	0.0001	0.070436	0.003905	0.070925	0.9999	0.827357	13545.83
		SAMPE-Jaya	833.4549	0.00073	0.066835	0.001752	0.05	0.9699	**0.549425**	**15173.96**
	100	TLBO	1301	3.6634E-05	0.0393	9.7193E-05	0.1253	0.8506	1.4696	–
		GBA	–	–	–	–	–	–	–	–
		Jaya algorithm	3852	3.5210E-05	0.0394	8.9802E-05	0.0873	0.8746	1.3058	16097.92
		SAMP-Jaya	5278.286	0.0001	0.08	0.004669	0.098994	0.9999	1.094008	15705.21
		SAMPE-Jaya	792.7362	0.000741	0.077365	0.001778	0.058267	0.9699	**0.64553**	**16658.33**
0	25	TLBO	4271	3.0070E-05	0.0287	7.3560E-05	0.0668	0.9162	0.5084	–
		GBA	3700	2.5910E-04	0.0388	9.9270E-04	0.063	0.9661	0.2904	–
		Jaya algorithm	3908.3	2.8332E-04	0.0231	9.4144E-04	0.0637	0.9637	**0.1747**	2072.917
		SAMP-Jaya	1466.724	0.000415	0.030541	0.000995	0.05	0.9699	0.206984	14520.83

(continued)

Table 5.2 (continued)

T_{si} (°C)	Q (W)	Method	N	D (m)	d_v (m)	t_w (m)	L_e (m)	ε	M_{total} (kg)	MFE
		SAMPE-Jaya		0.0003	0.030541	0.001189	0.05	0.9999	0.205804	**6067.708**
	50	TLBO	1681	3.6390E-05	0.0329	9.0940E-04	0.1742	0.9469	0.6359	–
		GBA	1796	3.5470E-04	0.0403	8.8110E-04	0.1683	0.06913	0.5146	–
		Jaya algorithm	1960	3.7000E-04	0.0409	8.4900E-04	0.1657	0.7161	0.4239	24677.08
		SAMP-Jaya	1856.836	0.0003	0.042675	0.001481	0.062723	0.9999	0.311128	14794.79
		SAMPE-Jaya	1182.003	0.000503	0.039273	0.001207	0.050718	0.969899	**0.281858**	**6585.417**
	75	TLBO	6390	3.6730E-05	0.0392	8.8660E-05	0.1081	0.9315	0.8532	–
		GBA	7176	6.6020E-05	0.0588	1.2000E-03	0.1524	0.9524	0.6406	–
		Jaya algorithm	7190	6.3500E-05	0.0553	9.9500E-04	0.1552	0.9188	0.4973	17507.29
		SAMP-Jaya	830.447	0.0003	0.053588	0.001701	0.086192	0.9999	0.42307	15742.71
		SAMPE-Jaya	1052.186	0.000558	0.046271	0.00134	0.071298	0.9699	**0.354825**	**16243.75**
	100	TLBO	7551	2.5820E-05	0.0386	9.7020E-05	0.0686	0.6567	1.6575	–
		GBA	3576	1.8020E-04	0.064	7.5400E-04	0.3663	0.7156	0.9148	–
		Jaya algorithm	3511.6	1.7292E-04	0.0561	7.4621E-04	0.3285	0.7768	0.7801	20386.25
		SAMP-Jaya	1804.779	0.0003	0.064962	0.001837	0.10256	0.9999	0.537944	18340
		SAMPE-Jaya	969.927	0.000598	0.052377	0.001436	0.089722	0.9699	**0.422935**	**15824.17**
15	25	TLBO	8612	2.9660E-05	0.0297	6.9900E-05	0.0793	0.9811	0.5234	–
		GBA	10641	4.7410E-05	0.04218	1.5140E-04	0.1422	0.9664	0.2785	–
		Jaya algorithm	10700	4.5800E-05	0.0388	1.3200E-04	0.1522	0.9501	0.2499	2570.833
		SAMP-Jaya	1861.138	0.0003	0.030376	0.000924	0.059909	0.9999	0.191677	4749.583
		SAMPE-Jaya	1651.5	0.000343	0.030376	0.000823	0.05	0.9699	**0.19112**	**2539.583**
	50	TLBO	1401	3.4418E-05	0.0393	9.7385E-05	0.0785	0.8845	0.6593	–
		GBA	14680	6.2630E-05	0.04407	5.1850E-05	0.3169	0.9063	0.3566	–

(continued)

Table 5.2 (continued)

T_{si} (°C)	Q (W)	Method	N	D (m)	d_v (m)	t_w (m)	L_e (m)	ε	M_{total} (kg)	MFE
		Jaya algorithm	14680	6.2630E-05	0.04407	5.1850E-05	0.3169	0.9063	0.3566	20539.17
		SAMP-Jaya	1812.103	0.0003	0.043278	0.001255	0.114057	0.9999	0.320409	19358.33
		SAMPE-Jaya	1264.937	0.000445	0.038271	0.001068	0.100628	0.9699	**0.283697**	**13567.92**
	75	TLBO	1494	2.9830E-05	0.0398	8.9351E-05	0.1334	0.9604	0.914	–
		GBA	3158	5.3700E-05	0.06357	5.1000E-04	0.2814	0.9654	0.6037	–
		Jaya algorithm	3479.4	5.25E-05	0.0593	4.9955E-04	0.2635	0.9835	0.517	16752.5
		SAMP-Jaya	1792.296	0.0003	0.05811	0.00138	0.140743	0.9999	0.456159	15312.08
		SAMPE-Jaya	1080.305	0.000525	0.043833	0.00126	0.153837	0.969894	**0.371016**	**13147.5**
	100	TLBO	6819	2.7130E-05	0.0333	7.3836E-05	0.1347	0.9197	1.9254	–
		GBA	7333	9.3980E-05	0.05329	9.5430E-04	0.3316	0.8476	0.7017	–
		Jaya algorithm	7174.7	9.8130E-05	0.0503	9.3768E-04	0.3482	0.8698	0.5757	19254.17
		SAMP-Jaya	1800.054	0.0003	0.074203	0.001436	0.153567	0.9999	0.596119	16786.25
		SAMPE-Jaya	1003.844	0.000556	0.051095	0.001334	0.185976	0.969899	**0.459103**	**15543.33**
30	25	TLBO	5384	3.4309E-05	0.0326	7.4947E-05	0.0943	0.9202	0.5771	–
		GBA	4754	1.9240E-04	0.03636	1.3000E-03	0.1207	0.8534	0.5015	–
		Jaya algorithm	4223.3	1.6119E-04	0.0477	3.4295E-04	0.1207	0.8261	0.3818	15066.25
		SAMP-Jaya	1728.133	0.000315	0.030277	0.000755	0.09133	0.9699	0.200262	15818.33
		SAMPE-Jaya	1821.058	0.0003	0.030277	0.000796	0.102409	0.9999	**0.195901**	**10248.33**
	50	TLBO	4565	1.9240E-04	0.03636	1.3000E-03	0.1207	0.8534	0.9617	–
		GBA	11880	6.9080E-05	0.05331	1.1000E-03	0.2435	0.9538	0.6639	–
		Jaya algorithm	3538.8	2.8800E-05	0.0288	8.9380E-05	0.0857	0.9169	0.5175	**12467.5**
		SAMP-Jaya	1786.039	0.0003	0.046978	0.001026	0.170601	0.9999	0.35198	15487.92
		SAMPE-Jaya	1314.121	0.000416	0.038146	0.000997	0.186939	0.9699	**0.311893**	16736.25

(continued)

Table 5.2 (continued)

T_{si} (°C)	Q (W)	Method	N	D (m)	d_v (m)	t_w (m)	L_e (m)	ε	M_{total} (kg)	MFE
	75	TLBO	7450	2.6820E-05	0.0405	9.4628E-05	0.0676	0.9467	1.6426	–
		GBA	6474	1.2230E-04	0.05371	1.2000E-03	0.3422	0.828	0.9626	–
		Jaya algorithm	6876.7	1.2029E-04	0.0536	1.0000E-03	0.2693	0.8101	**0.9247**	15920.83
		SAMP-Jaya	1776.947	0.0003	0.067741	0.001062	0.184712	0.9999	0.516233	18272.5
		SAMPE-Jaya	1136.497	0.000479	0.045451	0.00115	0.269091	0.9699	**0.430415**	**14940.42**
	100	TLBO	4579	2.6571E-05	0.0382	9.0315E-05	0.1602	0.5889	2.6856	–
		GBA	7631	9.7270E-05	0.07796	1.3000E-03	0.1843	0.7727	1.4598	–
		Jaya algorithm	6615.5	1.5731E-04	0.0493	1.0000E-03	0.3988	0.7592	1.3104	14614.58
		SAMP-Jaya	6375.4	9.9375E-05	0.0657	1.1000E-03	0.2138	0.7646	0.691162	18797.92
		SAMPE-Jaya	1072.018	0.000504	0.054871	0.00121	0.312516	0.9699	**0.553047**	**14542.08**

Note Bold values show optimal solutions; '–': values are not available

Table 5.3 Optimal mass of heat pipe with Ammonia as working fluid at different values of T_{si} and Q

T_{si} (°C)	Q (W)	Method	N	D (m)	d_v (m)	t_w (m)	L_e (m)	ε	M_{total} (kg)	MFE
−15	25	Jaya algorithm	3020.9464	0.0001705	0.029814	0.0004091	0.05	0.51312748	0.236383	21185
		SAMP-Jaya	2871.1696	0.0001768	0.029814	0.0004242	0.129813555	0.96989985	**0.181055**	20196
		SAMPE-Jaya	2871.1696	0.0001768	0.029814	0.0004242	0.129813555	0.96989985	**0.181055**	**20185**
	50	Jaya algorithm	2462.5276	0.0002145	0.0375633	0.0005147	0.05	0.27779511	0.392641	20917.5
		SAMP-Jaya	2182.624	0.0002346	0.0375635	0.0005631	0.2627481	0.9698971	0.28451	20622.5
		SAMPE-Jaya	2185.1448	0.0002346	0.0375633	0.000563	0.262764301	0.96989949	**0.284507**	**16778**
	75	Jaya algorithm	1833.681	0.0002807	0.0434113	0.0006737	0.4	0.9699	0.397124	21955
		SAMP-Jaya	1833.681	0.0002807	0.0434113	0.0006737	0.4	0.9699	0.397124	19436
		SAMPE-Jaya	1834.979	0.0002807	0.0434118	0.0006737	0.4	0.9699	**0.397122**	**17321.67**
	100	Jaya algorithm	1833.626	0.0002807	0.0581048	0.0006737	0.4	0.9699	0.528833	17795.83
		SAMP-Jaya	1833.626	0.0002807	0.0581048	0.0006737	0.4	0.9699	0.528833	21199.33
		SAMPE-Jaya	1833.626	0.0002807	0.0581048	0.0006737	0.4	0.9699	0.528833	**19723.33**
0	25	Jaya algorithm	3149.126	0.0001661	0.0284093	0.0003985	0.05000176	0.3024126	0.256573	20935.83
		SAMP-Jaya	2830.5647	0.0001903	0.0284092	0.0004567	0.4	0.9699	0.245339	**19028.67**
		SAMPE-Jaya	2773.439	0.0001805	0.0284093	0.0004331	0.2401268	0.9699	**0.202301**	20259.17
	50	Jaya algorithm	2139.71	0.0002403	0.0357934	0.0005766	0.3999983	0.8635097	0.36876	20733.33
		SAMP-Jaya	2068.867	0.0002271	0.0414805	0.0005457	0.3998826	0.9699	0.366499	19346.66
		SAMPE-Jaya	2254.818	0.0002244	0.0412681	0.0005386	0.3999994	0.9698999	**0.363643**	**19279.17**
	75	Jaya algorithm	2252.898	0.0002245	0.0621692	0.0005387	0.4	0.9699	0.545868	21006.67

(continued)

Table 5.3 (continued)

T_{si} (°C)	Q (W)	Method	N	D (m)	d_v (m)	t_w (m)	L_e (m)	ε	M_{total} (kg)	MFE
		SAMP-Jaya	2289.15	0.0002358	0.04121	0.0005938	0.05	0.07480693	0.537763	20790
		SAMPE-Jaya	2232.228	0.0002393	0.0409732	0.0005744	0.05	0.07985223	**0.521519**	**20661.67**
	100	Jaya algorithm	2219.414	0.0002286	0.08	0.0005486	0.4	0.9413235	0.732674	20646.67
		SAMP-Jaya	2033.962	0.0002634	0.0450969	0.0006321	0.05	0.04644479	**0.623851**	20604
		SAMPE-Jaya	2033.962	0.0002634	0.0450969	0.0006321	0.05	0.04644479	**0.623851**	**19110.83**
15	25	Jaya algorithm	2709.074	0.0001867	0.0280469	0.0004481	0.4	0.7898598	0.292401	20910
		SAMP-Jaya	3317.4393	0.0001628	0.0280661	0.000391	0.05	0.11718371	0.28055	19544.66
		SAMPE-Jaya	3270.684	0.0001622	0.0280469	0.0003893	0.05	0.1194038	**0.279287**	**19138.67**
	50	Jaya algorithm	2619.438	0.0002043	0.0353369	0.0004904	0.05	0.03665674	0.420318	20666.67
		SAMP-Jaya	2619.438	0.0002043	0.0353369	0.0004904	0.05	0.03665674	0.420318	20430.66
		SAMPE-Jaya	2619.438	0.0002043	0.0353369	0.0004904	0.05	0.03665674	0.420318	**19217.33**
	75	Jaya algorithm	3179.1198	0.0001671	0.08	0.000401	0.05	0.10610671	0.815624	19545.83
		SAMP-Jaya	2294.466	0.0002339	0.0404506	0.0005614	0.05	0.00627305	**0.533083**	**20538**
		SAMPE-Jaya	2294.466	0.0002339	0.0404506	0.0005614	0.05	0.00627305	**0.533083**	20550.67
	100	Jaya algorithm	2063.202	0.0002586	0.0445217	0.0006207	0.05986843	0.0001	0.63957	20900.83
		SAMP-Jaya	2592.306	0.0002054	0.0349811	0.0004931	0.05202193	0.0001	0.429932	19545.33
		SAMPE-Jaya	2592.306	0.0002054	0.0349811	0.0004931	0.05202193	0.0001	**0.429932**	**16596.5**

(continued)

Table 5.3 (continued)

T_{si} (°C)	Q (W)	Method	N	D (m)	d_v (m)	t_w (m)	L_e (m)	ε	M_{total} (kg)	MFE
30	25	Jaya algorithm	3306.901	0.0001629	0.0277645	0.0003909	0.05	0.0001	0.296561	18137.71
		SAMP-Jaya	3277.512	0.0001629	0.0277645	0.000391	0.05	0.04641671	**0.289545**	17237.71
		SAMPE-Jaya	3277.512	0.0001629	0.0277645	0.000391	0.05	0.04641671	**0.289545**	**17143.83**
	50	Jaya algorithm	2591.8229	0.0002054	0.0349811	0.0004931	0.052020674	0.0001	0.429933	16412.71
		SAMP-Jaya	2592.306	0.0002054	0.0349811	0.0004931	0.05202193	0.0001	**0.429932**	20150.3
		SAMPE-Jaya	2592.306	0.0002054	0.0349811	0.0004931	0.05202193	0.0001	**0.429932**	**15949.03**
	75	Jaya algorithm	2236.2494	0.0002378	0.0400434	0.0005708	0.077325811	0.0001	0.562958	15322.08
		SAMP-Jaya	2236.2494	0.0002378	0.0400434	0.0005708	0.077325811	0.0001	0.562958	18788.96
		SAMPE-Jaya	2236.2494	0.0002378	0.0400434	0.0005708	0.077325811	0.0001	0.562958	**15291.88**
	100	Jaya algorithm	2006.3586	0.0002647	0.0440735	0.0006354	0.102911968	0.00010006	0.692561	16364.88
		SAMP-Jaya	2006.3586	0.0002647	0.0440735	0.0006354	0.102911968	0.00010006	0.692561	16151.67
		SAMPE-Jaya	2006.3586	0.0002647	0.0440735	0.0006354	0.102911968	0.00010006	0.692561	**12505.50**

Note Bold values show optimal solutions

Table 5.4 Optimal mass of heat pipe with Ethanol as working fluid at different values of T_{si} and Q

^0C	Q	*Method*	N	D (m)	d_v (m)	t_w (m)	L_e (m)	ε	M_{total} (kg)	*MFE*
-15	25	Jaya algorithm	977.79059	0.0006258	0.045545421	0.001501867	0.05	0.969899005	0.354952	18192.5
		SAMP-algorithm	976.9594	0.0006253	0.04556215	0.001500675	0.05	0.9698991	0.354951	**17794.69**
		SAMPE-Jaya	980.0708	0.000625	0.04557043	0.001500052	0.05	0.9699	**0.354947**	**16555.17**
	50	Jaya algorithm	681.1676	0.001	0.05358831	0.0024	0.05	0.9699	0.522173	17835.42
		SAME-Jaya	788.25602	0.0007691	0.058074451	0.001845898	0.05	0.9699	0.491295	17006
		SAMPE-Jaya	787.1024	0.0007691	0.0580747	0.001845841	0.05	0.9699	**0.491288**	**15895**
	75	Jaya algorithm	693.0486	0.0008695	0.066911196	0.002086771	0.05	0.9699	0.597511	16966.67
		SAMP-Jaya	694.813	0.0008665	0.0670321	0.002079726	0.05000031	0.9699	0.597504	17322.5
		SAMPE-Jaya	694.6743	0.0008668	0.06702162	0.002080303	0.05	0.9699	**0.597499**	**16016.67**
	100	Jaya algorithm	637.66954	0.0009431	0.074235647	0.002263527	0.05	0.9699	0.688264	19574.58
		SAMP-Jaya	634.9009	0.0009449	0.0741551	0.002267867	0.05	0.9699	0.68825	14780.83
		SAMPE-Jaya	634.0668	0.0009459	0.07411445	0.002270107	0.05	0.9699	**0.688251**	**14249.58**
0	25	Jaya algorithm	1048.4496	0.0005859	0.040925717	0.001406063	0.05	0.969899995	0.310691	19983.54
		SAMP-Jaya	1046.718	0.0005871	0.04088569	0.001409142	0.05	0.9699	**0.310689**	**15796.46**
		SAMPE-Jaya	1046.718	0.0005871	0.04088569	0.001409142	0.05	0.9699	**0.310689**	17130.83
	50	Jaya algorithm	1135.8947	0.0004802	0.08	0.001152434	0.05	0.9699	0.548163	18412.5
		SAMP-Jaya	841.6491	0.0007218	0.05210715	0.001732442	0.05	0.9699	0.428183	19453.54
		SAMPE-Jaya	841.7701	0.0007222	0.0520959	0.001733192	0.05	0.9699	**0.428182**	**17645.17**
	75	Jaya algorithm	740.76254	0.0008141	0.060100729	0.001953805	0.05	0.96989969	0.519423	18966.46
		SAMP-Jaya	742.4007	0.0008132	0.0601326	0.001951755	0.05000038	0.9699	0.519416	18446.46
		SAMPE-Jaya	742.9384	0.0008145	0.06008425	0.001954762	0.05	0.9699	**0.519412**	**14180**
	100	Jaya algorithm	678.67743	0.0008878	0.066471178	0.002130802	0.05000102	0.969899694	0.597206	19252.92
		SAMP-Jaya	677.8856	0.0008875	0.06648659	0.002129882	0.05	0.9699	0.597203	16443.75
		SAMPE-Jaya	677.8856	0.0008875	0.06648659	0.002129882	0.05	0.9699	**0.597203**	**15069.17**

(continued)

Table 5.4 (continued)

^0C	Q	*Method*	N	D (m)	d_v (m)	t_w (m)	L_e (m)	ε	M_{total} (kg)	*MFE*
15	25	Jaya algorithm	1094.0086	0.0005197	0.033145357	0.001247323	0.055943	0.9699	0.24421	16015.14
		SAMP-Jaya	1094.0086	0.0005197	0.033145357	0.001247323	0.055943	0.9699	0.24421	16315.21
		SAMPE-Jaya	1094.0086	0.0005197	0.033145357	0.001247323	0.055943	0.9699	**0.24421**	**14930**
	50	Jaya algorithm	314	0.0009304	0.046767859	0.002233391	0.1350799	0.9699	0.499491	14987.29
		SAMP-Jaya	849.0016	0.0006741	0.04198606	0.001618321	0.1130586	0.9699	0.372448	13954.17
		SAMPE-Jaya	844.3783	0.0006752	0.04193639	0.001620587	0.1132918	0.9699	**0.372401**	**13224.79**
	75	Jaya algorithm	314	0.000988	0.05463844	0.002371428	0.1850261	0.9699	0.640274	19083.47
		SAMP-Jaya	1008.43	0.0005331	0.08	0.001279332	0.07248382	0.9699	0.588768	16383.96
		SAMPE-Jaya	760.1577	0.000737	0.05139369	0.001768762	0.1512963	0.9698998	**0.496607**	**13301.25**
	100	Jaya algorithm	314	0.001	0.064070101	0.0024	0.21395275	0.9699	0.778196	16868.96
		SAMP-Jaya	699.5579	0.0007948	0.05917906	0.001907752	0.1887848	0.9699	0.620035	18742.92
		SAMPE-Jaya	715.3008	0.0007837	0.06000446	0.001880824	0.183829	0.9699	**0.619772**	**16402.5**
30	25	Jaya algorithm	1652.7585	0.0003151	0.050453223	0.000756178	0.05	0.9699	0.306361	16471.6
		SAMP-Jaya	1329.616	0.00041	0.03302082	0.000984007	0.09630751	0.9699	0.237382	15892.71
		SAMPE-Jaya	1329.616	0.00041	0.03302082	0.000984007	0.09630751	0.9699	**0.237382**	**15599.88**
	50	Jaya algorithm	1008.842	0.0005424	0.04160365	0.001301653	0.1985322	0.9699	0.378634	19879.58
		SAMP-Jaya	1008.842	0.0005424	0.04160365	0.001301653	0.1985322	0.9699	0.378634	18711.88
		SAMPE-Jaya	1008.842	0.0005424	0.04160365	0.001301653	0.1985322	0.9699	**0.378634**	**17873.33**
	75	Jaya algorithm	949.37464	0.0005777	0.047625359	0.001386648	0.05	0.276170401	0.986993	14860
		SAMP-Jaya	824.8618	0.0006377	0.04887396	0.001530425	0.2952345	0.9699	0.532401	16447.92
		SAMPE-Jaya	898.5314	0.0006101	0.05112846	0.001464446	0.2717625	0.9698452	**0.530439**	**15820.97**
	100	Jaya algorithm	759.2649	0.0007247	0.05443068	0.001739208	0.4	0.9699	0.701614	17936.04
		SAMP-Jaya	852.08436	0.0006432	0.061953031	0.001544812	0.31583895	0.9699	0.686814	18441.67
		SAMPE-Jaya	840.4394	0.0006388	0.06246325	0.001533289	0.3107852	0.9699	**0.68603**	**17594.83**

Note Bold values show optimal solutions

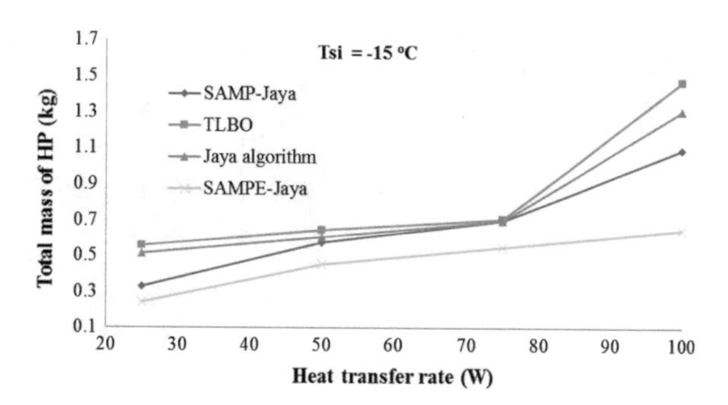

Fig. 5.5 Optimal mass of HP at−15 °C and different values of Q filled Methanol

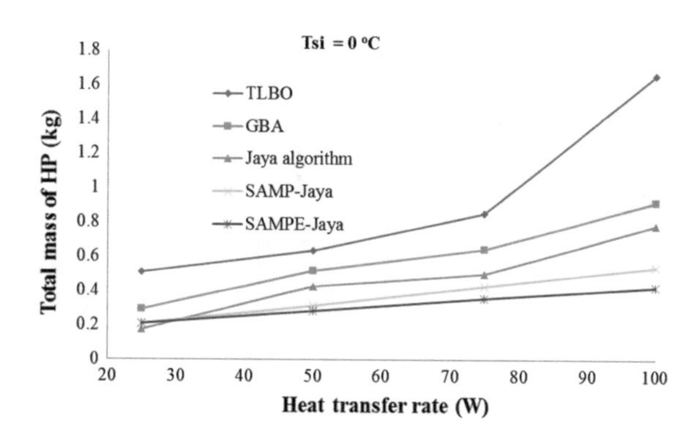

Fig. 5.6 Optimal mass of HP at 0 °C and different values of Q filled Methanol

optimum value of total mass of HP obtained by SAMPE-Jaya algorithm is less in each case as compared to the other methods (Figs. 5.6, 5.7 and 5.8).

Figures 5.9, 5.10, 5.11 and 5.12 present the convergence for SAMPE-Jaya algorithm with different operational conditions and filled with methanol. Figure 5.9 presents the convergence graph of SAMPE-Jaya algorithm at −15 °C condenser temperature with different values of heat transfer rate. It can be observed from Fig. 5.9 that the convergence of the SAMPE-Jaya algorithm takes place before the

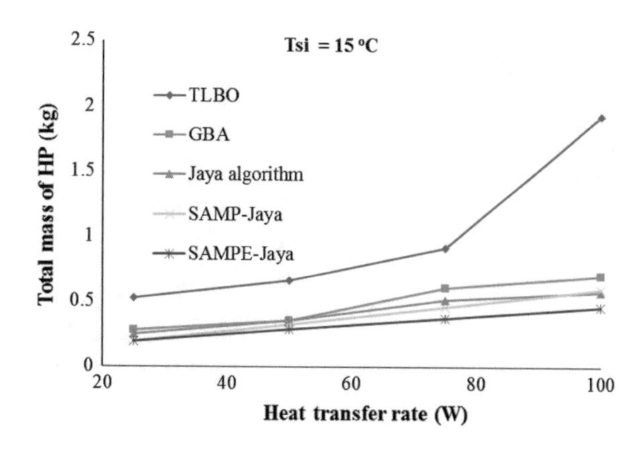

Fig. 5.7 Optimal mass of HP at 15 °C and different values of Q filled Methanol

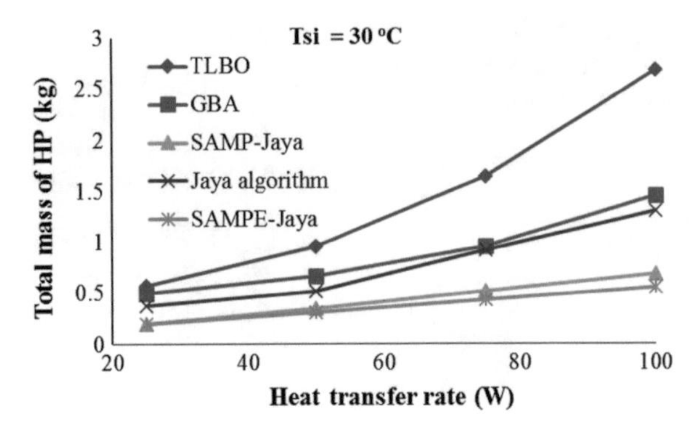

Fig. 5.8 Optimal mass of HP at 30 °C and different values of Q filled Methanol

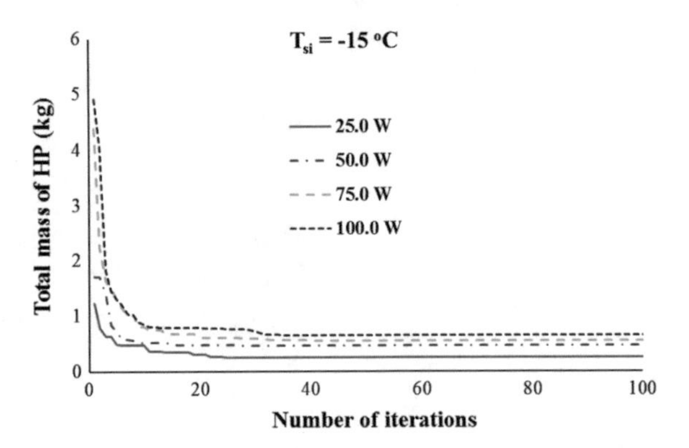

Fig. 5.9 Convergence of SAMPE-Jaya algorithm for the optimal mass of HP at −15 °C and different values of Q filled Methanol

40th iteration in each case. Figure 5.10 presents the convergence graph of SAMPE-Jaya algorithm at 0 °C condenser temperature with different values of heat transfer rate. It can be observed from Fig. 5.10 that the convergence of the SAMPE-Jaya algorithm takes place before the 25th iteration in each case. Figure 5.11 presents the convergence graph of SAMPE-Jaya algorithm at 15 °C condenser temperature with different values of heat transfer rate. It can be observed

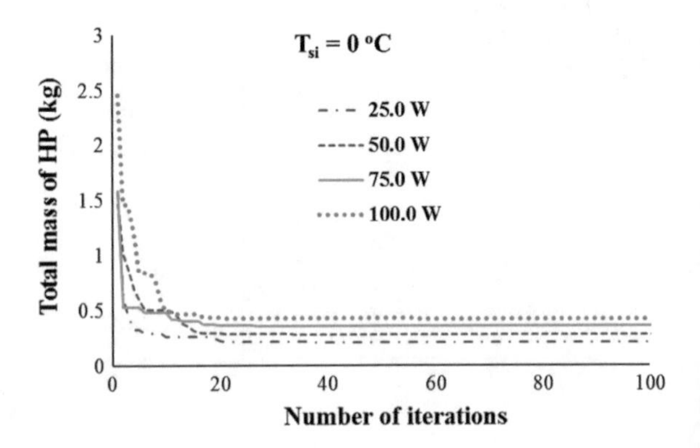

Fig. 5.10 Convergence of SAMPE-Jaya algorithm for the optimal mass of HP at 0 °C and different values of Q filled Methanol

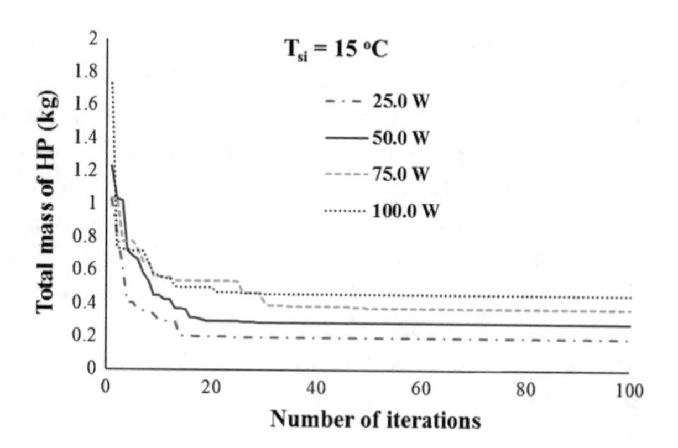

Fig. 5.11 Convergence of SAMPE-Jaya algorithm for the optimal mass of HP at 15 °C and different values of Q filled Methanol

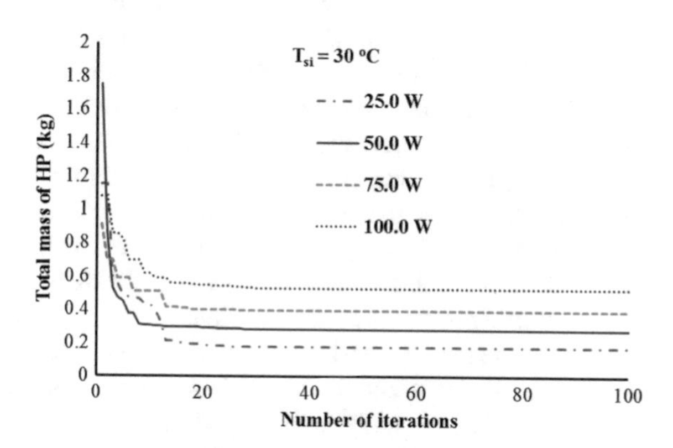

Fig. 5.12 Convergence of SAMPE-Jaya algorithm for the optimal mass of HP at 30 °C and different values of Q filled Methanol

from Fig. 5.11 that the convergence of the SAMPE-Jaya algorithm takes place before the 35th iteration in each case. Figure 5.12 presents the convergence graph of SAMPE-Jaya algorithm at 30 °C condenser temperature with different values of heat transfer rate. It can be observed from Fig. 5.12 that the convergence of the SAMPE-Jaya algorithm takes place before the 20th iteration in each case.

5.1.2.2 *Case Study 2: Multi-objective Optimization*

Equation (5.20) is used for the multi-objective optimization. In this equation, equal weights ($w_1 = w_2 = 0.5$) are considered for both the objectives for making fair comparison with the results of the previous researchers. However, the weights of the objective functions can be decided by using the analytic hierarchy process (Rao 2007). The values of the objective functions obtained by SAMP-Jaya algorithm with its comparison to other optimization algorithms for the design optimization of HP are presented in Table 5.4.

It can be observed from Table 5.5 that the performance of the SAMPE-Jaya algorithm is better as compared to the performance of NPGA, GEM, TLBO, Jaya and self-adaptive Jaya algorithms. The value of thermal resistance obtained by using Jaya algorithm is reduced by 55, 22.38, 15.44, 14.74 and 9.56% as compared to the results of NPGA, GEM, TLBO, Jaya and self-adaptive Jaya algorithms respectively. It is due to the significant increment in the value of N and reduction in the value of W which reduces the value of thermal resistance. Similarly, the value of maximum heat transfer rate obtained by using the proposed SAMPE-Jaya algorithm is enhanced by 34.94, 25.02, 19.37, 16.09 and 12.93% as compared to the results of NPGA, GEM, TLBO, Jaya and self-adaptive Jaya algorithms respectively. It is due to the increment in the value of N and δ which increases the heat transfer rate of HP significantly. Furthermore, average number of population size required by the SAMPE-Jaya algorithm is less as compared to the other algorithms.

It can be concluded based on these computational experiments that the performance of the proposed SAMPE-Jaya algorithm is better as compared to the GBA, GEM, TLBO and Jaya and SAMP-Jaya algorithms for the design optimization of HP. The SAMPE-Jaya algorithm achieves better results within less computational cost as compared to the other algorithms. In case study 1, the designs suggested by SAMPE-Jaya algorithm are better as compared to the results of GBA, TLBO, Jaya and SAMPE-Jaya algorithms when Methanol was considered as the working fluid. Similarly, the SAMPE-Jaya algorithm performed better as compared to the Jaya and SAMP-Jaya algorithms while Ammonia and Ethanol were considered as working fluids. Furthermore, the computational experiments suggested that Ammonia should be used as the working fluid when condenser section temperature is below 0 °C and Methanol should be used when the condenser section temperature is above 0 °C. In case study 2, the objective function values of heat transfer rate and thermal resistance obtained by SAMPE-Jaya algorithm are better as compared to the designs suggested by NPGA, GEM,TLBO, Jaya and self-adaptive Jaya algorithms.

5.2 Heat Sink

In electronic systems, a heat sink is a passive heat exchanger that cools a device by dissipating heat into the surrounding medium. In computers, heat sinks are used to cool central processing unit of graphics processor. Heat sinks are used with high-power semiconductor devices such as power transistor and optoelectronics such as lasers and light emitting diodes (LEDs), where the heat dissipation ability of the basic device is insufficient to moderate its temperature (Incropera and DeWitt 1996).

A heat sink is designed to maximize its surface area in contact with the cooling medium surrounding it, such as the air. Air velocity, choice of material, protrusion design and surface treatment are factors that affect the performance of a heat sink. Heat sink attachment methods and thermal interface materials also affect the die temperature of the integrated circuit. Thermal adhesive or thermal

Table 5.5 Compromised values of the objective functions for case study 2

Method	d_v (m)	W (m)	d (m)	δ (m)	N	Q_{max} (W)	R_{min} (C/W)	Avg. no. of populations
NPGA	0.0073	0.000370	0.001548	0.000652	16	486.1	0.02317	NA
GEM	0.0081	0.0014	0.0011	0.0012	22	524.69	0.0134	50
TLBO	0.0078	0.000714	0.0015	0.000638	24	549.53	0.0123	50
Jaya algorithm	0.0076	0.007246	0.00151	0.0006984	24	565.0168	0.0122	50
Self-adaptive Jaya	0.0076	0.00735	0.00158	0.0006984	25	580.71	0.0115	20.065
SAMPE-Jaya	0.007186	0.0007807	0.0010704	0.0012032	26.211	655.978	0.0104	15

NPGA (Zhang et al. 2009); TLBO (Rao and More 2015); Jaya algorithm (Rao and More 2017); Self-adaptive Jaya (Rao and More 2017)

grease improve the heat sink's performance by filling air gaps between the heat sink and the heat spreader on the device.

Several investigations have focused on the development and optimization of heat sink. Micro-channel heat sink (MCHS) was proposed first by Tuckerman and Pease (1981) who had designed and tested a very compact water-cooled integral MCHS for silicon integrated circuits. Jeevan et al. (2004) used a genetic algorithm and a box optimization method by considering fixed pumping powers for minimizing the thermal resistance by considering the optimum values of channel height, channel width and rib width. The authors had shown that the double-layer MCHS created a smaller amount of thermal resistance compared to the single-layer MCHS. Park et al. (2004) adopted a commercial finite-volume computational fluid dynamic code with the sequential linear programming method and weighting method to obtain an optimal design of plate heat exchangers with staggered pin arrays for a fixed volume. Kobus and Oshio (2005a) carried out a theoretical and experimental study on the performance of pin fin heat sinks. A theoretical model was proposed to predict the influence of various geometrical, thermal and flow parameters on the effective thermal resistance of heat sinks. Subsequently, Kobus and Oshio (2005b) investigated the effect of thermal radiation on the heat transfer of pin fin heat sinks and presented an overall heat transfer coefficient that was the sum of an effective radiation and a convective heat transfer coefficient.

Husain and Kim (2008) carried out the optimization of a MCHS with temperature dependent fluid properties using surrogate analysis and hybrid multi-objective evolutionary (MOE) algorithm. Two design variables related to the micro-channel depth, width and fin width were chosen and their ranges were decided through preliminary calculations of three-dimensional Navier–Stokes and energy equations. Objective functions related to the heat transfer and pressure drop i.e., thermal resistance and pumping power were formulated to analyze the performance of the heat sink. In another work, Husain and Kim (2010) optimized a liquid flow micro-channel heat sink with the help of three-dimensional numerical analysis and multiple surrogate methods. Two objective functions, thermal resistance and pumping power were selected to assess the performance of the micro-channel heat sink.

Hu and Xu (2009) proposed minimum thermal resistance as an objective function and a nonlinear, single objective and multi-constrained optimization model was proposed for the micro-channel heat sink in electronic chips cooling. The sequential quadratic programming (SQP) method was used to do the design optimization of the structure size of the micro-channel. The numerical simulation results showed that the heat transfer performance of micro-channel heat sink was affected intensively by its dimension. U-shaped heat sinks were studied by Liang and Hung (2010). Chong et al. (2002) developed a thermal resistance model and applied an algorithm called multiple variables constrained direct search for optimizing the performance of a single and double layer MCHSs at fixed pressure drops.

Karathanassis et al. (2013) used multi-objective optimization of MCHS for concentrating photovoltaic/thermal (CPVT) system by applying genetic algorithm.

Two different micro-channel configurations were considered, fixed and stepwise variable-width micro-channels respectively. Xie et al. (2013) designed three types of water-cooled MCHS, a rectangular straight, mono and bi-layer wavy MCHS and then evaluated using computational fluid dynamics (CFD). The authors had considered parameters such as amplitude of heat transfer, pressure drop and thermal resistance to monitor the effects on the MCHS. Results showed that for removing an identical heat load, the overall thermal resistance of the single-layer wavy micro-channel heat sink decreased with increasing volumetric flow rate, but the pressure drop was increased greatly. At the same flow rate, the double-layer wavy micro-channel heat sinks reduced not only the pressure drop but also the overall thermal resistance compared to the single-layer wavy micro-channel heat sinks.

Lin et al. (2014) developed a combined optimization procedure to look for optimal design for a water-cooled, silicon-based double layer MCHS. The geometry and flow rate distribution were considered for optimization. By using this method the authors had enhanced the performance of MCHS by optimizing the geometry and flow rate allocation for two-layer heat sink. Subhashi et al. (2016) found best values of design variables of a heat sink having honeycomb fins by using the RSM. Rao et al. (2016) used Jaya algorithm and Rao and More (2017) attempted self-adaptive Jaya algorithm for the design optimization of MCHS.

5.2.1 *Multi-objective Design Optimization of a Micro-Channel Heat Sink*

This case study was introduced by Husain and Kim (2010). A 10 mm * 10 mm * 0.42 mm silicon based micro-channel heat sink (MCHS) considered by Husain and Kim (2010) is shown in Fig. 5.13. Water was used as coolant liquid and it flowed into the micro-channel and left at the outlet. The silicon substrate occupied the remaining portion of heat sink. No slip condition was assumed at the inner walls of the channel, i.e. u = 0.

The thermal condition in the z-direction was given as:

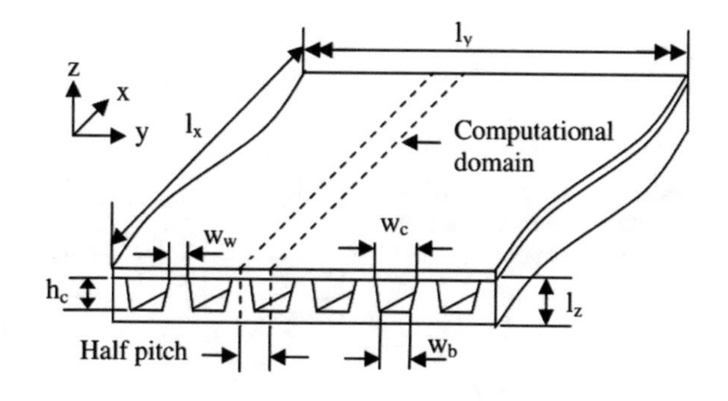

Fig. 5.13 Conventional diagram of trapezoidal MCHS (Husain and Kim 2010; Reprinted with permission from Elsevier)

$$-k_s((\partial T_s)/(\partial x_i)) = q \text{ at } z = 0 \text{ and } k_s((\partial T_s)/(\partial x_i)) = 0 \text{ at } z = l_z \quad (5.42)$$

The design variables considered by Husain and Kim (2010) were $\alpha = w_c/h_c$, $\beta = w_w/h_c$, and $\gamma = w_b/w_c$. where, w_c is the micro-channel width at bottom; w_b is the micro-channel width at top; w_w is the fin width and h_c is the micro-channel depth. h_c is kept 400 μm during the whole optimization procedure.

In this case study, two objective functions are considered and those are (i) thermal resistance associated with heat transfer performance and (ii) the pumping power to drive the coolant or to pass the coolant through the micro-channel. Table 5.6 shows design variables α, β and γ and their limits for both rectangular ($w_b/w_c = 1$) and trapezoidal ($0.5 < w_b/w_c < 1$) cross sections of MCHS.

The two objective functions considered are, thermal resistance and pumping power. The thermal resistance is given by:

$$R_{th} = \Delta Tmax/(As * q) \quad (5.43)$$

where, A_s is area of the substrate subjected to heat flux, and ΔT_{max} is the maximum temperature in MCHS, which is given as:

$$\Delta T_{max} = T_{s,o} - T_{f,i} \quad (5.44)$$

The pumping power to move the coolant (water) through MCHS is calculated as:

$$\bar{P} = n * u_{avg} * A_c * \Delta p \quad (5.45)$$

where, Δp was the pressure drop and u_{avg} was the mean velocity.

Pumping power and thermal resistance compete with each other because a decrease in pumping power contributes to increase in thermal resistance. Husain and Kim (2010) calculated the objectives by using Navier-stokes and heat conduction equations at specific design points. The Response Surface Approximation (RSA) was then used to obtain the functional forms of the two objective functions. The polynomial responses are expressed as:

$$R_{th} = 0.096 + 0.31 * \alpha - 0.019 * \beta - 0.003 * \gamma - 0.007 * \theta * \beta + 0.031 * \alpha * \gamma - 0.039 * \beta * \gamma + 0.008 * \alpha^2 + 0.027 * \beta^2 + 0.029 * \gamma^2 \quad (5.46)$$

Table 5.6 Design variables and their ranges for the case study considered

Limits	Variables		
	α (w_c/h_c)	*β (w_w/h_c)*	*γ (w_b/w_c)*
Upper	0.10	0.02	0.50
Lower	2.50	1.0	1.00

$$\bar{P} = 0.94 - 1.695 * \alpha - 0.387 * \beta - 0.649 * \gamma - 0.35 * \alpha * \beta + 0.557 * \alpha * \gamma - 0.132 * \beta * \gamma + 0.903 * \alpha^2 + 0.016 * \beta^2 + 0.135 * \gamma^2 \quad (5.47)$$

The design variables α, β and γ are in terms of the ratios of the micro-channel width at bottom to depth (i.e. w_c/h_c), fin width to the micro-channel depth (i.e. w_w/h_c) and micro-channel width at top to width at bottom (w_b/w_c) respectively. Solving equations for α, β and γ will give the optimum values of the dimensions of the micro-channel, i.e. w_c, w_w, w_b and h_c. The three design variables α, β and γ have significant effect on the thermal performance of micro-channel heat sink. Design and manufacturing constraints can be handled in a better way, and Pareto optimal solutions can be spread over the whole range of variables. The Pareto optimal analysis provides information about the active design space and relative sensitivity of the design variables to each objective function which is helpful in comprehensive design optimization. Thus, Eqs. (5.46) and (5.47) have the physical meaning.

The solution obtained by a priori approach depends on the weights assigned to various objective functions by designer or decision maker. By changing the weights of importance of different objective functions a dense spread of the Pareto points can be obtained. Following a priori approach in the present work, the two objective functions are combined into a single objective function. The combined objective function Z, is formed as:

$$\text{Minimize;} \quad Z = w_1\left(\frac{Z_1}{Z_{1min.}}\right) + w_2\left(\frac{Z_2}{Z_{2min.}}\right) \quad (5.48)$$
$$Z_1 = R_{th} \text{ and } Z_2 = \bar{P}$$

where, w_1 and w_2 are the weighs assigned to the objective functions Z_1 and Z_2 respectively between 0 and 1. These weights can be assigned to the objective functions according to the designer's/decision maker's priorities. Z_{1min} and Z_{2min} are the optimum values of the Z_1 and Z_2 respectively, obtained by solving the optimization problem when only one objective is considered at a time and ignoring the other. Now, the Eq. (5.48) can be used to optimize both the objectives simultaneously.

Husain and Kim (2010) used these surrogate models and a hybrid multi-objective evolutionary algorithm (MOEA) involving NSGA-II and Sequential Quadratic Programming (SQP) method to find out the Pareto optimal solutions. Husain and Kim (2010) used NSGA-II algorithm to obtain Pareto optimal solutions and the solutions were refined by selecting local optimal solutions for each objective function using a Sequential Quadratic Programming (SQP) method with NSGA-II solutions as initial solutions. Then K-means clustering method was then used to group the global Pareto optimal solutions into five clusters. The whole procedure was termed as a hybrid multi-objective optimization evolutionary algorithm (MOEA).

Now, the model considered by Husain and Kim (2010) is attempted using SAMPE-Jaya algorithm. Husain and Kim (2010) used a hybrid MOEA coupled with surrogate models to obtain the Pareto optimal solutions. Rao et al. (2016) used TLBO and Jaya algorithms to obtain the Pareto optimal solutions. The values of the design variables given by the SAMPE-Jaya algorithm, Jaya algorithm, TLBO algorithm, hybrid MOEA and numerical analysis are shown in Table 5.7. Table 5.8 shows the results comparison of SAMPE-Jaya algorithm, Jaya algorithm, TLBO algorithm, hybrid MOEA and numerical analysis. It can be observed from Table 5.8 that the SAMPE-Jaya algorithm has performed better as compared with hybrid MOEA, Numerical analysis, TLBO and Jaya algorithms for different weights of the objective functions for the bi-objective optimization problem considered. The performance of TLBO algorithm comes next to Jaya algorithm. Figure 5.14 presents the convergence of Jaya and SAMPE-Jaya algorithms for MCHS problem with equal weights. Figure 5.15 shows Pareto fronts obtained by using SAMPE-Jaya algorithm, Jaya algorithm, TLBO algorithm and hybrid MOE algorithm representing five clusters. Also, it can be observed that the SAMPE-Jaya algorithm has provided better results than the hybrid MOEA proposed by Husain and Kim (2010). Every peak end of the Pareto curve represents the higher value of one objective and lower value of another.

In order to make a fair comparison between the performances of the algorithms for the multi-objective optimization problems a quantity measure index known as hypervolume is calculated. Hypervolume is defined as the n-dimensional space that is enclosed by a set of points. It encapsulates in a single unary value a measure of the spread of the solutions along the Pareto front, as well as the closeness to the Pareto-optimal front. Table 5.9 presents the value of hypervolume obtained by the various algorithms for this case study. An observation can be made from Table 5.9 that value of the hypervolume obtained by the SAMPE-Jaya for the heat sink design optimization is better than the MOGA, Numerical method, TLBO and Jaya algorithm. Hence, it can be concluded that the performance of the SAMPE-Jaya algorithm is better than the hybrid MOEA, TLBO and Jaya algorithms.

In the case of MCHS, the proposed SAMPE-Jaya algorithm has obtained better pareto-optimal solutions as compared to hybrid-MOEA, numerical analysis, TLBO and Jaya algorithms. The concept of SAMPE-Jaya algorithm is simple and it is not having any algorithmic-specific parameters to be tuned and, therefore, it can be easily implemented to the engineering problems where the problems are usually complicated and have a number of design parameters and having the discontinuity in the objective function.

Table 5.7 Design variables of objective functions by using SAMPE-Jaya, Jaya, TLBO and NSGA-II of case study 3

S. No	Design variables											
	α				β				γ			
	TLBO	Jaya	SAMPE-Jaya	Hybrid MOEA	TLBO	Jaya	SAMPE-Jaya	Hybrid MOEA	TLBO	Jaya	SAMPE-Jaya	Hybrid MOEA
1	–	–	0.63399	–	–	–	0.02	–	–	–	1	–
2	0.7952	0.448167	0.7541864	0.994	0.040	0.879132	0.8945767	0.140	0.534	0.9999	1	0.528
3	0.595	0.068346	0.5704825	0.459	0.735	0.526466	0.6177118	0.693	0.5912	0.55973	0.5903021	0.991
4	0.325	0.324	0.2725593	0.096	0.745	0.721	0.706121	0.638	0.601	0.59701	0.7248777	0.982
5	0.132	0.0324	0.1	0.000	0.7601	0.857001	0.7258657	0.886	0.6299	0.9692	0.68915	0.971
6	0.1067	0.124	0.1	0.000	0.69	0.67	0.7480976	0.609	0.528	0.5692	0.6818574	0.456
7	–	–	0.0324	–	–	–	0.857001	–	–	–	0.9692	–

Table 5.8 Comparison of results of SAMPE-Jaya, Jaya, TLBO algorithm, hybrid MOEA and numerical analysis of case study 3

S. No	Hybrid MOEA (Husain and Kim 2010)		Numerical analysis (Husain and Kim 2010)		TLBO (Rao et al. 2016)		Jaya (Rao et al. 2016)		SAMPE-Jaya	
	R_{TH}	$\bar{P}$	R_{TH}	$\bar{P}$	R_{TH}	$\bar{P}$	R_{TH}	$\bar{P}$	R_{TH}	$\bar{P}$
1	–	–	–	–	–	–	–	–	0.163280	0.068143
2	0.145	0.097	0.143	0.094	0.143	0.0931	0.138308	0.086174	**0.13830**	**0.086144**
3	0.118	0.195	0.119	0.175	0.1172	0.1933	0.116959	0.192433	**0.11694**	**0.19211**
4	0.100	0.455	0.100	0.410	0.1033	0.4054	0.103345	0.402558	**0.10297**	**0.40195**
5	0.094	0.633	0.094	0.634	0.094	**0.6282**	0.093279	**0.6282**	0.**09225**	0.63266
6	0.093	0.828	0.094	0.821	0.0927	0.6966	0.093779	0.6444	**0.09321**	**0.6427**
7	–	–	–	–	–	–	–	–	0.09327	0.6253

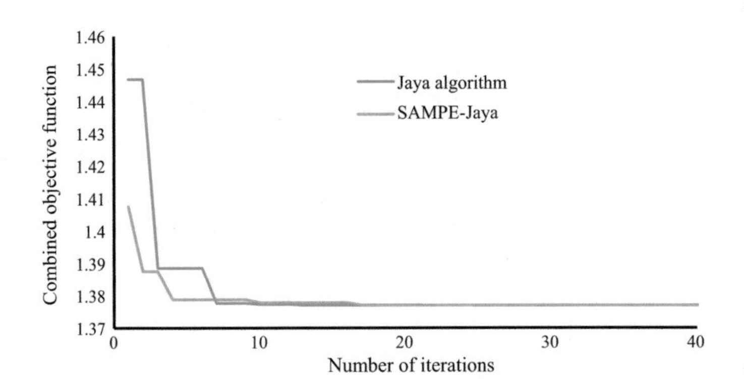

Fig. 5.14 Convergence of Jaya and SAMPE-Jaya algorithms for MCHS problem with equal weights of the objective functions

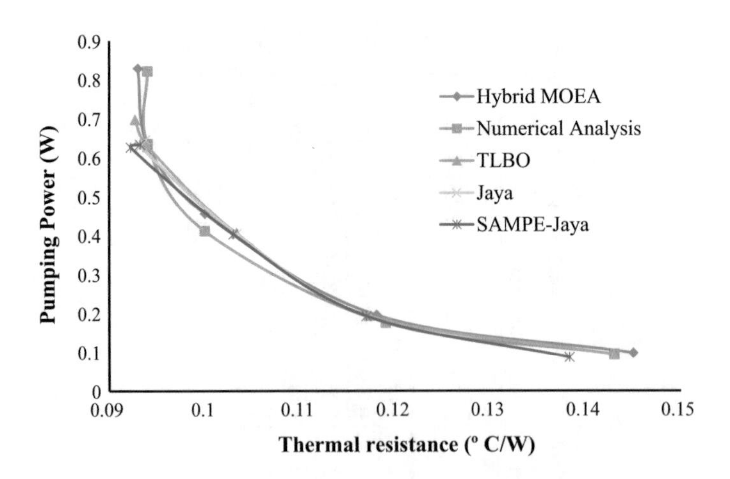

Fig. 5.15 Pareto optimal curves for MCHS problem

Table 5.9 *Hypervolume for case study of heat sink*

	Hybrid MOEA	Numerical method	TLBO	Jaya	SAMPE-Jaya
Hypervolume	0.072884	0.074248	0.07389326	0.074665546463	**0.0750687665**

References

Agha, S. R. (2011). Heat pipe performance optimization: A taguchi approach. *Journal of Research in Mechanical Engineering and Technology, 31,* 3410–3419.

Bertossi, R., Guilhem, N., Ayel, V., Romestant, C., & Bertin, Y. (2012). Modeling of heat and mass transfer in the liquid film of rotating heat pipes. *International Journal of Thermal Sciences, 52,* 40–49.

Buksa, J. J., & Hillianus, K. A. (1989). Sprite: a computer code for the optimization of space based heat pipe radiator systems. In: *Energy Conversion Engineering Conference 1989; Proceeding of the 24th Intersociety*. vol. 1, 39–44.

Chong, S. H., Ooi, K. T., & Wong, T. N. (2002). Optimization of single and double layer counters flow microchannel heat sinks. *Applied Thermal Engineering, 22,* 1569–1585.

Cui, X., Zhu, Y., Li, Z., & Shun, S. (2014). Combination study of operation characterstics and heat transfer mechanism for pulsating heat pipe. *Applied Thermal Engineering, 65,* 394–402.

Dong, Q. X., Zhen, L., JiAn, M., & ZhiXin, L. (2012). Entransy dissipation analysis and optimization of separated heat pipe system. *Science China, 55*(8), 2126–2131.

Faghri, A. (2014). Heat pipes: Review, opportunities and challenges. *Frontiers in Heat Pipes (FHP), 5,* 1–48.

Hu G., & Xu, S. (2009). Optimization design of microchannel heat sink based on SQP method and numerical simulation. In *Proceedings of IEEE*, 89–92.

Husain, V., & Kim, K. Y. (2008). Optimization of a micro-channel heat sink with temperature dependent fluid properties. *Applied Thermal Engineering, 28,* 1101–1107.

Husain, V., & Kim, K. Y. (2010). Enhanced multi-objective optimization of a micro-channel heat sink through evolutionary algorithm coupled with multiple surrogate models. *Applied Thermal Engineering, 30,* 1683–1691.

Incropera, F. P., & DeWitt, D. P. (1996). *Fundamentals of heat and mass transfer*. New York: Wiley.

Jeevan, K. Azid, I.A., & Seetharamu, K.N. (2004). Optimization of double layer counter flow (DLCF) micro-channel heat sink used for cooling chips directly, In *Proceedings of the Eectronics Packaging Technology Conference, Singapore,* 553–558.

Jeong, M. J., Kobayami, T., & Yoshimura, S. (2007). Multidimensional visualization and clustering for multiobjective optimization of artificial satellite heat pipe design. *Journal of Mechanical Science and Technology, 21,* 1964–1972.

Karathanassis, I. K., Papanicolaou, E., Belessiotis, V., & Bergeles, G. C. (2013). Multi-objective design optimization of a micro heat sink for concentrating photovoltaic/thermal (CPVT) systems using a genetic algorithm. *Applied Thermal Engineering, 59,* 733–744.

Kim, S. J., Seo, J. K., & Do, K. H. (2003). Analytical and experimental investigation on the operational characteristics and thermal optimization of a miniature heat pipe with a grooved structure. *International Journal of Heat and Mass Transfer, 46,* 2051–2063.

Kiseev, V. M., Vlassov, V. V., & Muraoka, I. (2010a). Experimental optimization of capillary structured for loop heat pipes and heat switches. *Applied Thermal Engineering, 30,* 1312–1329.

Kiseev, V. M., Vlassov, V. V., & Muraoka, I. (2010b). Optimization of capillary structures for inverted meniscus evaporators of loop heat pipes and heat switches. *International Journal of Heat and Mass Transfer, 53,* 2143–2148.

Kobus, C. J., & Oshio, T. (2005a). Development of a theoretical model for predicting the thermal performance characteristics of a vertical pin-fin array heat sink under combined forced and natural convection with impinging flow. *International Journal of Heat Mass Transfer, 48*(6), 1053–1063.

Kobus, C. J., & Oshio, T. (2005b). Predicting the thermal performance characteristics of staggered vertical pin fin array heat sinks under combined mode radiation and mixed convection with impinging flow. *International Journal of Heat Mass Transfer, 48*(13), 2684–2696.

Liang, T. S., & Hung, Y. M. (2010). Experimental investigation of thermal performance and optimization of heat sink U-shape heat pipes. *Energy Conversion and Management, 51,* 2109–2116.

Lin, L., Chang, Z., Zang, X., & Wang, X. (2014). Optimization of geometry and flow rate distribution for double-layer microchannel heat sink. *International Journal of Thermal Sciences, 78,* 158–168.

Lips, S., & Lefevre, F. (2011). A general analytical model for the design conventional heat pipes. *International Journal of Heat and Mass Transfer, 72,* 288–298.

Maheshkumar, P., & Muraleedharan, C. (2011). Minimization of entropy generation in flat heat pipe. *International Journal of Heat and Mass Transfer, 54,* 645–648.

Maziuk, V., Kulakov, A., Rabetsky, M., Vasiliev, L., & Vukovic, M. (2009). Miniature heat-pipe thermal performance prediction tool-software development. *Applied Thermal Engineering, 21,* 559–571.

Morawietz, K., & Hermann, M. (2014). Integrated development and modeling of heat pipe solar collectors. *Energy Procedia, 48,* 157–162.

Nithiynandam, K., & Pitchumani, R. (2011). Analysis and optimization of latent thermal energy storage system with embedded heat pipes. *International Journal of Heat and Mass Transfer, 54,* 4596–4610.

Park, K., Choi, D. H., & Lee, K. S. (2004). Numerical shape optimization for high-performance of a heat sink with pin–fins. *Numerical Heat Transfer Part A, 46,* 909–927.

Rao, R. V. (2007). Vendor selection in a supply chain using analytic hierarchy process and genetic algorithm methods. *International Journal of Services and Operations Management, 3,* 355–369.

Rao, R. V., & More, K. C. (2015). Optimal design of heat pipe using TLBO (teaching-learning-based-optimization) algorithm. *Energy, 80,* 535–544.

Rao, R. V. (2016). *Teaching learning based optimization algorithm and its engineering applications*. Switzerland: Springer International Publishing.

Rao, R. V., More, K. C., Taler, J., & Oclon, P. (2016). Dimensional optimization of a micro-channel heat sink using Jaya algorithm. *Applied Thermal Engineering, 103,* 572–582.

Rao, R. V., & More, C. (2017). Design optimization and analysis of selected thermal devices using self-adaptive Jaya algorithm. *Energy Conversion and Management, 140,* 24–35.

Rao, R. V., & Rakhade, R. D. (2011). Multi-objective optimization of axial "U" shaped micro grooves heat pipe using grenade explosion method (GEM). *International Journal of Advances in thermal Engineering, 2*(2), 61–66.

Riegler, R. L. (2003). *Heat transfer optimization of grooved heat pipe*. Columbia: University of Missouri.

Roper, C. S. (2011). Multi-objective optimization for design of multifunctional sandwich panel heat pipes with micro-architected truss cores. *International Journal of Heat and Fluid Flow, 32,* 239–248.

Said, S. A., & Akash, B. A. (1999). Experimental performance of a heat pipe. *International Communications in Heat and Mass Transfer, 26,* 679–684.

Senthilkumar, R. (2010). Thermal analysis of heat pipe using Taguchi method. *International Journal of Engineering Science and Technology, 2*(4), 564–569.

Shabgard, H., & Faghri, A. (2011). Performance characteristics of cylindrical heat pipes with multiple heat sources. *Applied Thermal Engineering, 31,* 3410–3419.

Shi, P. Z., Chua, K. M., Wong, Y. M., & Tan, Y. M. (2006). Design and performance optimization of miniature heat pipes in LTCC. *Journal of Physics: Conference Series, 34,* 142–147.

Sousa, F. L., Vlassov, V., & Ramos, F. M. (2004). Generalized extremal optimization: An application in heat pipe design. *Applied Thermal Engineering, 28,* 911–931.

Subhashi, S., Sahin, B., & Kaymaz, I. (2016). Multi-objective optimization of a honeycomb heat sink using Response Surface Method. *International Journal of Heat and Mass Transfer, 101,* 295–302.

Tuckerman, D. B., & Pease, R. F. W. (1981). High-performance heat sinking for VLSI. *IEEE Electron Devices Letters, 5,* 126–129.

Turgut, O. E., & Çoban, M. T. (2016). Thermal design of spiral heat exchangers and heat pipes. *Heat Mass Transfer, 53,* 899–916.

Vlassov, V. V., Souza, F. L., & Takahashi, W. K. (2006). Comprehensive optimization of a heat pipe radiator assembly filled with ammonia or acetone. *International Journal of Heat and Mass Transfer, 49,* 4584–4595.

Wang, Z., Wang, X., & Tang, Y. (2012). Condenser design optimization and operation characteristics of a novel miniature loop heat pipe. *Energy Conversion and Management, 64,* 35–42.

Wang, J. C. (2014). U and L-shaped heat pipes heat sinks for cooling electronic components employed a least square smoothing method. *Microelectronics and Reliability, 54,* 1344–1354.

Xie, G., Chen, Z., Sunden, B., & Zhang, W. (2013). Numerical predictions of the flow and thermal performance of water-cooled single-layer and double-layer wavy microchannel heat sinks. *Numerical Heat Transfer, Part A: Applications, 63,* 201–225.

Yang, X., Karamanoglu, M., Luan, T., & Koziel, S. (2014). Mathematical modeling and parameter optimization of pulsating heat pipes. *Journal of Computational Science, 5,* 119–125.

Yau, Y. H., & Ahmadzadehtalpatapeh, M. (2010). A review on the application of horizontal heat pipe heat exchangers in air conditioning systems in the tropics. *Applied Thermal Engineering, 30,* 77–84.

Zhang, C., Chen, Y., Shi, M., & Peterson, G. P. (2009). Optimization of heat pipe with axial "U" shaped micro grooves based on a niched Pareto genetic algorithm (NPGA). *Applied Thermal Engineering, 29,* 3340–3345.

Chapter 6
Multi-objective Design Optimization of Ice Thermal Energy Storage System Using Jaya Algorithm and Its Variants

Abstract This chapter presents the details of the performance optimization of an Ice Thermal Energy Storage (ITES) system carried out using TLBO algorithm, Jaya and self-adaptive Jaya algorithms. The results achieved by using Jaya and self-adaptive Jaya algorithms are compared with those obtained by using the GA and TLBO techniques for ITES system with phase change material (PCM). In ITES system, two objective functions including exergy efficiency (to be maximized) and total cost rate (to be minimized) of the whole system are considered. The Jaya and self-adaptive Jaya algorithms are proved superior to GA and TLBO optimization algorithms in terms of robustness of the results. The self-adaptive Jaya takes less computational time and the function evaluations as compared to the other algorithms.

6.1 Ice Thermal Energy Storage System

Thermal energy storage (TES) is achieved with greatly differing technologies that collectively accommodate a wide range of needs. It allows excess thermal energy to be collected for later use, hours, days or many months later, at individual building, multi-user building, district, town or even regional scale depending on the specific technology. The examples: energy demand can be balanced between daytime and night time; summer heat from solar collectors can be stored inter-seasonally for use in winter; and cold obtained from winter air can be provided for summer air conditioning. Storage media include: water or ice-slush tanks ranging from small to massive, masses of native earth or bedrock accessed with heat exchangers in clusters of small-diameter boreholes (sometimes quite deep); deep aquifers contained between impermeable strata; shallow, lined pits filled with gravel and water and top-insulated; and eutectic, phase-change materials.

Other sources of thermal energy for storage include heat or cold produced with heat pumps from off-peak, lower cost electric power, a practice called peak shaving; heat from combined heat and power (CHP) power plants; heat produced by renewable electrical energy that exceeds grid demand and waste heat from

R. Venkata Rao, *Jaya: An Advanced Optimization Algorithm and its Engineering Applications*, https://doi.org/10.1007/978-3-319-78922-4_6

industrial processes. Heat storage, both seasonal and short term, is considered an important means for cheaply balancing high shares of variable renewable electricity production and integration of electricity and heating sectors in energy systems almost or completely fed by renewable energy.

Phase Change Material (PCM) Phase change materials (PCM) can be divided into three major categories: organic, eutectics and salt hydrates materials. Among these materials, salt hydrates are the most common materials for use in air conditioning (A/C) applications. Hydrated salts are specific salts (e.g. potassium fluoride tetrahydrate ($KF{\cdot}4H_2O$)) and calcium chloride hexahydrate ($CaCl_2{\cdot}6H_2O$) that are able to store cooling energy during their solidification (charging time). When the environment temperature is lower than the PCM solidification temperature, PCMs are in solid state. On the other hand, they release the stored cooling energy during melting (discharging time) when the environment temperature is greater than their melting temperature.

There is an increasing interest among researchers in the design, development and optimization of TES system. Saito (2001) compared various types of TES systems and described their merits and demerits. Dincer (2002) performed analysis of cold thermal energy system design and optimization. The basic types of thermal energy storage (TES) techniques as sensible and latent heat storage was illustrated by Dincer (2002). Khudhair and Farid (2004) investigated thermal energy systems including PCMs for use in building cooling applications. The thermal energy systems is changing the electricity utilization of building cooling from on day time (during peak hours) to night time (during off peak hours). Stritih and Butala (2007) carried out experimental and numerical analysis of cooling buildings using night-time cold accumulation in PCM. Hence from last two decades the study of ice thermal energy systems and PCMs is an interesting research subject. The performance of ice storage charging and discharging processes on the basis of exergy and energy analysis was evaluated by MacPhee and Dincer (2009) and Koca et al. (2008).

In latent heat storage, energy is stored by changing the phase of energy storage material at a constant temperature. For storing a certain amount of energy, the lesser weight and volume of material is required in the latent heat storage technique (Sanaye and Shirazi 2013a). In another work, Sanaye and Shirazi (2013b) used multi-objective optimization and 4E analysis of an ITES for air conditioning application. Navidbakhsh et al. (2013) also used multi-objective optimization and 4E analysis of ice thermal energy storage (ITES) with PCM for air conditioning application. Palacio (2014) described a thermal energy system for reducing power system cost.

Ooka and Ikeda (2015) reviewed the optimization techniques for TES control systems. Rahdar et al. (2016a) used two different thermal energy systems including ITES and PCM for a heating, ventilation and air conditioning (HVAC) application. In another work, Rahdar et al. (2016b) used NSGA-II and MOPSO for optimization of R-717 and R-134a ITES air conditioning systems. Lefebvre and Tezel (2017) focused on adsorption TES processes for heating applications. Shu et al. (2017) designed a system for a permanent refuge chamber in an underground mine with an accommodation capacity of 50 persons.

6.2 Case Study of an Ice Thermal Energy Storage (ITES) System

This case study covers the energy, exergy, economic and environmental (4E) analyses as well as multi-objective optimization of an Ice thermal energy storage (ITES) system incorporating PCM material as the partial cold storage for air conditioning applications. To implement this job, an ITES system including charging and discharging processes as well as a PCM cold storage is modelled and analyzed. Then the multi-objective optimization of the system is performed in order to obtain optimal design parameters using genetic algorithm technique. The objective functions are the exergy efficiency (to be maximized) and the total cost rate (including capital and maintenance costs, operational cost, and the penalty cost due to CO_2 emissions of the whole system, which is to be minimized), subjected to a list of constraints (Navidbaksh et al. 2013).

An ITES system incorporating PCM as the partial cold storage is modelled for air conditioning applications. The schematic diagram of the modelled system is shown in Fig. 6.1. The whole system includes two main parts: ITES system and PCM cold storage. The ITES system itself includes two cycles: Charging cycle including evaporator, compressor, condenser, cooling tower, pump and expansion valve; Discharging cycle including air handling unit (AHU), discharging pump and ice storage tank.

In charging cycle (vapor compression refrigeration system), R134a is used as refrigerant, and water/glycol solution (chilled water) is the secondary cooling fluid in discharging process. The charging cycle is used to make ice during off-peak hours when the electricity price is low (during night-time). During on peak hours when the electricity price is high, chilled water inside the tubes passes through the ice storage tank and is pumped into AHU to provide the building cooling load (discharging process). Therefore, the whole charging cycle is turned off during on-peak hours.

PCM as the partial cold storage consists of parallel plates filled with calcium chloride hexahydrate ($CaCl_2{\cdot}6H_2O$), Where ambient air is passed through gaps between them. During night-time, when the ambient temperature is lower than the PCM solidification temperature, the PCM releases heat into the atmosphere and then solidify (charging process). During the day, when the ambient temperature is greater than the PCM melting temperature, the PCM begins to melt and absorbs heat from the flow of warm air at the AHU inlet. Therefore, the air flow temperature is decreased at the AHU inlet (discharging process).

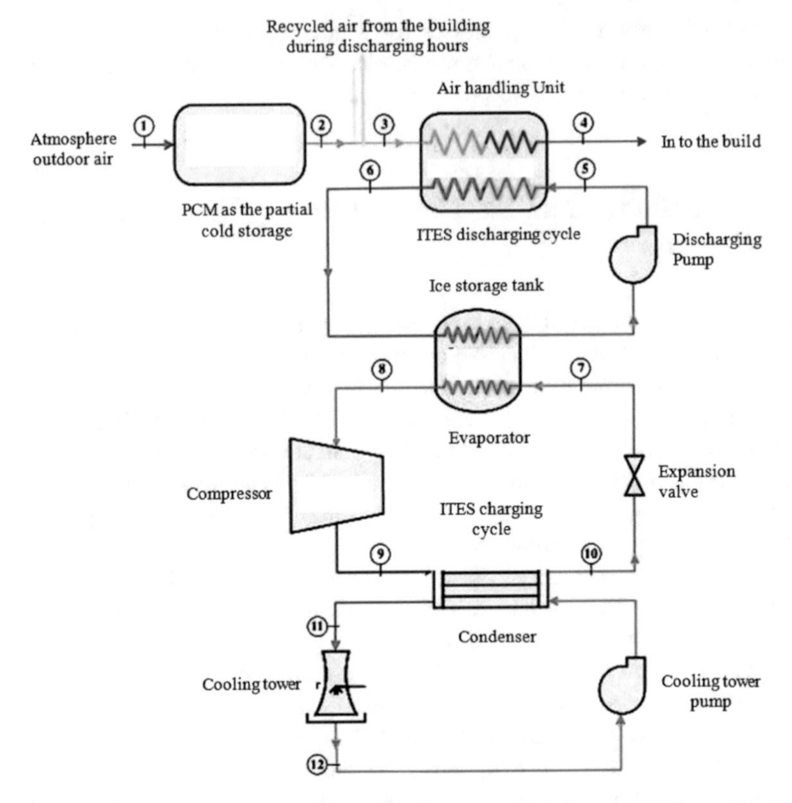

Fig. 6.1 The schematic diagram of the ITES system (Navidbaksh et al. 2013; reprinted with permission from Elsevier)

6.2.1 Energy Analysis

The following includes energy analysis of the whole system for both charging and discharging processes. The following assumptions are made to simplify the modelling of the system (Navidbaksh et al. 2013):

1. Expansion valve heat losses and pressure drop within the pipes were assumed negligible.
2. Thermal capacity of the unit casing was assumed negligible. All kinetic and potential effects were assumed negligible.
3. The states of the refrigerant at evaporator and condenser outlets were considered as saturated vapor and saturated liquid, respectively.
4. The storage tank temperature distribution was assumed constant.
5. The specific heat and density of PCM are different for solid and liquid phases.
6. The heat leakage of PCM storage is neglected (well-insulated unit's casing).

The required cooling energy of a building (Q_c, kWh) can be estimated from its cooling load.

$$Q_c = \int_{t_{dc}} \dot{Q}_c(t)dt \tag{6.1}$$

where, t_{dc} is the discharging time (h). Since the whole building cooling load is supplied via ice and PCM storage systems, the coefficient α (between 0 and 0.1) is defined to allocate the building required cooling load to both of them. Therefore, the cooling capacity of each system can be obtained as:

$$Q_{c,PCM} = \alpha \times \mathrm{Q_c} \tag{6.2}$$

$$Q_{c,ITES} = (1 - \alpha) \times \mathrm{Q_c} \tag{6.3}$$

The mass flow rate of air within AHU can be obtained as follows:

$$\dot{m}_a = \frac{Q_{c,ITES}}{(h_3 - h_4)} = \frac{Q_{c,ITES}}{c_{p,a}(T_3 - T_4) + (\omega_3 h_{g3} - \omega_4 h_{g4})} \tag{6.4}$$

where, $Q_{c,ITES}$ is the rate of cooling energy supplied by ice storage tank, h is the specific enthalpy of moist air, ω is absolute humidity, and h_g is the specific enthalpy of water vapor.

In addition, the required mass of PCM is obtained by:

$$m_{pcm} = \frac{3600_{C,PCM}}{Cp_{PCM,L}(T_{amb,d} - T_{m,PCM}) + i_{ph,PCM} + Cp_{PCM,S}(T_{m,PCM} - T_{amb,n})} \tag{6.5}$$

where, $Cp_{PCM,L}$ and $Cp_{PCM,S}$ are specific heats for the PCM in liquid and solid phases, $T_{amb,d}$ and $T_{amb,n}$ are ambient temperatures during the day and night, $T_{m,PCM}$ and $i_{ph,PCM}$ are the melting temperature and melting latent heat of PCM, respectively.

The cooling energy stored in the ice storage tank can be obtained as:

$$Q_{ST} = \frac{Q_{C,ITES}}{\eta_{ST}} \tag{6.6}$$

where,

$$\eta_{ST} = \frac{Q_{ST} - Q_{1,ch} - Q_{1,dc}}{Q_{ST}} = 1 - \left(\frac{Q_{1,ch} + Q_{1,dc}}{Q_{ST}}\right) \tag{6.7}$$

where, $Q_{1,ch}$ and $Q_{1,dc}$ are the amounts of heat leakage of storage tank during charging and discharging processes.

6.2.1.1 Charging Process

During charging process in ITES system, a vapor compression refrigeration system is used to make and store ice in the ice storage tank. The evaporator heat transfer rate is defined as:

$$\dot{Q}_{EV} = \frac{Q_{ST}}{t_{ch}} \tag{6.8}$$

where, t_{ch} is the system charging time (h).

The amount of refrigerant mass flow rate is:

$$\dot{m}_r = \frac{\dot{Q}_{EV}}{(h_8 - h_7)} \tag{6.9}$$

The compressor electricity consumption is:

$$\dot{W}_{Comp} = \dot{m}_r(h_9 - h_8) \tag{6.10}$$

where, h_9 is obtained from the following relation:

$$\dot{\eta}_{Comp} = \frac{(h_{9s} - h_8)}{(h_9 - h_8)} \tag{6.11}$$

The condenser heat transfer rate can be obtained as follows:

$$Q_{Cond} = \dot{m}_r(h_9 - h_{10}) \tag{6.12}$$

The coefficient of performance (COP) of refrigeration system is:

$$COP = \frac{\dot{Q}_{EV}}{\dot{W}_{Comp}} \tag{6.13}$$

In addition, the heat transfer surface area of evaporator and condenser can be estimated as (Navidbakhsh et al. 2013):

$$A_{EV} = \frac{NTU \times \left(\dot{m}C_p\right)_{min}}{U_{EV}} \tag{6.14}$$

$$A_{Cond} = \frac{\dot{Q}_{EV}}{U_{Cond} \times F \times \Delta T_{LMPTD}} \tag{6.15}$$

The amount of heat leakage of the ice storage tank during charging process ($Q_{1,ch}$) is expressed as:

$$Q_{1,ch} = A_{ST} \frac{T_{amb,n} - T_{ST}}{R_{th}} t_{ch} \tag{6.16}$$

where, the terms A_{ST}, R_{th}, and t_{ch} are the ice storage tank heat transfer surface area, thermal resistance of the storage tank, and the system charging time, respectively. The following relations are used to find the ice storage tank volume and its surface area (Navidbakhsh et al. 2013):

$$V_{ST} = \frac{3600 Q_{ST}}{\rho_w C_{p,w} \left(T_{dc} - T_{FP,w}\right) + \rho_{ice} C_{p,ice} \left(T_{FP,w} - T_{ST}\right)} \tag{6.17}$$

$$A_{ST} = 6\pi \left(\frac{V_{ST}}{2\pi}\right)^{2/3} \tag{6.18}$$

The rate of stored cooling energy at the PCM cold storage during night-time is:

$$\dot{Q}_{PCM,ch} = \frac{Q_{C,PCM}}{t_{ch}} \tag{6.19}$$

6.2.1.2 Discharging Process

In discharging cycle, the ice storage tank is used for cooling water/glycol solution (chilled water). Then, the chilled water is pumped into the AHU in order to supply the cooling load of the building. The amount of heat leakage of the ice storage tank during discharging process is:

$$Q_{1,dc} = A_{ST} \frac{T_{amb,d} - T_{dc}}{R_{th}} t_{dc} \tag{6.20}$$

where, T_{dc} and t_{dc} are discharging temperature and discharging hours respectively.

During the day, when the ambient temperature is greater than the PCM melting temperature, the warm atmospheric air is passed through gaps between PCM slabs. At this time, the PCM starts to melt and absorbs heat from warm air flow to decrease the temperature of air flow into the AHU inlet. The rate of stored cooling energy at the PCM cold storage which is released during daytime is:

$$\dot{Q}_{PCM,dc} = \frac{Q_{C,PCM}}{t_{dc}} \tag{6.21}$$

6.2.2 Exergy Analysis

Exergy or available work is defined as the maximum theoretical useful work that a system can perform in a given state when it comes to the ambient conditions. The exergy balance equation for a closed system can be written as:

$$E_f - E_i = E^Q - E^w - E_D \tag{6.22}$$

where, the terms E^Q, E^w and E_D are the exergy transfers associated with heat transfer, the exergy transfer accompanying net useful work, and the exergy destruction, respectively. The term $(E_f - E_i)$ is the exergy change in the closed system which can be obtained by:

$$E_f - E_i = m\left[\left(u_f - u_i\right) + p_0\left(v_f - v_i\right) - T_0\left(s_f - s_i\right)\right] \tag{6.23}$$

Furthermore, the steady state form of exergy rate balance for a control volume can be expressed as follows:

$$\frac{dE_{cv}}{dt} = \sum_j \dot{E}_j^Q - \dot{E}_j^Q + \sum_{.} \dot{E}_i - \sum_e \dot{E}_e - \dot{E}_D = 0 \tag{6.24}$$

where, $\dot{E}_j^Q$ and $\dot{E}_j^Q$ are the time rate of exergy transfer by heat transfer and work, $\dot{E}_i$ and $\dot{E}_e$ are the exergy transfer rate at control volume inlet and outlet, and finally $\dot{E}_D$ is the rate of exergy destruction due to irreversibilities, respectively. The exergy flow rate of the system is divided into two parts of physical and chemical exergy.

$$\dot{E} = \dot{E}^{PH} + \dot{E}^{CH}(4.172)$$

Physical exergy in the general form is defined by:

$$\dot{E}^{PH} = \dot{m}[(h - h_0) - T_0(s - s_0)] \tag{6.25}$$

where, for liquids (incompressible fluids):

$$\dot{E}^{PH} = \dot{m}c_p T_0 \left[\frac{T}{T_0} - 1 - \ln\frac{T}{T_0}\right] \tag{6.26}$$

While for ideal gases:

$$\dot{E}^{PH} = \dot{m}c_p T_0 \left[\frac{T}{T_0} - 1 - \ln\frac{T}{T_0} + \frac{(k-1)}{k}\ln\frac{p}{p_0}\right] \tag{6.27}$$

Since no chemical reaction does not occur, the chemical exergy is not used in the modelling of the system. Equations (6.22)–(6.27) show the calculation of exergy destruction rate in each system component:

$$\dot{E}_{D,AHU} = \left(\dot{E}_3 + \dot{E}_5\right) - \left(\dot{E}_4 + \dot{E}_6\right) \tag{6.28}$$

$$\dot{E}_{D,Comp} = \dot{E}_8 - \dot{E}_9 + \dot{W}_{Comp} \tag{6.29}$$

$$\dot{E}_{D,EX} = \dot{E}_{10} - \dot{E}_7 \tag{6.30}$$

$$\dot{E}_{D,Cond+CT} = \left(\dot{E}_9 + \dot{E}_{12}\right) - \left(\dot{E}_{10} + \dot{E}_{11}\right) \tag{6.31}$$

$$\dot{E}_{D,EV} = \left(\dot{E}_7 + \dot{E}_8\right) + \dot{E}^{Q}_{EV} \tag{6.32}$$

The exergy destruction rate at the ice storage tank $\dot{E}_{D,ST}$ can be expressed as follows (Navidbakhsh et al. 2013):

$$\dot{E}_{D,ST} = \dot{E}_{D,ST,ch} + \dot{E}_{D,ST,dc} \tag{6.33}$$

where, $\dot{E}_{D,ST,ch}$ and $\dot{E}_{D,ST,dc}$ are the exergy destruction rate for ice storage tank during charging and discharging processes respectively. Applying Eq. (4.182) for the ice storage tank $\dot{E}_{D,ST,ch}$ is calculated as:

$$\dot{E}_{D,ST,ch} = \frac{E_{D,ST,ch}}{t_{ch}} = \frac{E^{Q}_{ST,ch} - \left(E_f - E_i\right)_{ST,ch}}{t_{ch}} \tag{6.34}$$

where,

$$E^{Q}_{ST,ch} = \left(Q_{1,ch} - Q_{ST}\right) - \left(1 - \frac{T_0}{T_{ST}}\right) \tag{6.35}$$

In charging cycle, initial state of the ice storage tank (i) is water at temperature T_{dc} and its final state (f) is ice at temperature T_{ST}. The exergy change in ice storage tank during charging process $\left(E_f - E_i\right)_{ST,ch}$ is obtained as:

$$\begin{aligned}\left(E_f - E_i\right)_{ST,ch} &= m_{w,ST}\left[\left(u_f - u_i\right) - T_o\left(s_f - s_i\right)\right]_{St,ch} \\ &= \left(-Q_{ST} + Q_{1,ch}\right) m_{w,ST} T_0 \left[c_{p,w} ln\left(\frac{T_{FP,w}}{T_{dc}}\right) \right. \\ &\quad \left. - \frac{i_{ph,ice}}{T_{FP,w}} + c_{p,ice} ln\left(\frac{T_{ST}}{T_{FP,w}}\right)\right]\end{aligned} \tag{6.36}$$

The exergy destruction rate of ice storage tank during discharging (melting) hours $\dot{E}_{D,ST,dc}$ is defined as:

$$\dot{E}_{D,ST,dc} = \frac{E_{D,ST,dc}}{t_{dc}} = \frac{E^{Q}_{ST,dc} - \left(E_f - E_i\right)_{ST,dc}}{t_{dc}} \tag{6.37}$$

where, indexes i and f denote initial state of ice at temperature T_{ST} and final state of water at temperature T_{dc}. The terms $E^{Q}_{ST,dc}$ and $\left(E_f - E_i\right)_{ST,dc}$ are obtained by the following relations respectively:

$$E^{Q}_{ST,dc} = \left(Q_{c,ITES} - Q_{1,dc}\right)\left(1 - \frac{T_o}{T_{dc}}\right) \tag{6.38}$$

$$\begin{aligned}\left(E_f - E_i\right)_{ST,dc} &= m_{w,ST}\left[\left(u_f - u_i\right) - T_o\left(s_f - s_i\right)\right]_{ST,dc}\\ \left(E_f - E_i\right)_{ST,dc} &= \left(Q_{C,ITES} + Q_{1,ch}\right) - m_{w,ST}T_0\left[c_{p,ice}\,ln\left(\frac{T_{FP,w}}{T_{ST}}\right) - \frac{i_{ph,ice}}{T_{FP,w}} + ln\left(\frac{T_{dc}}{T_{FP,w}}\right)\right]\end{aligned} \tag{6.39}$$

There are the same explanations for terms in the bracket as are for the charging mode in the ice storage vessel. The exergy destruction rate of PCM $\dot{E}_{D,PCM}$ can be expressed as:

$$\dot{E}_{D,PCM} = \dot{E}_{D,PCM,ch} + \dot{E}_{D,PCM,dc} \tag{6.40}$$

where, $\dot{E}_{D,PCM,ch}$ and $\dot{E}_{D,PCM,dc}$ are the exergy destruction rate of the PCM during charging (solidification) and discharging (melting) processes, respectively. $\dot{E}_{D,PCM,ch}$ is obtained from:

$$\dot{E}_{D,PCM,ch} = \frac{E_{D,PCM,ch}}{t_{ch}} = \frac{E^{Q}_{PCM,ch} - \left(E_f - E_i\right)_{PCM,ch}}{t_{ch}} \tag{6.41}$$

where, $E^{Q}_{PCM,ch} = \left(-Q_{C,PCM}\right)\left(1 - \frac{T_o}{T_{PCM}}\right)$

In charging process, initial and final states of the PCM (i and f) are liquid (at temperature ($T_{amb,d}$) and solid (at temperature $T_{amb,n}$), respectively. The exergy change for PCM storage during charging process $\left(E_f - E_i\right)_{PCM,ch}$ is estimated as follows:

$$\begin{aligned}\left(E_f - E_i\right)_{PCM,ch} &= m_{PCM}\left[\left(u_f - u_i\right) - T_o\left(s_f - s_i\right)\right]_{PCM,ch} = \left(-Q_{C,PCM}\right)\\ &\quad - m_{PCM}T_0\left[c_{pPCM,L}ln\left(\frac{T_{m,PCM}}{T_{amb,d}}\right) - \frac{i_{ph,PCM}}{T_{m,PCM}} + c_{pPCM,s}ln\left(\frac{T_{amb,n}}{T_{m,PCM}}\right)\right]\end{aligned} \tag{6.42}$$

The first and last terms in the bracket in Eq. (6.42) specify the PCM specific entropy change in liquid and solid states respectively (sensible entropy change), and the second term specifies the entropy of phase change of PCM (solidification entropy change). Furthermore, the exergy destruction rate of PCM during discharging (melting) process $\dot{E}_{D,PCM,dc}$ is defined as:

$$\dot{E}_{D,PCM,dc} = \frac{E_{D,PCM,dc}}{t_{dc}} = \frac{E^{Q}_{PCM,dc} - \left(E_f - E_i\right)_{PCM,dc}}{t_{dc}} \tag{6.43}$$

where, the initial state of PCM (i) is solid at temperature $T_{amb,n}$, and its final state (f) is liquid at temperature $T_{amb,d}$. The terms $E^{Q}_{PCM,dc}$ and $\left(E_f - E_i\right)_{PCM,dc}$ are obtained by:

$$E^{Q}_{PCM,dc} = \left(Q_{C,PCM}\right)\left(1 - \frac{T_o}{T_{PCM}}\right) \tag{6.44}$$

$$\begin{aligned}\left(E_f - E_i\right)_{PCM,dc} &= m_{PCM}\left[\left(u_f - u_i\right) - T_o\left(s_f - s_i\right)\right]_{PCM,dc} = \left(-Q_{C,PCM}\right)\\ &- m_{PCM}T_0\left[c_{pPCM,s}\ln\left(\frac{T_{m,PCM}}{T_{amb,n}}\right) - \frac{i_{ph,PCM}}{T_{m,PCM}} + c_{pPCM,l}\ln\left(\frac{T_{amb,d}}{T_{m,PCM}}\right)\right]\end{aligned} \tag{6.45}$$

6.2.3 Economic Analysis

The presented economic analysis takes into account the capital and maintenance costs of system components and the operational cost of the system (including the cost of electricity consumption). The capital cost of each component (Z_k) is estimated based on the cost functions which are listed in Table 6.1. To convert the capital cost (in terms of US dollar) into the cost per unit of time $\dot{Z}_k$, one may write:

$$\dot{Z}_k = \frac{Z_k \times CRF \times \emptyset}{N \times 3600} \tag{6.46}$$

where, N, $\emptyset$, and CRF are the annual operational hours of the system, maintenance factor, and the capital recovery factor, respectively. CRF is determined through the following relation:

$$CRF = \frac{i(1+i)^n}{(1+i)^n - 1} \tag{6.47}$$

where, the terms i and n are the interest rate and system life time, respectively.

Table 6.1 The cost functions of various equipments in ITES system

System component	Capital cost function
PCM	30 US$ per kilogram of PCM material
Air handling unit (AHU)	$Z_{AHU} = 24202\, xA_{AHU}^{0.4162}$
Pump	$Z_{pump} = 705.48 \times W_{pump}^{0.71}\left(1 + \frac{0.2}{1-\eta_{pump}}\right)$
Ice storage tank	$Z_{ST} = 8.67 \times 10^{[2.9211\exp(0.1416 \times log V_{ST})]}$
Evaporator	$Z_{EV} = 16648.3 \times A_{EV}^{0.6123}$
Compressor	$Z_{comp} = \frac{39.5 \times \dot{m}_r}{0.9 \times \eta_{pcomp}}\left(\frac{P_{dc}}{P_{suc}}\right) In\left(\frac{P_{dc}}{P_{suc}}\right)$
Expansion valve	$Z_{EX} = 114.5 \times \dot{m}_r$
Condenser	$Z_{cond} = (516.621 \times A_{contd}) + 268.45$
Cooling tower	$Z_{CT} = 746.749\,(\dot{m}_{CT})^{0.79}(\Delta T_{CT})^{0.57}\left(T_{in,CT} - T_{WT,OUT}\right)^{-0.9924} * \left(0.022 T_{WB,OUT} + 0.39\right)^{2.447}$
Compression chiller	$Z_{chiller} = 150.2 \times \dot{Q}_{chiller}$

Moreover, the operational cost of the whole system including the cost of electricity consumption $\dot{C}elec$ can be determined as follows:

$$\begin{aligned}\dot{C}elec = & \left[\left(\dot{W}_{fan,PCM} + \dot{W}_{Comp} + \dot{W}_{pump,CT} + \dot{W}_{fan,CT}\right) \times \frac{C_{ele,off-peak}}{3600}\right] \\ & + \left[\left(\dot{W}_{fan\,,PCM} + \dot{W}_{fan,PCM} + \dot{W}_{fan,AHU} + \dot{W}_{pump,dc}\right) \times \frac{C_{ele,on-peak}}{3600}\right]\end{aligned} \quad (6.48)$$

where, $C_{ele,off-peak}$ and $C_{ele,on-peak}$ are the electricity unit cost during off-peak and on-peak hours, respectively.

6.2.4 Environmental Analysis

Global warming is one of the most serious environmental problems that the world is facing today. The amount of CO_2 emission from modelled system can be determined using the emission conversion factor as follows:

$$m_{CO_2}[\mathrm{kg}] = \mu_{CO_2}\left[\mathrm{kg\,kWh}^{-1}\right] \times annual\, electricity\, consumption\,[\mathrm{kWh}] \quad (6.49)$$

where, μ_{CO_2} is the emission conversion factor of electricity from grid. The value of this parameter is 0.968 kg kWh^{-1}. The penalty cost of CO_2 emission c_{co_2} is considered as 90 US dollars per ton of CO_2 emissions. Thus, the rate of penalty cost of CO_2 emission $\dot{C}_{env}$ is defined as:

$$\dot{C}_{env} = \frac{(m_{co_2}/1000) \times C_{CO_2}}{N \times 3600} \tag{6.50}$$

6.3 Optimization of ITES System with Phase Change Material (PCM)

Navidbaksh et al. (2013) presented a case study of optimal design of ITES system incorporating PCM as the partial cold storage installed in a commercial building in Ahwaz, a city in south of Iran. The working hours of the building were 7 a.m. to 7 p.m. The chiller cooling load capacity for simple ITES (without PCM as the partial cold storage ($\alpha = 0$)) and conventional (system with capability of load change steps equal to 25% nominal cooling load) systems. Thermo-physical properties of used PCM ($CaCl_2{\cdot}6H_2O$) were as listed in Table 6.2. The thermodynamic properties of refrigerant used in the modelling of vapor compression refrigeration cycle (R134a) were obtained from an in-house developed database. The thermal resistance (R_{th}) of the storage tank is 1980 m^2K/kW. The comfort temperature and the relative humidity of the room as well as the ambient pressure were assumed to be 21 °C, 0.55, and 1 atm, respectively. The electricity cost during on-peak hours is 0.09 US$/kWh and during off-peak hours was 0.06 US$/kWh. To determine CRF (Eq. (4.195)), the annual interest rate, approximate life time of the system, and the maintenance factor were considered as 14%, 15 years, and 1.06, respectively. The salvage value (ΔZ_{sv}) was considered to be 10% of the difference between the capital costs of two systems. The annual operational hours of the whole system were 2100 h and for charging (12 p.m.–7 a.m.) and 3600 h discharging (7 a.m.–7 p.m.) cycles respectively (from March to December).

In this study, two objective functions including exergy efficiency (to be maximized) and total cost rate (to be minimized) of the whole system are considered for multi-objective optimization. The mathematical formulation of the aforementioned objective functions is as follows:

Table 6.2 Thermo physical properties of used PCM ($CaCl_2{\cdot}6H_2O$)

Property	Value
Melting point (°C)	29
Density (Solid) (Kg m^{-3})	1800
Density (Liquid) (Kg m^{-3})	1560
Specific heat (solid) (kJ kg^{-1} K^{-1})	1.46
Specific heat (liquid) (kJ kg^{-1} K^{-1})	2.13
Melton latent heat (kJ kg^{-1})	187.49
Toxic effect	No

- Exergy efficiency (objective function I)

$$\varphi = \frac{\dot{E}_{D,tot}}{\dot{E}_{in}} \tag{6.51}$$

where $\dot{E}_{D,tot}$ represents the sum of exergy destruction rate of system components, and $\dot{E}_{in}$ is the electricity consumption of the whole system (including compressor, fans, and pumps) respectively.

- Total cost rate (objective function II)

$$\dot{C}_{tot} = \sum\nolimits_k Z_k + \dot{C}_{elec} + \dot{C}_{eenv} \tag{6.52}$$

Moreover, the following design parameters are selected for optimization of the system: chilled water temperature at air handling unit (AHU) inlet (T_5) and outlet (T_6), storage temperature within the ice storage tank (T_{ST}), refrigerant saturated temperature at evaporator (T_{EV}) and at condenser (T_{Cond}), and cooling load coefficient (α). The list of mentioned design parameters and their ranges of variation as well as system constraints are summarized in Table 6.3.

Navidbaksh et al. (2013) applied GA for an ITES system incorporating PCM as the partial cold storage was modeled for A/C applications and the results are give in Tables 6.4, 6.5, 6.6 and 6.7. The basic parameters of the refrigeration system including the refrigerant mass flow rate $\dot{m}_r$, compressor power consumption $\dot{W}_{comp}$, and the coefficient of performance (COP) are obtained from the modeling results and compared with the corresponding values reported in Table 6.4 to ensure the accuracy of modelling.

Table 6.3 Design variables and their ranges for ITES system

Design Variable	Ranges		Design variable	Ranges	
	Min	Max		Min	Max
T_5	3	5	T_6	11	13
T_{ST}	−10	0	T_{EV}	−30	0
T_{Cond}	$(T_{WB,out} + 5)$	60	α	0	0.1

Table 6.4 The comparison of computed values of system operating parameters

Input		Output	TLBO	(Navidbakhsh et al. 2013)	Jaya	Self-adaptive Jaya
$T_{EV}\,(oC)$	20	$\dot{m}_r$	0.2	0.2001	0.2	0.2
$T_{cond}\,(oC)$	40	$\dot{W}_{comp}$	9	9.1102	9.12	9.12
$Q_{EV}\,(kW)$	25.9	COP	2.87	2.8430	2.912	2.912

Table 6.5 The optimum values of ITES system design parameters with different methods of optimization

Design Parameters	Maximum exergy efficiency				Minimum total annual cost				Combined objective function			
	GA (Navidbaksh et al. 2013)	TLBO	Jaya	Self-adaptive Jaya	GA (Navidbaksh et al. 2013)	TLBO	Jaya	Self-adaptive Jaya	GA (Navidbaksh et al. 2013	TLBO	Jaya	Self-adaptive Jaya
$T_5(oC)$	3.36	3.52	3.65	3.65	4.53	4.89	4.91	4.91	4.19	4.20	4.45	4.45
$T_6(oC)$	12.45	13.02	12.95	12.95	11.21	10.95	10.82	10.82	11.53	12.14	12.18	12.18
$T_{ST}(oC)$	−3.93	−4.14	−4.14	−4.14	−1.34	−1.17	−1.19	−1.19	−2.05	−2.47	−2.54	−2.54
$T_{EV}(oC)$	−5.38	−5.73	−5.84	−5.84	−3.70	−4.18	−3.97	−3.97	−4.37	−4.82	−4.75	−4.75
$T_{cond}(oC)$	36.29	35.4	35.2	35.2	38.69	40.0	39.97	39.97	37.93	38.65	38.67	38.67
α	0.1	0.1	0.1	0.1	0.011	0.013	0.011	0.011	0.063	0.058	0.062	0.062

Table 6.6 The results of exergy and energy analysis with GA, TLBO, Jaya and self-adaptive Jaya algorithms

	Maximize exergy efficiency				Minimize total annual cost				Combined objective function			
	GA (Navidbakhsh et al. 2013)	TLBO	Jaya	Self adaptive Jaya	GA (Navidbakhsh et al. 2013)	TLBO	Jaya	Self adaptive Jaya	GA (Navidbakhsh et al. 2013)	TLLBO	Jaya	Self-adaptive Jaya
Required PCM (kg)	1.8258 *E*04	2.0367*E*04	2.0145E04	2.0145E04	2.0493 E03	1.9723E03	19569E03	19569E03	1.1737 E04	1.1715E04	1.1547E03	1.1547E03
Annual CO_2 emmission	1.62*E*06	1.58E06	1.56E06	1.56E06	1.87*E*06	1.79*E06*	1.78E06	1.78E06	1.73*E*06	1.69*E06*	1.67E06	1.67E06
AHU (kW)	101.55	100.33	100.01	100.01	125.86	124.23	122.46	122.46	113.71	109.45	107.54	107.54
ST (kW)	81.94	78.98	77.42	77.42	96.98	95.22	94.85	94.85	90.93	87.23	85.97	85.97
EV (kW)	68.87	68.32	65.92	65.92	83.88	82.91	82.12	82.12	75.83	70.97	68.54	68.54
Comp (kW)	153.75	150.89	150.01	150.01	177.21	173.35	170.35	170.35	164.23	160.35	158.69	158.69
EX (kW)	17.64	16.44	15.45	15.45	28.24	26.57	24.78	24.78	23.32	22.34	21.56	21.56
Cond+CT (kW)	247.59	240.35	238.78	238.78	278.46	276.76	275.34	275.34	258.49	250.09	249.87	249.87
PCM storage (kW)	55.72	52.47	49.98	49.98	25.67	24.45	23.58	23.58	38.55	34.76	33.91	33.91
Total exergy efficiency (%)	42.17	47.66	**48.54**	**48.54**	31.68	35.85	**37.57**	**37.57**	38.47	41.22	**42.05**	**42.05**

Bold value indicates the best results

Table 6.7 The result of economic analysis with GA, TLBO, Jaya and self-adaptive Jaya algorithms

	Maximize exergy efficiency				Minimize total annual cost				Combined objective function			
	GA (Navidbakhsh et al. 2013)	TLBO	Jaya	Self-adaptive Jaya	GA (Navidbakhsh et al. 2013)	TLBO	Jaya	Self-adaptive Jaya	GA (Navidbakhsh et al. 2013)	TLBO	Jaya	Self-adaptive Jaya
PCM cost (MUS$)	0.54773	0.64101	0.6112	0.6112	0.06148	0.05916	0.05245	0.05245	0.35211	0.35145	0.34781	0.34781
AHU cost (MUS$)	0.2191	0.2067	0.1994	0.1994	0.2304	0.2289	0.2281	0.2281	0.2217	0.2190	0.2184	0.2184
ST cost (MUS$)	0.0273	0.0263	0.0215	0.0215	0.0319	0.0296	0.0216	0.0216	0.0309	0.0289	0.0246	0.0246
Discharging pump cost	0.0921	0.0917	0.0901	0.0901	0.1005	0.09772	0.0924	0.0924	0.0961	0.0923	0.0894	0.0894
EV cost (MUS$)	0.1428	0.1389	0.1316	0.1316	0.1471	0.1419	0.13974	0.13974	0.1449	0.1402	0.1364	0.1364
EX cost	0.2792	0.2759	0.2714	0.27145	0.2863	0.2842	0.2801	0.2801	0.2824	0.2792	0.2731	0.2731
Comp cost (MUS$)	0.00075	0.0007	0.00069	0.000694	0.0007681	0.00073	0.00073	0.00073	0.0007548	0.00073	0.00073	0.00073
(Cond + CZ) cost (MUS$)	0.2579	0.2519	0.2464	0.2464	0.2701	0.2684	0.2614	0.2614	0.2612	0.2595	0.2534	0.2534
CT pump cost (MUS$ year)	0.0324	0.0319	0.03274	0.03274	0.0368	0.0352	0.0307	0.0307	0.0353	0.0332	0.03145	0.03145
CO_2 penalty cost	0.1458	0.1324	0.1287	0.1287	0.1683	0.1641	0.1606	0.1606	0.1557	0.1529	0.1488	0.1488
Total annual cost (MUS$)	1.9769	1.7974	**1.617**	**1.617**	1.6392	1.3121	**1.2677**	**1.2677**	1.7938	1.5574	**1.5240**	**1.5240**

Bold value indicates the best results

Table 6.5 indicates the results of hybrid ITES system optimization at the optimum design points with single-objective and multi-objective optimization methods. The results show that the highest exergy destruction occurs in condenser and cooling tower for maximizing exergy efficiency, minimizing annual cost and optimization with multi-objective optimization respectively. Compressor and AHU have the next highest exergy destruction, respectively. For optimum results GA required 30,000 function evaluations and for the fare comparison TLBO, Jaya and self-adaptive Jaya algorithms also used the same function evaluations.

The system exergy efficiency after maximizing exergy efficiency, minimizing annual cost and optimization with multi-objective optimization are shown in Table 6.6. Moreover, the economic parameters of the system at the maximizing exergy efficiency, minimizing annual cost and optimization with multi-objective optimization are shown in Table 6.7. Therefore, in multi-objective optimization technique (4E analysis), simultaneous analysis of economic and thermodynamic aspects is performed by using TLBO, Jaya and self-adaptive Jaya algorithm. Thus, this approach gives a reasonable trade-off between the first and the second objectives.

Navidbakhsh et al. (2013) used the genetic algorithm for optimization of ITES system and used 150 generations and population size of 200 and thus 30,000 function evaluations were used. Figure 6.2 shows the effect of number of generations on convergence using TLBO, Jaya and self-adaptive Jaya algorithms. The convergence takes place after 15th, 12th and 09th generations by using TLBO, Jaya and self-adaptive Jaya algorithms respectively for maximizing Exergy efficiency of ITES system. Figure 6.3 shows the effect of number of generations on the convergence rate using TLBO, Jaya and self-adaptive Jaya algorithms. The convergence takes place after 17th, 13th and 11th generations using TLBO, Jaya and self-adaptive Jaya algorithms respectively for minimizing the total annual cost of ITES system. Figure 6.4 shows the effect of number of generations on convergence rate using TLBO, Jaya and self-adaptive Jaya algorithms. The convergence takes place after 18th, 15th and 13th generations by using TLBO, Jaya and self-adaptive

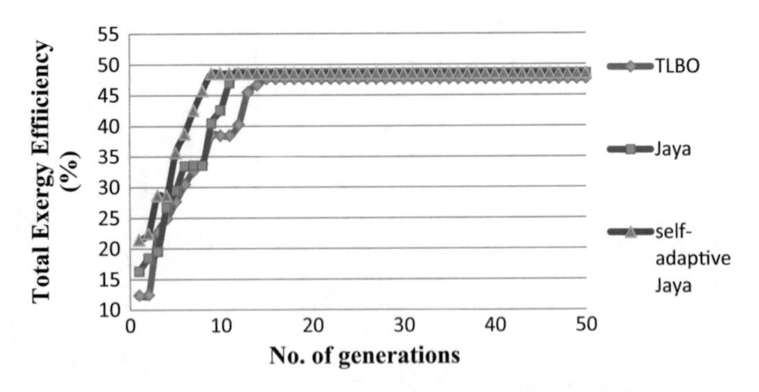

Fig. 6.2 Convergence of TLBO, Jaya and self-adaptive Jaya algorithms for maximizing exergy efficiency of ITES system

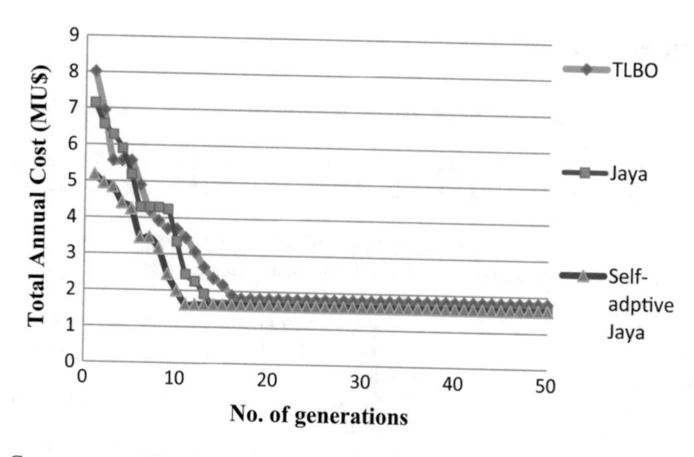

Fig. 6.3 Convergence of TLBO, Jaya and self-adaptive Jaya algorithms for minimizing the total annual cost of ITES system

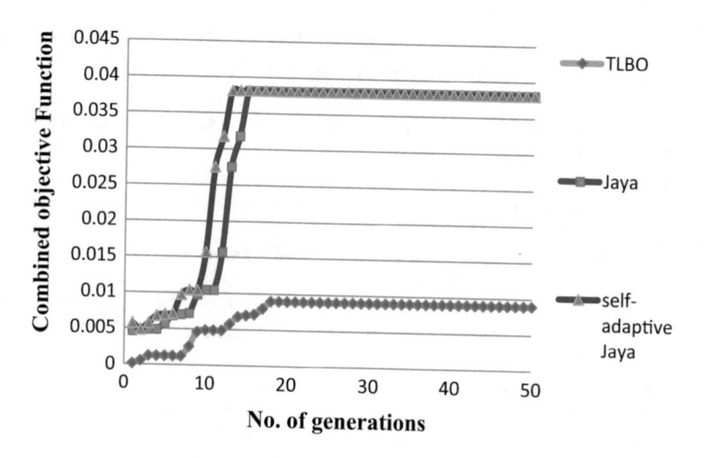

Fig. 6.4 Convergence of TLBO, Jaya and self-adaptive Jaya algorithms for maximizing the combined objective function of ITES system

Jaya algorithms respectively for maximizing the combined objective function of ITES system.

The Jaya and self-adaptive Jaya algorithms are proved superior to GA and TLBO optimization algorithms in terms of robustness of the results. The self-adaptive Jaya takes less computational time and the function evaluations as compared to the other algorithms. These algorithms can be applied with the same ease and effectiveness to the ITES systems optimization considering any number of objectives. Other variants of Jaya algorithm such as SAMP-Jaya and SAMPE-Jaya algorithms can also be used for the same purpose and the readers are encouraged to try.

References

Dincer, I. (2002). On thermal energy storage systems and applications in buildings. *Energy Buildings, 34,* 377–388.

Khudhiar, A. M., & Farid, M. M. (2004). A review on energy conservation in building applications with thermal storage by latent heat using phase change materials. *Energy Conversion and Management, 45,* 263–275.

Koca, A., Oztop, H. F., Koyun, T., & Varol, Y. (2008). Energy and exergy analysis of a latent heat storage system with phase change material for a solar collector. *Renewable Energy, 33,* 567–574.

Lefebvre, D., & Tezel, F. H. (2017). A review of energy storage technologies with a focus on adsorption thermal energy storage processes for heating applications. *Renewable and Sustainable Energy Reviews, 67,* 116–125.

MacPhee, D., & Dincer, I. (2009). Performance assessment of some ice TES systems. *International Journal of Thermal Sciences, 48,* 2288–2299.

Navidbakhsh, M., Shirazi, A., & Sanaye, S. (2013). Four E analysis and multi-objective optimization of an ice storage system incorporating PCM as the partial cold storage for air conditioning applications. *Applied Thermal Engineering, 58,* 30–41.

Ooka, R., & Ikeda, S. (2015). A review on optimization techniques for active thermal energy storage control. *Energy and Buildings, 106,* 225–233.

Palacio, S. N., Valentine, K. F., Wong, M., & Zhang, K. M. (2014). Reducing power system costs with thermal energy storage. *Applied Energy, 129,* 228–237.

Rahdar, M. H., Emamzadeh, A., & Ataei, A. A. (2016a). Comparative study on PCM and ice thermal energy storage tank for air-conditioning systems in office buildings. *Applied Thermal Engineering, 96,* 391–399.

Rahdar, M. H., Heidari, M., Ataei, A., & Choi, J. K. (2016b). Modeling and optimization of R-717 and R-134a ice thermal energy storage air conditioning systems using NSGA-II and MOPSO algorithms. *Applied Thermal Engineering, 96,* 217–227.

Saito, A. (2001). Recent advances in research on cold thermal energy storage. *International Journal of Refrigeration, 25,* 177–189.

Sanaye, S., & Shirazi, A. (2013a). Four E analysis and multi-objective optimization of an ice thermal energy storage for air-conditioning applications. *International Journal of Refrigeration, 36,* 828–841.

Sanaye, S., & Shirazi, A. (2013b). Thermo-economic optimization of an ice thermal energy storage system for air-conditioning applications. *Energy and Buildings, 60,* 100–109.

Shu, W., Longzhe, J., Zhonglong, H., Yage, L., Shengnan, O., Na, G., et al. (2017). Discharging performance of a forced-circulation ice thermal storage system for a permanent refuge chamber in an underground mine. *Applied Thermal Engineering, 110,* 703–709.

Stritih, U., & Butala, V. (2007). Energy saving in building with PCM cold storage. *International Journal of Energy Research, 31,* 1532–1544.

Chapter 7
Single- and Multi-objective Optimization of Traditional and Modern Machining Processes Using Jaya Algorithm and Its Variants

Abstract This chapter describes the formulation of process parameters optimization models for traditional machining processes of turning, surface grinding and modern machining processes of wire electric discharge machining (wire EDM), electro-discharge machining (EDM), micro-electric discharge machining, electro-chemical machining (ECM), abrasive waterjet machining (AWJM), focused ion beam (FIB) micro-milling, laser cutting and plasma arc machining. The TLBO and NSTLBO algorithms, Jaya algorithm and its variants such as Quasi-oppositional (QO) Jaya, multi-objective (MO) Jaya, and multi-objective quasi-oppositional (MOQO) Jaya are applied to solve the single and multi-objective optimization problems of the selected traditional and modern machining processes. The results are found better as compared to those given by the other advanced optimization algorithms.

7.1 Parameters Optimization Models for Machining Processes

In this work the output parameters of manufacturing processes are considered as objectives. These objectives are expressed as functions of input process parameters.

7.1.1 Process Parameters Optimization Model for Turning Process

In this work process parameters optimization of a widely used conventional machining process known as turning is considered. The optimization problem formulated in this work is based on the empirical models developed by Palanikumar et al. (2009). The objectives considered are: minimization of tool flank wear 'V_b' (mm), minimization of surface roughness 'R_a' (μm) and maximization of material

R. Venkata Rao, *Jaya: An Advanced Optimization Algorithm and its Engineering Applications*, https://doi.org/10.1007/978-3-319-78922-4_7

removal rate '*MRR*' (mm^3/min). The process input parameters considered are: cutting speed '*v*' (m/min), feed '*f*' (mm/rev) and depth of cut '*d*' (mm).

Objective functions

The objective functions are expressed by Eqs. (7.1)–(7.3)

$$\begin{aligned} minimize\ V_b = {} & 0.09981 + 0.00069v + 1.41111f - 0.17944d + 0.000001v^2 \\ & - 3.11111f^2 + 0.00222d^2 - 0.00267vf + 0.00007vd \\ & + 0.96667fd \end{aligned} \tag{7.1}$$

$$\begin{aligned} minimize\ R_a = {} & 1.9065 - 0.0103v + 11.1889f + 0.3283d \\ & + 0.000001v^2 - 7.1111f^2 + 0.0022d^2 + 0.0340vf \\ & - 0.0015vd - 4.433fd \end{aligned} \tag{7.2}$$

$$maximize\ MRR = 1000vfd \tag{7.3}$$

Process input parameter bounds

The bounds on process input parameters are expressed as follows

$$50 \le v \le 150 \tag{7.4}$$

$$0.1 \le f \le 0.20 \tag{7.5}$$

$$0.5 \le d \le 1.5 \tag{7.6}$$

7.1.2 Process Parameters Optimization Model for Surface Grinding Process

In this work the process parameters optimization of a widely used conventional machining process known as surface grinding is considered. The optimization aspects of rough grinding process and finish grinding process are discussed separately in two different case studies. The analytical models developed by Wen et al. (1992) for cost of production 'C_T' (\$/pc.), workpiece removal parameter '*WRP*' (mm^3/min N) and surface roughness 'R_a' (μm) are considered to formulate the optimization problem. The process input parameters considered are: work-piece speed 'V_s' (mm/s), wheel speed 'V_w' (mm/s), depth of dressing '*doc*' (mm) and lead of dressing '*L*' (mm).

7.1.2.1 Case Study 1

This case study describes the optimization aspects of rough grinding process. The objective functions for cost of production and workpiece removal parameter are expressed by Eqs. (7.7) and (7.8)

$$\begin{aligned} \text{minimize } C_T = & \frac{M_c}{60p}\left(\frac{L_w+L_e}{V_w 1000}\right)\left(\frac{b_w+b_e}{f_b}\right)\left(\frac{a_w}{a_p}+S_p+\frac{a_w b_w L_w}{\pi D_e b_s a_p G}\right)+\frac{M_c}{60p}\left(\frac{S_d}{V_r}+t_1\right) \\ & +\frac{M_c t_{ch}}{60N_t}+\frac{M_c \pi b_s D_e}{60pN_d L V_s 1000}+C_s\left(\frac{a_w b_w L_w}{pG}+\frac{\pi(doc)b_s D_e}{pN_d}\right)+\frac{C_d}{pN_{td}} \end{aligned} \tag{7.7}$$

Due to space limitation the details of all the constants considered in this work are not provided here.

$$\textit{maximize } \mathrm{WRP} = 94.4\frac{(1+(2\mathrm{doc}/3\mathrm{L}))\mathrm{L}^{11/19}(\mathrm{V_w}/\mathrm{V_s})^{3/19}\mathrm{V_s}}{\mathrm{D_e^{43/304}VOL^{0.47}d_g^{5/38}R_c^{27/19}}} \tag{7.8}$$

In order to satisfy both the objectives simultaneously, a combined objective function (*COF*) is formed by assigning equal importance to both the objectives. The *COF* for rough grinding process is expressed by Eq. (7.9)

$$\text{minimize } COF = w1\frac{CT}{CT^*} - w2\frac{WRP}{WRP^*} \tag{7.9}$$

where, $w1 = w2 = 0.5$; $CT^* = 10$ (\$/pc); $WRP^* = 20$ (mm^3/min N)

Constraints

(a) Thermal damage constraint (Wen et al. 1992)

$$\begin{aligned} U = 13.8 + & \frac{9.64\times10^{-4}V_s}{a_p V_w}+\left(6.9\times10^{-3}\frac{2102.4V_w}{D_e V_s}\right) \\ & \times\left(A_0+\frac{K_u V_s L_w a_w}{V_w D_e^{1/2} a_p^{1/2}}\right)\frac{V_s D_e^{1/2}}{V_w a_p^{1/2}} \end{aligned} \tag{7.10}$$

$$U^* = 6.2 + 1.76\left(\frac{D_e^{1/4}}{a_p^{3/4} V_w^{1/2}}\right) \tag{7.11}$$

The thermal damage constraint is expressed by Eq. (7.12)

$$U^* - U \geq 0 \tag{7.12}$$

(b) Wheel wear parameter (*WWP*) constraint (Wen et al. 1992)

$$WWP = \left(\frac{k_p a_p d_g^{5/38} R_c^{27/29}}{D_c^{1.2/VOL-43/304} VOL^{0.38}} \right) \times \frac{(1 + (doc/L)) L^{27/19} (V_s/V_w)^{3/19} V_w}{(1 + (2doc/3L))} \quad (7.13)$$

The *WWP* constraint is expressed by Eq. (7.14)

$$\frac{WRP}{WWP} - G \geq 0 \quad (7.14)$$

(c) Machine tool stiffness constraint (*MSC*) (Wen et al. 1992)

$$K_c = \frac{1000 V_w f_b}{WRP} \quad (7.15)$$

$$K_s = \frac{1000 V_s f_b}{WWP} \quad (7.16)$$

$$MSC - \frac{|R_{em}|}{K_m} \geq 0 \quad (7.17)$$

where,

$$MSC = \frac{1}{2K_c} \left(1 + \frac{V_w}{V_s G} \right) + \frac{1}{K_s} \quad (7.18)$$

(d) Surface roughness constraint (Wen et al. 1992)

The surface roughness constraint is as given by,

$$R_a = 0.4587 T_{ave}^{0.30} \quad \text{for } 0 < T_{avg} < 0.254 \text{ else,} \quad (7.19)$$

$$R_a = 0.7866 T_{ave}^{0.72} \quad \text{for } 0.254 < T_{avg} < 2.54 \quad (7.20)$$

$$T_{avg} = 12.5 \times 10^3 \frac{d_g^{16/27} a_p^{19/27}}{D_e^{8/27}} \left(1 + \frac{doc}{L} \right) L^{16/27} \left(\frac{V_w}{V_s} \right)^{16/27} \quad (7.21)$$

$$R_a \leq 1.8\ \mu\text{m} \quad (7.22)$$

Process input parameter bounds

The bounds on the process input parameters are expressed as follows

$$1000 \le V_s \le 2023 \quad (7.23)$$

$$10 \le V_w \le 27.7 \quad (7.24)$$

$$0.01 \le doc \le 0.137 \quad (7.25)$$

$$0.01 \le L \le 0.137 \quad (7.26)$$

7.1.2.2 Case Study 2

In the case of finish grinding process, minimization of production cost and minimization of surface roughness are considered as objectives. While the constraint is on the limiting value of workpiece removal parameter along with other constraints such as wheel wear, machine tool stiffness and thermal damage to the workpiece.

Objective functions

The objective functions for production cost and surface roughness for finish grinding process are same as those expressed by Eqs. (7.7) and (7.19)–(7.22), respectively. In order to satisfy both the objectives simultaneously a combined objective function for finish grinding process is formulated and is expressed by Eq. (7.27).

$$\text{minimize } Z_R = w1\frac{CT}{CT^*} + w2\frac{R_a}{R_a^*} \quad (7.27)$$

where, $w1 = 0.3$ and $w2 = 0.7$; $CT^* = 10$ (\$/pc); $R_a^* = 1.8$ (μm)

Constraints

The thermal damage constraint, wheel wear parameter constraint, machine tool stiffness constraint is same as given by Eqs. (7.12), (7.14) and (7.17), respectively. In addition, the constraint on the *WRP* is considered i.e. $WRP \ge 20$. The parameter bounds in case of finish grinding process are same as those considered for rough grinding process. The values of constants considered in the case of finish grinding process are same as those considered in rough grinding process except the value of $a_w = 0.055$ mm and $a_p = 0.0105$ mm/pass is considered in case of finish grinding process (Wen et al. 1992).

7.1.3 Process Parameters Optimization Model for Wire-EDM Process

7.1.3.1 Case Study 1

The optimization problem formulated in this work is based on the analysis given by Kuriakose and Shunmugam (2005). The process input parameters considered are: applied voltage '*V*', ignition pulse current '*IAL*' (A), pulse-off time 'T_B' (μs), pulse duration 'T_A' (μs), servo reference mean voltage 'A_j'(V), servo speed variation '*S*' (mm/min), wire speed 'W_s' (m/min), wire tension 'W_b' (kg) and injection pressure '*Inj*' (bar). The objectives considered are: maximization of cutting velocity '*CV*' (mm/min) and minimization of surface roughness 'R_a' (μm).

Objective functions

The objective functions are expressed by Eqs. (7.28) and (7.29).

$$\begin{aligned} \mathit{Maximize}\ (CV) = {} & 1.662 + 0.002375 IAL - 0.0639 T_B + 0.628 T_A - 0.01441 A_j \\ & + 0.008313 S - 0.001792 W_S - 0.673 W_b - 0.0294 Inj \end{aligned} \quad (7.28)$$

$$\begin{aligned} \mathit{Minimize}\ (R_a) = {} & 2.017 - 0.01236 IAL + 0.0075\, T_B + 1.792\, T_A \\ & - 0.006056 A_j + 0.01\, S - 0.009583 W_s + 0.258\, W_b \\ & - 0.0683 Inj \end{aligned} \quad (7.29)$$

Process input parameter bounds

The bounds on the process parameters are expressed as follows

$$8 \le IAL \le 16 \quad (7.30)$$

$$4 \le T_B \le 8 \quad (7.31)$$

$$0.6 \le T_A \le 1.2 \quad (7.32)$$

$$30 \le A_j \le 60 \quad (7.33)$$

$$4 \le S \le 12 \quad (7.34)$$

$$4 \le W_s \le 8 \quad (7.35)$$

$$0.8 \le W_b \le 1 \quad (7.36)$$

$$2 \le Inj \le 4 \quad (7.37)$$

7.1.3.2 Case Study 2

The optimization problem formulated in this work is based on the mathematical models developed by Garg et al. (2012). The objectives considered are: maximization of cutting speed '*CS*' (mm/min) and minimization surface roughness '*SR*' (μm). The process input parameters considered are: pulse on time 'T_{on}' (μs), pulse off time 'T_{off}' (μs), peak current '*IP*' (A), spark set voltage '*SV*' (V), wire feed '*WF*' (m/min) and wire tension '*WT*' (g).

Objective functions

The objective functions are expressed by Eqs. (7.38) and (7.39).

$$\begin{aligned} \text{maximize } CS = {} & -24.85563 + 0.29637 \times T_{on} + 0.12237 \times T_{off} \\ & + (6.53472E-4) \times IP + 0.1454 \times SV + 0.060880 \times WT \\ & + (1.52323E-3) \times T_{off}^2 - (3.15625E-3) \times T_{on} \times T_{off} \\ & - (1.66667E-3) \times T_{on} \times SV + (7.84375E-4) \times T_{off} \times SV \\ & - (1.30312E-3) \times SV \times WT \end{aligned} \tag{7.38}$$

$$\begin{aligned} \text{minimize } SR = {} & 2.28046 + (0.014514 \times T_{on}) - 0.01175 \times T_{off} \\ & - (7.54444E-3) \times IP - (4.466E-3) \times SV - 0.19140 \times WF \\ & - 0.8279 \times WT + (7.35417E3) \times T_{on} \times WT \\ & + (1.08333E-3) \times IP \times WF \end{aligned} \tag{7.39}$$

Process input parameter bounds

The bounds on process input parameters are expressed as,

$$112 \leq T_{\text{on}} \leq 118 \tag{7.40}$$

$$48 \leq T_{\text{off}} \leq 56 \tag{7.41}$$

$$140 \leq IP \leq 200 \tag{7.42}$$

$$35 \leq SV \leq 55 \tag{7.43}$$

$$6 \leq WF \leq 10 \tag{7.44}$$

$$4 \leq WT \leq 8 \tag{7.45}$$

7.1.3.3 Case Study 3

In this work the optimization aspects of micro-WEDM process are considered. The optimization problem formulated in this work is based on the empirical models developed by Kuriachen et al. (2015). The objectives considered are: maximization of material removal rate '*MRR*' (mm^3/min) and minimization of surface roughness '*SR*' (μm). The process input parameters considered are: gap voltage '*A*' (V), capacitance '*B*' (μF), feed rate '*C*' (μm/s) and wire tension '*D*' (gm).

Objective functions

The objective functions in terms of coded values of process input parameters are expressed by Eqs. (7.46) and (7.47). The process input parameters are coded between −1 and 1 and the bounds on the process parameters in terms of actual values are expressed by Eqs. (7.46) and (7.47).

$$\begin{aligned} maximize\ \sqrt{MRR} = {} & 0.14 + 0.006812A + 0.024B + 0.014C - 0.007979AB \\ & + 0.00385BC - 0.039B^2 \end{aligned} \tag{7.46}$$

$$\text{minimize } SR = 1.13 - 0.11A + 0.080B - 0.17C - 0.16BC + 0.60A^2 + 0.28C^2 \tag{7.47}$$

Process input parameter bounds

The bounds on the process input parameters are expressed as follows:

$$100 < A < 150 \tag{7.48}$$

$$0.01 < B < 0.4 \tag{7.49}$$

$$3 < C < 9 \tag{7.50}$$

$$4.125 < D < 12.375 \tag{7.51}$$

7.1.4 Process Parameters Optimization Model for EDM Process

7.1.4.1 Case Study 1

In this work the objectives considered are: maximization of material removal rate '*MRR*' (mg/min), minimization of tool wear rate '*TWR*' (mg/min), minimization of taper angle '*Θ*' (degree) and minimization of delamination factor '*DF*' in electro-discharge machining (EDM) process. For this purpose, experiments are performed and the data collected is used to develop regression models for *MRR*, *TWR*, Θ and *DF* are developed and the same as used as objective functions to formulate an optimization problem.

The process input parameters considered are: gap voltage 'V_g' (V), pulse current 'I_p' (A), pulse-on time 'T_{on}' (μs) and tool rotation speed '*N*' (rpm). Design of experiments is used as a tool to generate the experimental procedure. The experiments are planned according to the rotational-central composite design (RCCD) and regression models for *MRR*, *TWR*, Θ and *DF* are developed. The experiments are conducted with 4 process parameters considering each at 5 levels. The values of other process parameters are maintained as constant such as duty cycle 40%, *Z* depth 15 mm, sensitivity 8, anti-arc sensitivity 7, work time 8.0 s, lift time 0.2 s, prepulse sparking current 0 A and straight polarity.

The experiments are performed in the Manufacturing Sciences laboratory of IIT Kanpur, India (Rao et al. 2016). ZNC Electronica EDM machine with a copper tool of 3 mm diameter is used for the purpose of experimentation. Carbon-carbon composite materials with 6% grade with approximate dimensions as 155 mm 75 mm × 3.5 mm is used as the workpiece material. A copper rod of 3 mm diameter and 7 mm length is used as tool. The tool is given negative polarity while the workpiece is given positive polarity. 30 experiments with 6 replicates of centre point are performed. Table 7.1 gives the values of *MRR, TWR*, Θ and *DF* measured after experimentation.

The regression models for *MRR*, *TWR*, Θ and *DF* are developed using a logarithmic scale with uncoded values of machining parameters and are expressed by Eqs. (7.52) and (7.55).

$$\begin{aligned} MRR = \exp(&-264.7311 + 14.62835 \times (\log V_g) + 0.63896 \times (\log I_p) \\ &+ 8.67444 \times (\log T_{on}) + 74.465 \times (\log N) - 1.0053 \times (\log V_g)^2 \\ &+ 0.2317 \times (\log I_p)^2 - 0.3459 \times (\log T_{on})^2 - 5.83289 \times (\log N)^2 \\ &- 0.63041 \times (\log V_g \times \log I_p) + 0.16643 \times (\log V_g \times \log T_{on}) \\ &- 0.87394 \times (\log V_g \times \log N) + 0.12709 \times (\log I_p \times \log T_{on}) \\ &- 0.94153 \times (\log T_{on} \times \log N)) \\ &(R^2 = 0.855) \end{aligned} \tag{7.52}$$

$$\begin{aligned} TWR = \exp(&-264.7887 + 16.8 \times (\log V_g) + 7.1385 \times (\log I_p) \\ &- 1.206 \times (\log T_{on}) + 79.1385 \times (\log N) - 0.9912 \times (\log V_g)^2 \\ &- 0.5355 \times (\log I_p)^2 + 0.03906 \times (\log T_{on})^2 - 6.3355 \times (\log N)^2 \\ &- 0.3342 \times (\log V_g \times \log I_p) + 0.4552 \times (\log V_g \times \log T_{on}) \\ &- 1.6904 \times (\log V_g \times \log N) + 0.06969 \times (\log I_p \times \log T_{on}) \\ &- 0.2617(\log T_{on} \log N)) \\ &(R^2 = 0.926) \end{aligned} \tag{7.53}$$

Table 7.1 Values of *MRR*, *TWR*, Θ and *DF* in EDM process measured after experimentation (Rao et al. 2016a)

S. No.	V_g	I_p	T_{on}	N	*MRR*	*TWR*	Θ (°)	*DF*
1	40	20	500	250	11.774	15.56	0.49896	1.13765
2	80	20	500	250	18.429	28.071	0.94956	1.17407
3	40	40	500	250	17.984	87.5	2.14977	1.19477
4	80	40	500	250	23.818	169.129	2.93734	1.28756
5	40	20	1000	250	10.485	9.706	1.01307	1.1483
6	80	20	1000	250	21.755	35.145	1.44775	1.18842
7	40	40	1000	250	17.957	89.586	2.34904	1.20387
8	80	40	1000	250	25.836	189.062	3.03773	1.29591
9	40	20	500	350	9.409	12.994	1.21824	1.12567
10	80	20	500	350	14.73	22.858	1.45351	1.16211
11	40	40	500	350	14.251	87.553	2.47527	1.18693
12	80	40	500	350	17.152	141.074	2.73688	1.27143
13	40	20	1000	350	7.567	13.593	1.43657	1.13062
14	80	20	1000	350	12.739	26.774	1.58195	1.16556
15	40	40	1000	350	13.983	88.552	1.86935	1.18385
16	80	40	1000	350	16.643	145.738	2.11996	1.2753
17	25	30	750	300	7.135	13.344	1.56341	1.15768
18	95	30	750	300	17.791	97.223	2.05367	1.20508
19	60	10	750	300	16.462	2.123	0.609447	1.1168
20	60	45	750	300	23.262	189.031	1.93848	1.23504
21	60	30	300	300	19.481	90.738	1.2623	1.23478
22	60	30	2000	300	11.879	63.277	2.5361	1.20539
23	60	30	750	200	23.644	92.36	1.75735	1.24253
24	60	30	750	400	4.167	8.568	1.45485	1.21992
25	60	30	750	300	22.61	76.726	1.54156	1.21704
26	60	30	750	300	22.532	80.963	1.51356	1.21996
27	60	30	750	300	21.873	75.491	1.50756	1.22892
28	60	30	750	300	22.095	79.266	1.53274	1.22455
29	60	30	750	300	20.032	75.118	1.53498	1.2119
30	60	30	750	300	22.263	76.726	1.52827	1.2338

$$\begin{aligned} \theta = \exp(&-60.4654 + 3.7949 \times (\log V_g) + 6.7335 \times (\log I_p) \\ &+ 10.0673 \times (log\, T_{on}) + 1.58424 \times (\log N) + 0.6458 \times (\log V_g)^2 \\ &+ 0.18217 \times (\log I_p)^2 + 0.24652 \times (\log T_{on})^2 + 1.2749 \times (\log N)^2 \\ &- 0.2535 \times (\log V_g \times \log I_p) - 0.1392 \times (\log V_g \times \log T_{on}) \\ &- 1.20511 \times (\log V_g \times \log N) - 0.89575 \times (\log I_p \times \log T_{on}) \\ &- 1.6643 \times (\log T_{on} \times \log N) \end{aligned} \tag{7.54}$$

$$(R^2 = 0.892)$$

$$DF = \exp(-0.58509 + 0.15295 \times (\log V_g) - 0.14645 \times (\log I_p) + 0.27323 \times (\log T_{on}) - 0.12994 \times (\log N) - 0.02262 \times (\log V_g)^2 + 0.00873 \times (\log I_p)^2 + (7.9329E - 05) \times (\log T_{on})^2 + 0.032075 \times (\log N)^2 + 0.07168 \times (\log V_g \times \log I_p) - 0.00957 \times (\log V_g \times \log T_{on}) - 0.01534 \times (\log V_g \times \log N) - 0.01683 \times (\log I_p \times \log T_{on}) - 0.03149 \times (\log T_{on} \times \log N)) \quad (R^2 = 0.898) \tag{7.55}$$

Process input parameter bounds

The bounds on the process parameters are expressed as follows:

$$25 \leq V_g \leq 95 \tag{7.56}$$

$$10 \leq I_p \leq 45 \tag{7.57}$$

$$300 \leq T_{on} \leq 2000 \tag{7.58}$$

$$200 \leq N \leq 400 \tag{7.59}$$

7.1.4.2 Case Study2

In this work process parameters optimization of micro-electric discharge machining process is considered (micro-EDM). The objectives considered are: maximization of *MRR* (mm^3/min) and minimization of *TWR* (mm^3/min). The regression models for *MRR* and *TWR* are developed based on actual data collected by means of experimentation and the same as used as objective functions in order to formulate an optimization problem.

The experiments are performed at Manufacturing Science Laboratory of IIT Kanpur, India and DT110 high precision, CNC controlled, micro-machining setup with integrated multi-process machine tool was used for the purpose of experimentation. The workpiece is die material EN24, cylindrical tungsten electrode (dia. 500 μm) is used as tool and conventional EDM oil is used as die electric. The feature shape considered for the study is a μ-channel of width approximately equal to the diameter of the tool, length of cut 1700 μm, the depth of channel is considered as 1000 μm.

In the present study the bulk machining approach for μ-EDM milling is used. As the bulk machining approach results in excessive tool wear intermittent tool dressing with block electro-discharge grinding (EDG) process is used. Review of literature shows that there are a number of process parameters that affect the performance of μ-EDM milling process. Therefore prior to actual experimentation dimensional analysis is performed to identify the most influential parameters of the process such as Energy '*E*' (μJ), feed rate '*F*' (μm/s), tool rotation speed '*S*' (rpm) and aspect ratio '*A*'. The useful levels of these parameters is identified using one factor at a time (OFAT) analysis and 2 levels of energy, 4 levels of feed rate, 3 levels of rotational speed, and 4 levels of aspect ratio are identified.

The regression models for *MRR* and *TWR* are formulated by considering a full factorial experimental design, considering all combination of process parameter values a total number of 96 experiments are conducted. The values of *MRR* (mm^3/min) and *TWR* (mm^3/min) are measured and recorded as shown in Table 7.2.

The regression models for *MRR* and *TWR* are developed using the experimental data, using a logarithmic scale, and are expressed by Eqs. (7.60) and (7.61) in the uncoded form of process parameters.

Table 7.2 Values of *MRR* and *TWR* for micro-EDM process measured after experimentation (Rao et al. 2016a)

S. No.	*E* (μJ)	*F* (μm/s)	*S* (rpm)	*A*	*MRR* (10^{-3} mm^3/min)	*TWR* (10^{-3} mm^3/min)
1	2000	60	100	1	9.16	1.99
2	2000	60	800	1	23.48	5.16
3	500	60	800	1	12.88	3.2
4	500	60	100	1	6.26	0.92
5	500	10	100	1	4.53	0.74
6	2000	10	800	1	12.58	2.29
7	500	10	800	1	8.48	1.48
8	500	10	500	1.5	7.06	1.08
9	500	25	800	1.5	10.14	1.67
10	500	45	500	1.5	9.92	1.55
11	500	45	800	0.5	9.39	1.43
12	500	10	500	2	5.97	1.05
13	500	25	100	0.5	3.6	0.54
14	2000	45	500	2	16	3.66
15	2000	60	500	1	19.01	4.48
16	500	45	100	2	4.05	0.83
17	2000	10	100	2	6.05	1.11
18	500	60	500	1	9.91	2.15
19	500	60	100	1.5	6.57	1.01
20	500	10	800	0.5	7.34	1.19
21	2000	60	800	1.5	28.17	5.87
22	2000	25	800	2	19.68	3.83
23	500	25	800	2	11.69	1.61

(continued)

Table 7.2 (continued)

S. No.	E (μJ)	F (μm/s)	S (rpm)	A	MRR (10^{-3} mm^3/min)	TWR (10^{-3} mm^3/min)
24	500	25	500	0.5	6.32	0.81
25	2000	10	100	1	4.24	0.53
26	500	60	100	0.5	5.52	1.27
27	2000	60	100	2	12.08	3.56
28	2000	10	500	1	5.28	1.34
29	500	25	100	1	4.56	0.79
30	2000	25	500	1	10.15	2.45
31	500	60	800	0.5	12.71	2.84
32	500	45	500	1	9.2	1.92
33	2000	60	800	2	25.39	5.14
34	500	45	100	1.5	5.72	1.02
35	2000	10	500	2	7.66	1.87
36	500	10	500	1	6.69	1.23
37	500	25	800	1	9.86	1.64
38	2000	45	800	2	23.75	5.06
39	2000	25	800	1	21	3.33
40	500	60	800	2	12.02	2.1
41	500	60	100	2	5.62	0.93
42	2000	45	100	2	11.62	3.12
43	500	45	800	2	12.62	1.96
44	500	60	500	0.5	9.26	1.49
45	2000	25	500	1.5	15	3.32
46	2000	25	500	0.5	6.7	1.13
47	2000	10	800	1.5	13.32	2.77
48	500	10	800	1.5	9.03	1.36
49	2000	60	500	0.5	16.12	3.33
50	500	45	500	2	10.63	1.73
51	500	45	100	1	5.55	0.73
52	2000	45	500	1	17.65	4.49
53	2000	60	500	2	17.46	3.84
54	500	10	100	2	4.52	0.73
55	2000	45	800	1.5	29.84	6.4
56	500	60	500	1.5	10.08	2.06
57	2000	45	500	1.5	22.29	5.04
58	2000	60	100	0.5	7.32	1.13
59	2000	45	100	0.5	4.84	0.97
60	2000	45	800	0.5	23.2	3.84
61	500	45	800	1.5	12.7	1.97
62	500	45	100	0.5	5.3	1.02
63	2000	10	800	0.5	9.88	1.79

(continued)

Table 7.2 (continued)

S. No.	E (μJ)	F (μm/s)	S (rpm)	A	MRR (10^{-3} mm^3/min)	TWR (10^{-3} mm^3/min)
64	2000	25	100	1	6.13	1.21
65	2000	60	800	0.5	25.64	3.58
66	500	10	100	0.5	3.99	0.63
67	2000	25	800	1.5	25.55	4.23
68	2000	45	100	1	6.93	1.37
69	500	10	500	0.5	6.02	0.97
70	500	10	100	1.5	5.28	0.93
71	2000	25	100	1.5	7.1	1.44
72	2000	60	500	1.5	22.68	5.3
73	500	60	800	1.5	13.8	2.5
74	500	25	500	1.5	7.82	1.19
75	2000	10	500	1.5	7.9	1.73
76	2000	10	500	0.5	3.95	0.64
77	500	25	500	2	8.48	1.33
78	500	45	500	0.5	7.55	1.11
79	2000	60	100	1.5	11.41	3.82
80	2000	10	100	1.5	6.79	0.95
81	2000	45	800	1	24.61	5.82
82	500	25	100	2	4.75	0.78
83	500	45	800	1	11.25	2.33
84	2000	45	100	1.5	9.73	2.14
85	2000	10	800	2	15.46	3.47
86	500	10	800	2	11.33	1.43
87	500	60	500	2	10.34	1.83
88	2000	10	100	0.5	2.73	0.31
89	2000	25	500	2	13.21	3.46
90	500	25	100	1.5	5.31	0.82
91	2000	25	100	0.5	3	0.44
92	2000	25	100	2	7.88	1.86
93	500	25	800	0.5	8.54	1.07
94	2000	45	500	0.5	13.38	1.79
95	500	25	500	1	7.44	1.57
96	2000	25	800	0.5	17.29	1.79

$$
\begin{aligned}
MRR = \exp(&11.15134 - 1.79325 \times (\log F) - 3.20333 \times (\log S) \\
&- 0.114931 \times (\log A) - 0.072533 \times (\log E)^2 + 0.06657 \times (\log F)^2 \\
&+ 0.251122 \times (\log S)^2 - 0.16314 \times (\log A)^2 + 0.21496 \times (\log E \times \log F) \\
&+ 0.099501 \times (\log E \times \log S) + 0.16903 \times (\log E \times \log A) \\
&+ 0.040721 \times (\log F \times \log S) - 0.11206 \times (\log F \times \log A) \\
&- 0.07489 \times (\log S \times \log A)) \\
&\qquad (R^2 = 0.94)
\end{aligned} \tag{7.60}
$$

$$
\begin{aligned}
TWR = \exp(&5.68347 - 2.22795 \times \log F - 1.77173 \times \log S \\
&- 1.29611 \times \log A - 0.07152 \times (\log E)^2 + 0.175929 \times (\log F)^2 \\
&+ 0.13946 \times (\log S)^2 - 0.34761 \times (\log A)^2 + 0.23781 \times (\log E \times \log F) \\
&+ 0.1005 \times (\log E \times \log S) + 0.380612 \times (\log E \times \log A) \\
&- 0.015495 \times (\log F \times \log S) - 0.120799 \times (\log F \times \log A) \\
&- 0.096066 \times (\log S \times \log A)) \\
&\qquad (R^2 = 0.933)
\end{aligned} \tag{7.61}
$$

The regression models for *MRR* and *TWR* expressed by Eqs. (7.60) and (7.61), respectively are used as objective functions to formulate an optimization problem.

Process input parameter bounds

The bounds on the process input parameters are expressed as follows

$$500 \le E \le 2000 \tag{7.62}$$

$$10 \le F \le 60 \tag{7.63}$$

$$100 \le S \le 800 \tag{7.64}$$

$$0.5 \le A \le 2.0 \tag{7.65}$$

7.1.5 Process Parameters Optimization Models for ECM Process

7.1.5.1 Case Study 1

The optimization problem formulated in this work is based on the empirical models developed by Bhattacharyya and Sorkhel (1999). The objectives considered are: maximization of material removal rate '*MRR*' (gm/min) and minimization of overcut '*OC*' (mm). The process input parameters considered are: electrolyte concentration 'x_1' (g/l), electrolyte flow rate 'x_2' (l/min), applied voltage 'x_3' (V) and inter-electrode gap 'x_4' (mm).

Objective functions

The objective functions in terms of coded values of process parameters are expressed by Eqs. (7.66) and (7.67). The process parameters are coded between -2 to 2 and the bounds on the process parameters in their actual form are expressed by Eqs. (7.68) and (7.71).

$$\begin{aligned} maximize\ MRR = {} & 0.6244 + 0.1523x_1 + 0.0404x_2 + 0.1519x_3 - 0.1169x_4 + 0.0016x_1^2 \\ & + 0.0227x_2^2 + 0.0176x_3^2 - 0.0041x_4^2 + 0.0077x_1x_2 + 0.0119x_1x_3 \\ & - 0.0203x_1x_4 + 0.0103x_2x_3 - 0.0095x_2x_4 + 0.03x_3x_4 \end{aligned} \tag{7.66}$$

$$\begin{aligned} minimize\ OC = {} & 0.3228 + 0.0214x_1 - 0.0052x2 + 0.0164x3 + 0.0118x4 - 0.0041x_1^2 \\ & - 0.0122x_2^2 + 0.0027x_3^2 + 0.0034x_4^2 - 0.0059x_1x_2 - 0.0046x_1x_3 \\ & - 0.0059x_1x_4 + 0.0021x_2x_3 - 0.0053x_2x_4 - 0.0078x_3x_4 \end{aligned} \tag{7.67}$$

Process input parameter bounds

The bounds on the process input parameters are expressed as follows

$$15 < x_1 < 75 \tag{7.68}$$

$$10 < x_2 < 14 \tag{7.69}$$

$$10 < x_3 < 30 \tag{7.70}$$

$$0.4 < x_4 < 1.2 \tag{7.71}$$

7.1.5.2 Case Study 2

The optimization problem formulated in this work is based on the empirical models developed by Acharya et al. (1986). The objectives considered are: minimization of dimensional inaccuracy 'Z_1' (μm), minimization of tool wear tool wear in terms of number of sparks per mm 'Z_2' and maximization of material removal rate 'Z_3'. The process input parameters considered are: feed '*f*' (μm/s), electrolyte flow velocity '*U*' (cm/s) and voltage '*V*' (volts). The constraints are on temperature of electrolyte, passivity and choking of electrolyte flow.

Objective functions

The first objective is to minimize the dimensional inaccuracy which is expressed as follows

$$Z_1 = f^{0.381067} U^{-0.372623} V^{3.155414} e^{-3.128926} \tag{7.72}$$

The second objective is to minimize the tool wear which is expressed as follows

$$Z_2 = f^{3.528345} U^{0.000742} V^{-2.52255} e^{0.391436} \tag{7.73}$$

The third objective is to maximize the material removal rate which is expressed as follows

$$Z_3 = f \tag{7.74}$$

Constraints

The constraints consider in this work are same as those considered by Acharya et al. (1986) and they are as follows.

- *Temperature constraint*

To avoid boiling the electrolyte, the electrolyte temperature at the outlet should be less than the electrolyte boiling temperature. The temperature constraint is expressed by Eq. (7.75).

$$1 - \left(f^{2.133007} U^{-1.088937} V^{-0.351436} e^{0.321968}\right) \geq 0 \tag{7.75}$$

- *Passivity constraint*

Oxygen evolved during electrochemical machining forms an oxide film, which is the root cause of passivity. To avoid passivity, the thickness of the oxygen gas bubble layer must be greater than the passive layer thickness. This constraint is expressed by Eq. (7.76)

$$\left(f^{-0.844369} U^{-2.526076} V^{1.546257} e^{12.57697}\right) - 1 \geq 0 \tag{7.76}$$

- *Choking constraint*

Hydrogen evolved at the cathode during the ECM process can choke the electrolyte flow. To avoid choking the electrolyte flow, the maximum thickness of the hydrogen bubble layer should be less than the equilibrium inter-electrode gap. This constraint is expressed by Eq. (7.77)

$$1 - \left(f^{0.075213} U^{-2.488362} V^{0.240542} e^{11.75651}\right) \geq 0 \quad (7.77)$$

Process input parameter bounds

The bounds on the process input parameters are as follows

$$8 \leq f \leq 200 \quad (7.78)$$

$$300 \leq U \leq 5000 \quad (7.79)$$

$$3 \leq V \leq 21 \quad (7.80)$$

7.1.6 Process Parameters Optimization Model for FIB Micro-milling Process

The optimization problem formulated in this work is based on the empirical models developed by Bhavsar et al. (2015). The objectives considered are: maximization of material removal rate '*MRR*' (μm^3/s) and minimization of surface roughness 'R_a' (nm). in focused ion beam (FIB) micro-milling. The process input parameters considered are: extraction voltage 'x_1' (kV), angle of inclination, 'x_2' (degree), beam current 'x_3' (nA), dwell time 'x_4' (μs) overlap 'x_5' (%).

Objective functions

The objective functions are expressed by Eqs. (7.81) and (7.82).

$$\begin{aligned} \text{maximize } MRR = {} & 0.0514 - 0.00506x_1 - 0.0269x_3 - 0.000032x_2^2 \\ & - 0.00009x_5^2 - 0.000103x_1x_2 + 0.0036x_1x_3 \\ & + 0.000228x_1x_5 + 0.000625x_2x_3 + 0.0001x_2x_5 \\ & + 0.000514x_3x_5 \end{aligned} \quad (7.81)$$

$$\begin{aligned} \text{minimize } R_a = {} & 245 + 3.61x_2 - 5.38x_2 - 0.304x_1^2 + 0.0428x_2^2 \\ & + 0.0735x_5^2 + 0.863x_1x_3 + 0.144x_1x_5 - 0.17x_2x_3 \\ & - 0.139x_2x_5 + 1.5x_3x_4 \end{aligned} \quad (7.82)$$

Parameter input parameter bounds

The bounds on the process input parameters are as follows

$$15\ (\mathrm{kV}) \le x_1 \le 30\ (\mathrm{kV}) \tag{7.83}$$

$$10^{\circ} \le x_2 \le 70^{\circ} \tag{7.84}$$

$$0.03\ (\mathrm{nA}) \le x_3 \le 3.5\ (\mathrm{nA}) \tag{7.85}$$

$$1\ (\mu\mathrm{s}) \le x_4 \le 10\ (\mu\mathrm{s}) \tag{7.86}$$

$$30(\%) \le x_5 \le 75(\%) \tag{7.87}$$

7.1.7 Process Parameters Optimization Model for Laser Cutting Process

The optimization problem formulated in this work is based on the analysis given by Pandey and Dubey (2012). The objectives considered are: minimization of surface roughness 'R_a' (μm) and minimization kerf taper 'K_t' (°). The process input parameters considered are: gas pressure 'x_1' (kg/cm^2), pulse width 'x_2' (ms), pulse frequency 'x_3' (Hz) and cutting speed 'x_4' (mm/min).

Objective functions

The objective functions are expressed by Eqs. (7.88) and (7.89)

$$\begin{aligned} \textit{minimize } R_a = & -33.4550 + 7.2650x_1 + 12.1910x_2 + 1.8114x_3 \\ & - 0.2813x_2^2 - 0.0371x_3^2 - 0.7193x_1x_2 + 0.0108x_3x_4 \\ & + 0.0752x_1x_2 \end{aligned} \tag{7.88}$$

$$\begin{aligned} \textit{minimize } K_t = & -8.567 - 2.528x_1 + 0.2093x_1^2 + 2.1318x_2^2 - 0.0371x_3^2 \\ & - 0.7193x_1x_2 + 0.0108x_3x_4 + 0.0752x_1x_3 \end{aligned} \tag{7.89}$$

Process input parameter bounds

The bounds on the process input parameters are expressed as follows

$$5 \le x_1 \le 9 \tag{7.90}$$

$$1.4 \le x_2 \le 2.2 \tag{7.91}$$

$$6 \leq x_3 \leq 14 \quad (7.92)$$

$$15 \leq x_4 \leq 25 \quad (7.93)$$

7.1.8 Process Parameters Optimization Model for AWJM Process

In this work the objectives considered are: maximization of kerf top '*Ktw*' (mm) width and minimization of taper angle (degrees) in abrasive waterjet machining (AWJM) process. The optimization problem is formulated based on the regression models developed by Shukla and Singh (2016). The process input parameters considered are: traverse speed 'x_1' (mm/min), stand-off distance 'x_2' (mm), and mass flow rate 'x_3' (g/s).

Objective functions

The objective functions are expressed by Eqs. (7.94) and (7.95)

$$\begin{aligned} \textit{maximize } Ktw = {} & 12.7538 - 0.0974x_1 + 0.0913x_2 - 1.0424x_3 \\ & + 0.0002x_1^2 - 0.0289x_2^2 + 0.12011x_3^2 + 0.0016x_1x_2 \\ & + 0.0032x_1x_3 \end{aligned} \quad (7.94)$$

$$\begin{aligned} \textit{minimize taperangle} = {} & 26.0879 - 0.2401x_1 + 0.5024x_2 - 2.2168x_3 \\ & + 0.0006x_1^2 - 0.0884x_2^2 - 0.3657x_3^2 + 0.0024x_1x_2 \\ & + 0.0028x_1x_3 \end{aligned} \quad (7.95)$$

Process input parameter bounds

The bounds on the process input parameters are expressed as follows

$$160 \leq x_1 \leq 200 \quad (7.96)$$

$$1 \leq x_2 \leq 5 \quad (7.97)$$

$$1.5 \leq x_3 \leq 3.5 \quad (7.98)$$

7.1.9 *Process Parameters Optimization Model for PAM Process*

This work aims to improve the performance of plasma arc machining process (PAM) process by means of process parameters optimization. The objectives considered are: maximization of material removal rate '*MRR*' (g/s) and minimization of dross formation rate '*DFR*' (g/s). The regression models for are developed using data collected by means of actual experimentation, and the same as used as fitness functions for MO-Jaya algorithm in order to obtain multiple trade-off solutions. The experiments are performed at Manufacturing Science Laboratory of IIT Kanpur, India and AISI 4340 steel (0.16–0.18% of C) is used as work material (Rao et al. 2016a). The experiments are planned according to the central composite design (CCD) and 4 process input parameters such as thickness of workpiece '*T*' (mm), current '*I*' (A), arc gap voltage 'V_g' (V) and speed '*S*' (mm/min) are considered each at 5 levels. Table 7.3 gives the plan of experiments based on CCD.

Thirty experimental runs are performed and *MRR* and *DFR* are measured and recorded. Thereafter, regression models for *MRR* and *DFR* are developed using a logarithmic scale.

$$\begin{aligned} MRR = \exp(&202.0963939 + 26.97654873 \times (\log T) - 115.7823 \times (\log I) \\ &+ 36.5388 \times (\log V_g) - 32.2698 \times (\log S) - 2.3015 \times (\log T)^2 \\ &+ 3.07499 \times (\log I)^2 - 10.03049 \times (\log V_g)^2 + 2.5766 \times (\log S)^2 \\ &+ 0.70759 \times (\log T \times \log I) - 0.25221(\log T \times \log V_g) \\ &- 3.92965 \times (\log T \times \log S) + 17.92577(\log I \times \log V_g) \\ &+ 0.91766 \times (\log I \times \log S) - 0.07549 \times (\log V_g \times \log S)) \\ &\qquad (R^2 = 0.95) \end{aligned} \tag{7.99}$$

$$\begin{aligned} DFR = \exp(&-310.030243 - 7.0437 \times (\log T) + 311.642 \times (\log I) \\ &- 169.3030 \times (\log V_g) + 56.3056 \times (\log S) - 0.5839 \times (\log T)^2 \\ &- 16.1736 \times (\log I)^2 + 17.4766 \times (\log V_g)^2 - 8.15487(\log S)^2 \\ &- 4.90491 \times (\log T \times \log I) + 4.68153 \times (\log T \times \log V_g) \\ &+ 0.17082 \times (\log T \times \log S) - 28.2996 \times (\log I \times \log V_g) \\ &- 8.91918 \times (\log I \times \log S) + 15.42233 \times (\log V_g \times \log S)) \\ &\qquad (R^2 = 0.7) \end{aligned} \tag{7.100}$$

Table 7.3 Values of *MRR* and *DFR* measured after experimentation (Rao et al. 2016a)

S. No.	T (mm)	I (A)	V_g (V)	S (mm/min)	MRR (g/s)	DFR (g/s)
1	2	40	135	500	0.1514	0.1164
2	1.5	35	145	600	0.1269	0.1623
3	1.5	45	145	600	0.3088	0.0058
4	2	30	135	700	0.2476	0.02
5	2	40	155	700	0.3529	0.0217
6	1	40	135	700	0.1495	0.0428
7	1.5	25	145	600	0.1367	0.0894
8	1.5	35	145	600	0.2537	0.0204
9	2	40	135	700	0.3206	0.0065
10	2.5	35	145	600	0.2939	0.0643
11	1	40	155	500	0.0696	0.1120
12	1	40	155	700	0.1958	0.0427
13	1	30	155	700	0.1571	0.0495
14	1.5	35	145	400	0.1230	0.0799
15	2	40	155	500	0.2530	0.0484
16	2	30	135	500	0.1791	0.0330
17	1	40	135	500	0.0447	0.0804
18	0.5	35	145	600	0.0023	0.0858
19	1.5	35	145	600	0.15	0.1260
20	2	30	155	500	0.1516	0.1095
21	1.5	35	145	800	0.3106	0.0235
22	1.5	35	125	600	0.1389	0.0899
23	2	30	155	700	0.1351	0.1921
24	1.5	35	145	600	0.1693	0.1144
25	1	30	135	700	0.1330	0.0218
26	1.5	3555	145	600	0.1440	0.1274
27	1.5	35	165	600	0.1308	0.1885
28	1	30	135	500	0.0580	0.0679
29	1.5	35	145	600	0.1711	0.1084
30	1	30	155	500	0.0236	0.1363

Process input parameter bounds

The bounds on the process input parameters are expressed as follows:

$$0.5 \leq T \leq 2.5 \tag{7.101}$$

$$25 \leq I \leq 45 \tag{7.102}$$

$$125 \leq V_g \leq 165 \tag{7.103}$$

$$400 \leq S \leq 800 \tag{7.104}$$

7.2 Process Parameters Optimization of Traditional and Modern Machining Processes

7.2.1 Turning Process

This problem described in Sect. 7.1.1 was previously attempted by Palanikumar et al. (2009) using NSGA-II, considering a population size of 100 and maximum number of generations equal to 100 (i.e. maximum number of function evaluations equal to 10,000).The algorithm-specific parameters used by NSGA-II are: crossover probability equal to 0.9, distribution index equal to 20 and mutation probability equal to 0.25. Palanikumar et al. (2009) solved the multi-objective optimization problem using a posteriori approach. Therefore, for a fair comparison of results a posteriori approach is used to solve the same multi-objective optimization problem. Rao et al. (2016b) used NSTLBO algorithm. The problem is solved using MOQO-Jaya, MO-Jaya and NSTLBO algorithms in order to see whether any improvement in the results can be achieved. The results of MOQO-Jaya, MO-Jaya and NSTLBO algorithms are compared with the results obtained by MOTLBO (Zou et al. 2014) algorithm and NSGA-II (Rao et al. 2017a). For the purpose of fair comparison of results, the maximum number of function evaluations for MOQO-Jaya, MO-Jaya and NSTLBO algorithms is maintained as 10,000. For this purpose a population size equal to 50 and maximum number of generations equal to 200 (i.e. maximum number of function evaluations = 50 × 200 = 10,000) is considered by the MOQO-Jaya, MO-Jaya algorithms, and a population size of 50 and maximum number of generations equal to 100 (i.e. maximum number of function evaluations = 2 × 50 × 100 = 10,000) is chosen for NSTLBO algorithm.

Now the performance of MOQO-Jaya, MO-Jaya, NSTLBO, MOTLBO and NSGA-II algorithms are compared on the basis of coverage and spacing. The MOQO-Jaya, MO-Jaya, NSTLBO and MOTLBO algorithms are executed 30 times independently and the best, mean and standard deviation of values of coverage and spacing obtained over 30 independent runs are reported in Table 7.4. The value Cov(A, B) = 0.02 implies that, 2% of the solutions obtained by MO-Jaya algorithm are dominated by the solutions obtained using MOQO-Jaya algorithm. The values Cov(A, C) = 0.1, Cov(A, D) = 0.06 and Cov(A, E) = 0.25 imply that, the solutions obtained by MOQO-Jaya algorithm dominate 10% of the solutions obtained by NSTLBO algorithm, 6% of the solutions obtained MOTLBO algorithm and 25% of the solutions obtained using NSGA-II. The value of spacing obtained by NSGA-II is $S(E)$ = 0.1993. It is observed that the MOQO-Jaya algorithm obtained a better value of spacing as compared to the other algorithms.

Table 7.4 The values of coverage and spacing for the non-dominated solutions obtained by MOQO-Jaya, MO-Jaya, NSTLBO, MOTLBO and NSGA-II algorithms for turning process

	Best	Mean	SD
Cov(*A*, *B*)	0.02	0.01	0.0110
Cov(*B*, *A*)	0	0.01	0.0113
Cov(*A*, *C*)	0.1	0.09	0.0115
Cov(*C*, *A*)	0	0.01	0.0113
Cov(*A*, *D*)	0.06	0.04	0.0138
Cov(*D*,*A*)	0.02	0.02	0
Cov(*A*, *E*)	0.25	0.2417	0.0129
Cov(*E*, *A*)	0.04	0.046	0.0175
S(*A*)	0.0221	0.0268	0.0330
S(*B*)	0.0221	0.0251	0.0312
S(*C*)	0.0349	0.0441	0.0412
S(*D*)	0.0321	0.0423	0.0436

A, *B*, *C*, *D* and *E* are the non-dominated set of solutions obtained using MOQO-Jaya, MO-Jaya, NSTLBO, MOTLBO and NSGA-II (Palanikumar et al. 2009) algorithms
SD is the standard deviation

Now the performances of MOQO-Jaya, MO-Jaya, NSTLBO, MOTLBO and NSGA-II algorithms are compared on the basis of hypervolume performance measure and the values of hypervolume are reported in Table 7.5. It is observed that the hypervolume of the Pareto-fronts obtained using MOQO-Jaya and MO-Jaya algorithms is higher than the hypervolume obtained by the other algorithms. This is mainly because, although the Pareto-fronts obtained using MOQO-Jaya, MO-Jaya, NSTLBO, MOTLBO algorithms and NSGA-II seem to overlap each other, but the solutions obtained by MOQO-Jaya and MO-Jaya algorithms are uniformly distributed along the Pareto-front. On the other hand, clustering is observed in the Pareto-optimal solutions obtained using NSGA-II. The number of function evaluations required by MOQO-Jaya, MO-Jaya, NSTLBO and MOTLBO are 850, 900, 1000 and 1150, respectively. Thus, the MOQO-Jaya algorithm has shown a faster convergence as compared to other algorithms.

Table 7.5 The values of hypervolume obtained using MOQO-Jaya, MO-Jaya, NSTLBO, MOTLBO and NSGA-II algorithms for turning process

Algorithm	Hypervolume	No. of function evaluations required	Computational time (s)
MOQO-Jaya	**9493.4**	**850**	7.91
MO-Jaya	**9493.4**	900	**5.68**
NSTLBO	9308.5	1000	6.07
MOTLBO	9294.3	1150	24.93
NSGA-II	9190.1	10000	NA

NA means not available in the literature
Values in bold indicate better performance of the algorithm

The computational time required by MOQO-Jaya, MO-Jaya, NSTLBO and MOTLBO algorithms for one complete simulation run is 7.91, 5.68, 6.07 and 24.93 s, respectively. However, the computational time required for complete one simulation run is lowest for MO-Jaya algorithm. The computational time required by NSGA-II for one complete simulation run was not reported by Palanikumar et al. (2009). The non-dominated set of solutions (Pareto-optimal solutions) obtained by MOQO-Jaya algorithm is shown in Table 7.6 for the purpose of demonstration.

Figure 7.1 shows the Pareto-fronts obtained by MOQO-Jaya algorithm, MO-Jaya algorithm, NSTLBO algorithm, MOTLBO algorithm and NSGA-II for turning process.

Table 7.6 The pareto-optimal set of solutions obtained by MOQO-Jaya algorithm for turning process

S. no.	*v* (m/min)	*f* (mm/rev)	*d* (mm)	V_b (mm)	R_a (μm)	*MRR* (mm^3/min)
1	54.2795	0.1	1.4994	0.1223	2.2931	8138.9121
2	58.3274	0.1	1.5	0.1248	2.2564	8749.1042
3	62.4693	0.1	1.5	0.1275	2.219	9370.3944
4	71.9101	0.1	1.4924	0.1343	2.1356	10732.096
5	75.3031	0.1004	1.4986	0.1369	2.1057	11,325.274
6	79.2396	0.1004	1.5	0.1396	2.0703	11,933.641
7	83.2911	0.1005	1.4991	0.1426	2.0346	12,545.298
8	88.5514	0.1003	1.4933	0.1464	1.9879	13,265.659
9	92.4951	0.1003	1.4992	0.1487	1.951	13906.875
10	102.2128	0.1	1.5	0.1551	1.8619	15331.917
11	109.2442	0.1	1.4713	0.1622	1.8069	16072.785
12	112.5882	0.1	1.5	0.1628	1.7692	16888.234
13	117.1517	0.1	1.4928	0.1667	1.7306	17488.753
14	121.8446	0.1	1.4921	0.1704	1.689	18180.313
15	125.5378	0.1	1.5	0.1727	1.6538	18830.675
16	130.8624	0.1	1.5	0.1769	1.6065	19629.367
17	134.3337	0.1	1.4946	0.18	1.5773	20077.852
18	136.3089	0.1	1.5	0.1812	1.5581	20446.334
19	150	0.1	1.5	0.1924	1.4371	22509.985
20	150	0.1029	1.5	0.1976	1.4601	23141.492
21	150	0.1055	1.5	0.2024	1.482	23745.633
22	150	0.108	1.4994	0.2069	1.5022	24289.652
23	150	0.1098	1.4977	0.2102	1.5173	24662.139
24	150	0.1136	1.5	0.2168	1.5472	25559.907
25	148.4627	0.1176	1.5	0.2226	1.5921	26196.464
26	150	0.1203	1.5	0.2284	1.6009	27075.545
27	150	0.1241	1.5	0.2349	1.6308	27926.109

(continued)

Table 7.6 (continued)

S. no.	v (m/min)	f (mm/rev)	d (mm)	V_b (mm)	R_a (μm)	MRR (mm^3/min)
28	150	0.1242	1.5	0.2349	1.6311	27934.667
29	150	0.1296	1.489	0.2444	1.6787	28943.341
30	150	0.1329	1.4992	0.2495	1.6998	29888.757
31	150	0.1368	1.5	0.2558	1.7295	30779.681
32	150	0.1395	1.5	0.2602	1.7506	31397.552
33	150	0.1431	1.5	0.2658	1.7777	32197.449
34	148.4572	0.1497	1.4984	0.2749	1.8392	33292.613
35	150	0.1496	1.5	0.2758	1.8265	33651.974
36	150	0.1545	1.4838	0.2835	1.873	34392.181
37	149.5388	0.1604	1.4985	0.2917	1.9109	35936.839
38	149.8541	0.1625	1.4976	0.2951	1.9252	36474.722
39	148.1898	0.1666	1.5	0.2999	1.9647	37042.167
40	149.8148	0.169	1.5	0.3042	1.9709	37970.197
41	150	0.172	1.5	0.3086	1.9916	38702.384
42	149.8909	0.175	1.494	0.3126	2.0174	39179.928
43	149.9526	0.1797	1.4978	0.319	2.0483	40360.413
44	150	0.1841	1.5	0.3249	2.0774	41420.598
45	150	0.1858	1.5	0.3272	2.0895	41806.908
46	149.1499	0.1888	1.4995	0.3305	2.1153	42215.482
47	149.4195	0.1928	1.4998	0.3358	2.1412	43195.272
48	150	0.1955	1.5	0.3396	2.1569	43996.501
49	150	0.197	1.5	0.3413	2.1665	44313.77
50	150	0.2	1.4997	0.3451	2.1876	44989.703

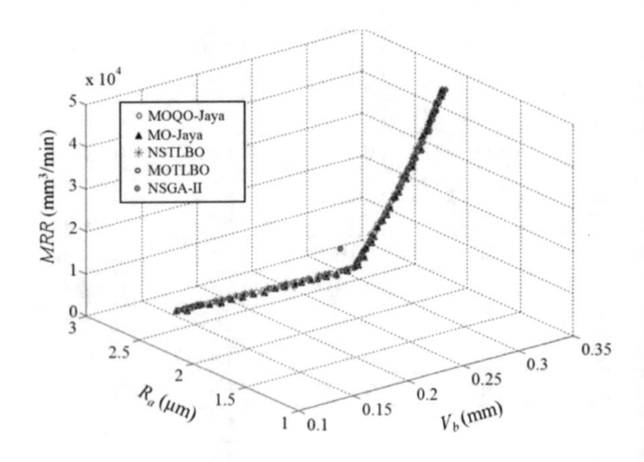

Fig. 7.1 The Pareto-fronts obtained by MOQO-Jaya, MO-Jaya, NSTLBO, MOTLBO and NSGA-II for turning process

7.2.2 Surface Grinding Process

7.2.2.1 Case Study 1

The problem described in Sect. 7.1.2.1 was previously solved by the other researchers using quadratic programming (QP) (Wen et al. 1992), GA (Saravanan et al. 2002), ACO algorithm (Baskar et al. 2004), PSO algorithm (Rao 2010), SA (Rao 2010), ABC algorithm (Rao 2010), HS (Rao 2010) and TLBO algorithm (Pawar and Rao 2013). The previous researchers had solved the multi-objective optimization problem using a priori approach. Therefore, for a purpose of fair comparison the same problem is now solved using Jaya and QO-Jaya algorithms using a priori approach.

A population size of 20 and maximum no. of generations equal to 100 (i.e. maximum number of function evaluations = 20 × 100 = 2000, which is same as that for GA) is considered for Jaya and QO-Jaya algorithms. The solution obtained using the Jaya and QO-Jaya algorithms for rough grinding process is reported in Table 7.7 along with the solution obtained by previous researchers using other advanced optimization algorithms.

Figure 7.2 shows the convergence graph of Jaya and QO-Jaya algorithms for rough grinding process. It is observed that the Jaya and QO-Jaya algorithms required 43 generations (i.e. 860 function evaluations) and 40 generations (i.e. 800 function evaluations) to converge at the minimum value of *COF* (i.e. −0.672), respectively. While the other algorithm such as GA (Sarvanan et al. 2002) required an initial-population size of 20 and no. of generations equal to 100, ACO algorithm (Baskar et al. 2004) considered a population size of 20. The no. of generations required by ACO algorithm was not reported by Baskar et al. (2004), PSO algorithm (Pawar et al. 2010) required 30 to 40 iterations for its convergence, SA (Rao 2010), ABC algorithm (Rao 2010) and HS (Rao 2010) required 75, 65 and 62 generations, respectively for convergence and TLBO algorithm (Pawar and Rao

Table 7.7 The results of Jaya and QO-Jaya algorithms and those other techniques rough grinding process (case study 1)

Technique	V_s	V_w	*doc*	*L*	C_T	*WRP*	*COF*
QP (Wen et al. 1992)	2000	19.96	0.055	0.044	6.2	17.47	−0.127
GA (Saravanan et al. 2002)	1998	11.30	0.101	0.065	7.1	21.68	−0.187
ACO (Baskar et al. 2004)	2010	10.19	0.118	0.081	7.5	24.20	−0.230
PSO (Pawar et al. 2010)	2023	10.00	0.110	0.137	8.33	25.63	−0.224
SA (Rao 2010)	2023	11.48	0.089	0.137	7.755	24.45	−0.223
HS (Rao 2010)	2019.3	12.455	0.079	0.136	7.455	23.89	−0.225
ABC (Rao 2010)	2023	10.973	0.097	0.137	7.942	25.00	−0.226
TLBO (Pawar and Rao 2013)	2023	11.537	0.0899	0.137	7.742	24.551	−0.226
Jaya	2023	22.7	0.137	0.01	**7.689**	**42.284**	**−0.672**
QO-Jaya	2023	22.7	0.137	0.01	**7.689**	**42.284**	**−0.672**

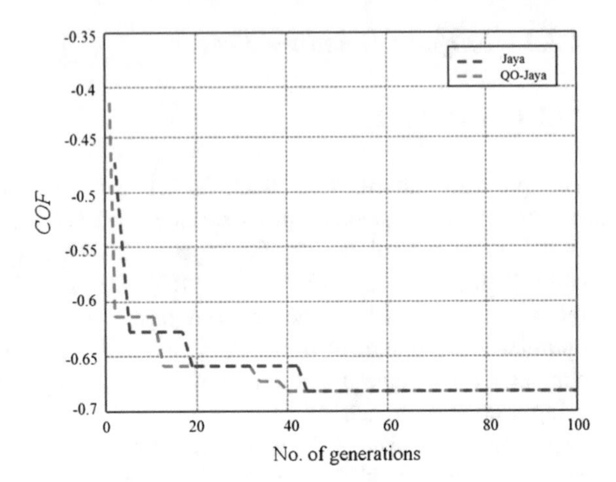

Fig. 7.2 The convergence graph for Jaya and QO-Jaya algorithm for rough grinding process (case study 1)

2013) required 30 generations to converge at the minimum value of the combined objective function. The Jaya and QO-Jaya algorithms required 0.97 and 1.76 s to perform 2000 function evaluations.

7.2.2.2 Case Study 2

In the case of finish grinding process the problem described in Sect. 7.1.2.2 was solved by the previous researchers using QP (Wen et al. 1992) and GA (Saravanan et al. 2002). The previous researchers had used a priori approach to solve the multi-objective optimization problem. Therefore, for a purpose of fair comparison, in the present work a priori approach is used to solve the multi-objective optimization problems. The problem is solved using Jaya and QO-Jaya algorithms considering a population size of 10 and maximum number of function evaluations equal to 2000 (which is same as that of GA). The solution obtained using the Jaya and QO-Jaya algorithms are reported in Table 7.8 along with the results of QP and GA. The Jaya and QO-Jaya algorithms required 0.83 s and 1.6 s to perform 2000 function evaluations.

Table 7.8 The results of Jaya and QO-Jaya algorithms and those other techniques for finish grinding process (case study 2)

Technique	V_s	V_w	*Doc*	*L*	C_T	R_a	*COF*
QP (Wen et al. 1992)	2000	19.99	0.052	0.091	7.7	0.83	0.554
GA (Sarvanan et al. 2002)	1986	21.40	0.024	0.136	7.37	0.827	0.542
Jaya	2023	22.7	0.011	0.137	**7.13**	**0.7937**	**0.522**
QO-Jaya	2023	22.7	0.011	0.137	**7.13**	**0.7937**	**0.522**

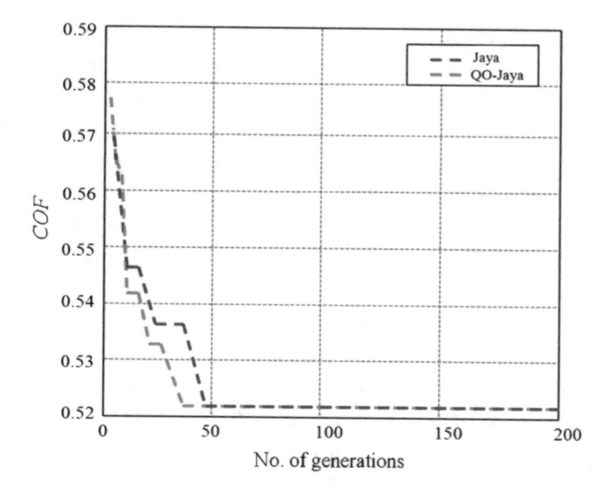

Fig. 7.3 The convergence graph for Jaya and QO-Jaya algorithms for finish grinding process (case study 2)

The convergence graph of Jaya and QO-Jaya algorithms for finish grinding process are shown in Fig. 7.3. Jaya and QO-Jaya algorithms achieved a lower value of *COF* (i.e. 0.522) compared to the other techniques. The Jaya algorithm required on 470 function evaluations (i.e. 47 generations) to converge at the minimum value of *COF*. The QO-Jaya algorithm required 430 function evaluations (i.e. 43 generations) to converge at the minimum value of *COF*. This shows that the Jaya algorithm is effective in solving the optimization problems of rough and finish grinding processes.

7.2.3 Wire Electro-discharge Machining Process

7.2.3.1 Case Study 1

This problem described in Sect. 7.1.3.1 was solved by Kuriakose and Shanmugam (2005) using NSGA, considering a population size of 50 and maximum number of generations equal to 250 (i.e. maximum number of function evaluations equal to $50 \times 250 = 12{,}500$, for a single simulation run). The algorithm-specific parameters for NSGA are as follows: sharing parameter equal to 1; dummy fitness value equal to 50; cross over probability equal to 0.8 and mutation probability equal to 0.1 (Kuriakose and Shanmugam 2005).

The previous researchers had solved the multi-objective optimization problem using a posteriori approach. Therefore, for a purpose of fair comparison, in the present work the multi-objective optimization problem is solved using a posteriori approach. The multi-objective optimization problem of WEDM process is now solved using MOQO-Jaya, MO-Jaya and NSTLBO algorithms, in order to see whether any improvement in the results can be achieved. For a fair comparison of results, the maximum number of function evaluations for MOQO-Jaya, MO-Jaya and NSTLBO algorithms are maintained as 12,500 (which is same as that of NSGA) for a single simulation run. For this purpose, a population size of 50 is chosen and number of iterations is set to 250 (i.e. maximum no. of function evaluations = 50 × 250 = 12,500 in a single simulation run) for the MOQO-Jaya and MO-Jaya algorithm. For NSTLBO algorithm a population size of 50 and maximum number of iterations equal to 125 are chosen (i.e. maximum number of function evaluations = 2 × 50 × 125 = 12,500). In addition, the same problem is solved using sequential quadratic programming (SQP) and Monte Carlo (MC) simulation and the results are compared.

Now, the performances of MOQO-Jaya, MO-Jaya, NSTLBO, NSGA and SQP algorithms are compared on the basis of coverage and spacing of the Pareto-fronts obtained by the respective algorithms. The values of coverage and spacing for a population size of 50 are reported in Table 7.9. It is observed that 12% of the non-dominated solutions obtained using NSTLBO algorithm are dominated by the non-dominated solutions obtained using MOQO-Jaya and MO-Jaya algorithms. On the other hand, 8% of the non-dominated solutions obtained using MOQO-Jaya and MO-Jaya algorithms are dominated by the non-dominated solutions obtained using NSTLBO algorithm. It is observed that, 80% of the non-dominated solutions obtained using NSGA are dominated by the non-dominated solutions obtained using MOQO-Jaya and MO-Jaya algorithms. On the other hand, only 4% of the non-dominated solutions obtained using MOQO-Jaya and MO-Jaya algorithms are dominated by the non-dominated solutions obtained using NSGA (Kuriakose and Shanmugam 2005). MOQO-Jaya and MO-Jaya algorithms dominate 21% of the solution obtained by SQP. The MOQO-Jaya algorithm and MO-Jaya algorithm obtained a better value of spacing as compared to the other algorithms.

Table 7.9 The values of coverage and spacing obtained by MOQO-Jaya, MO-Jaya, NSTLBO, NSGA and SQP algorithms for WEDM process (case study 1)

	Best	Mean	SD
Cov(A, B)	0	0.0162	0.0132
Cov(B, A)	0	0.0147	0.0099
Cov(A, C)	0.12	0.0686	0.0347
Cov(C, A)	0.08	0.0746	0.0582
Cov(A, D)	0.80	0.75	0.0297
Cov(D, A)	0.04	0.0227	0.0149
Cov(A, E)	0.21	0.1667	0.0349
Cov(E, A)	0.1	0.0587	0.0207
$S(A)$	0.0112	0.0139	0.0057
$S(B)$	0.0112	0.0142	0.0032
$S(C)$	0.0119	0.0207	0.0267
$S(D)$	0.1496	0.1589	0.0052
$S(E)$	0.0262	–	–

A, *B*, *C*, *D* and *E* are the non-dominated set of solutions obtained by MOQO-Jaya, MO-Jaya algorithm, NSTLBO, SQP and NSGA respectively

SD is the standard deviation

Table 7.10 The values of hypervolume obtained using MOQO-Jaya, MO-Jaya, NSTLBO, NSGA and SQP algorithms for WEDM process (case study 1)

Algorithm	Hypervolume	No. of FEs required for convergence	Computational time (s)
MOQO-Jaya	**1.1799**	**2440**	10.85
MO-Jaya	**1.1799**	2500	**7.89**
NSTLBO	1.1786	4380	11.63
NSGA-II	1.1081	12,500	NA
SQP	0.9551	2786	12.58

NA means not available in the literature
Values in bold indicate better performance of the algorithm

Now the performance of MOQO-Jaya, MO-Jaya, NSTLBO, NSGA and SQP are compared on the basis of hypervolume performance measure and the values of hypervolume obtained by the respective algorithms are reported in Table 7.10. The MOQO-Jaya and MO-Jaya algorithms obtained a higher hypervolume as compared to NSTLBO, NSGA and SQP algorithms. The number of function evaluations required by MOQO-Jaya, MO-Jaya, NSTLBO and SQP are 2440, 2500, 4380 and 2786, respectively. The MOQO-Jaya algorithm showed a faster convergence as compared to other algorithms. The computational time required by MOQO-Jaya, MO-Jaya, NSTLBO and SQP for one complete simulation run is 10.85, 7.89, 11.63 and 12.58 s. The computational time required for one complete simulation run is lowest for MO-Jaya algorithm. The computational time required by NSGA for one complete simulation run was not reported by Kuriakose and Shanmugam (2005).

All the solutions obtained using MC simulation lie in the region of the search space which is dominated by the solutions obtained using MO-Jaya algorithm. Figure 7.4 shows the Pareto-fronts obtained by MOQO-Jaya, MO-Jaya, NSTLBO, NSGA, SQP and MC simulation. The Pareto-optimal set of solution obtained by MOQO-Jaya algorithm is reported in Table 7.11 for the purpose of demonstration.

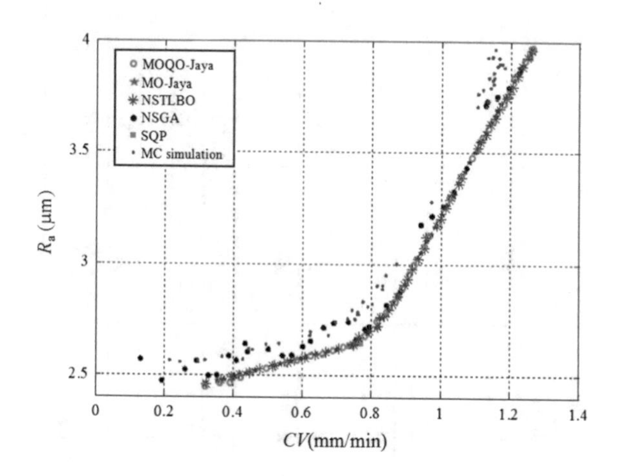

Fig. 7.4 The Pareto-front for WEDM process obtained by MOQO-Jaya, MO-Jaya, NSTLBO, NSGA, SQP and MC simulation for WEDM process (case study 1)

Table 7.11 The Pareto-optimal set of solutions obtained by MOQO-Jaya algorithm for WEDM process (case study 1)

S. No.	*IAL*	T_B	T_A	A_j	$S \times 7.32$	W_s	W_b	Inj	*CV*	R_a
1	15.9999	4	0.6	60	4	8	0.8	4	0.3195	2.4576
2	15.9994	4	0.6	58.6771	4	8	0.8	4	0.3386	2.4656
3	15.9991	4	0.6	57.6558	4	8	0.8	4	0.3533	2.4718
4	15.9994	4	0.6	55.9787	4	8	0.8	4	0.3775	2.482
5	15.9989	4.0016	0.6	52.7276	4	8	0.8	4	0.4242	2.5017
6	15.9992	4.0016	0.6	46.3391	4	8	0.8	4	0.5163	2.5404
7	15.9999	4.0001	0.6	44.5225	4.0364	8	0.8	4	0.5428	2.5517
8	15.999	4.0009	0.6	43.1861	4	8	0.8	4	0.5617	2.5595
9	16	4	0.6	39.6141	4.3858	8	0.8	4	0.6165	2.5849
10	15.9993	4.0007	0.6	38.0578	4	8	0.8	4	0.6357	2.5905
11	15.999	4.0026	0.6	37.2847	4	8	0.8	4	0.6467	2.5952
12	16	4	0.6	35.2711	4	8	0.8	4	0.6759	2.6074
13	15.9992	4.002	0.6	33.5746	4	8	0.8	4	0.7002	2.6177
14	15.9997	4.0002	0.6	32.3642	4.7592	8	0.8	4	0.724	2.6326
15	16	4	0.6	30	5.771	8	0.8	4	0.7665	2.657
16	15.9996	4.0009	0.6	30	7.9452	8	0.8	4	0.7846	2.6788
17	15.9998	4	0.6	30	12	8	0.8	4	0.8183	2.7193
18	15.9999	4	0.6	30	12	8	0.8	3.539	0.8319	2.7508
19	15.9999	4.0001	0.6	30	12	7.6084	0.8	3.0383	0.8473	2.7887
20	15.9998	4.0002	0.6	30	12	6.6765	0.8	2.5505	0.8633	2.831
21	15.9999	4.0004	0.6493	30	11.7809	7.0209	0.8	3.1277	0.8749	2.8745
22	16	4	0.6674	30	12	4.8731	0.8	3.3412	0.8856	2.9151
23	16	4	0.6349	30	12	5.6172	0.8	2	0.9033	2.9413
24	15.9995	4	0.6641	30	11.6623	7.8222	0.8	2	0.9149	2.9691
25	16	4.0001	0.6801	30	12	5.6963	0.8	2.1432	0.9273	3.0118

(continued)

Table 7.11 (continued)

S. No.	*IAL*	T_B	T_A	A_j	$S \times 7.32$	W_s	W_b	Inj	*CV*	R_a
26	16	4	0.7262	30	12	7.5329	0.8	2.5851	0.94	3.0466
27	16	4	0.7278	30	12	6.3916	0.8	2	0.9603	3.1003
28	16	4	0.7723	30.3431	12	8	0.8	2.26	0.9728	3.1449
29	16	4	0.786	30	12	8	0.8	2.6021	0.9762	3.148
30	16	4	0.7836	30	12	7.8292	0.8	2.0671	0.9908	3.182
31	15.9999	4.0001	0.8593	30	12	8	0.8	3.3241	1.001	3.2302
32	15.9999	4.0002	0.8399	30	12	5.9427	0.8	2.5979	1.0139	3.2646
33	16	4	0.8264	30	12	4	0.8	2	1.0265	3.2999
34	16	4	0.8314	30	12	4.3741	0.8	2	1.0289	3.3053
35	15.9997	4	0.8873	30	12	5.2076	0.8	2.3586	1.052	3.373
36	16	4	0.9051	30	12	5.1072	0.8	2.2958	1.0652	3.4101
37	16	4.0005	0.9606	30	12	6.9236	0.8	2.9885	1.0764	3.445
38	16	4	0.9377	30	11.9848	7.3347	0.8	2	1.0903	3.4673
39	16	4.0003	0.9605	30	12	6.6032	0.8	2.1963	1.1002	3.5019
40	16	4	0.9884	30	12	5.1802	0.8	2	1.1261	3.579
41	16	4	1.0198	30	12	7.6664	0.8	2	1.1414	3.6114
42	16	4.0003	1.0511	30	12	4.9615	0.8	2	1.1658	3.6934
43	16	4	1.0665	30	12	4.6318	0.8	2	1.1761	3.7241
44	16	4	1.0737	30	12	4	0.8	2	1.1818	3.7431
45	16	4	1.0998	30	12	7.1442	0.8	2	1.1925	3.7597
46	16	4.0001	1.1119	30	12	5.5681	0.8	2	1.2029	3.7965
47	16	4	1.1205	30	12	4.6854	0.8	2	1.2099	3.8204
48	16	4	1.1455	30	12	4.9096	0.8	2	1.2252	3.8631
49	16	4.0002	1.1774	30	12	4	0.8	2	1.2469	3.929
50	16	4	1.2	30	12	4.0334	0.8	2	1.261	3.9691

7.2.3.2 Case Study 2

This problem described in Sect. 7.1.3.2 was solved by Garg et al. (2012) using NSGA-II considering a population size equal to 100 and maximum number of generations equal to 1000 (i.e. maximum number of function evaluations equal to 100,000), the crossover probability and mutation probability were selected as 0.9 and 0.116, respectively. Garg et al. (2012) used a posteriori approach to solve the multi-objective optimization problem. Therefore, for a purpose of fair comparison, in the present work a posteriori approach is used to solve the multi-objective optimization problem of WEDM process. The multi-objective optimization problem is solved using MOQO-Jaya, MO-Jaya, NSTLBO and MOTLBO (Zou et al. 2014) algorithms to see whether any improvement in the results can be achieved. For a purpose of fair comparison of results the maximum number of function evaluations considered by MOQO-Jaya, MO-Jaya, NSTLBO and MOTLBO algorithms is maintained as 100,000.

For this purpose, a population size of 50 and maximum number of generations equal to 1000 (i.e. maximum number of function evaluations = 50 × 1000 = 50,000) are chosen for the MOQO-Jaya and MO-Jaya algorithms and a population size of 50 and maximum number of generations equal to 500 (i.e. maximum number of function evaluations = 2 × 50 × 500 = 50,000) is chosen for NSTLBO and MOTLBO algorithms. Now the performance of MOQO-Jaya, MO-Jaya, NSTLBO, MOTLBO and NSGA-II algorithms are compared on the basis of coverage and spacing performance measures. The MOQO-Jaya, MO-Jaya, NSTLBO and MOTLBO algorithms are executed 30 times independently and the best, mean and standard deviation of values of coverage and spacing obtained over 30 independent runs are reported in Table 7.12.

Table 7.12 Values of coverage and spacing for the non-dominated solutions obtained by MOQO-Jaya, MO-Jaya, NSTLBO, MOTLBO and NSGA-II for WEDM process (case study 2)

	Best	Mean	SD
Cov(*A*, *B*)	0.04	0.0218	0.0166
Cov(*B*, *A*)	0.02	0.02	0
Cov(*A*, *C*)	0.04	0.0223	0.0516
Cov(*C*, *A*)	0	0.02	0.0104
Cov(*A*, *D*)	0.06	0.044	0.0126
Cov(*D*, *A*)	0	0.02	0.0113
Cov(*A*, *E*)	0.3	0.231	0.056
Cov(*E*, *A*)	0.04	0.0393	0.0249
S(*A*)	0.0119	0.01202	0.0122
S(*B*)	0.0120	0.01320	0.0119
S(*C*)	0.0122	0.01331	0.0120
S(*D*)	0.0125	0.01346	0.0126
S(*E*)	0.0377	–	–

A, *B*, *C*, *D* and *E* are the non-dominated set of solutions obtained using MOQO-Jaya, MO-Jaya, NSTLBO, MOTLBO and NSGA-II algorithms

SD is the standard deviation

The value Cov(*A*, *B*) = 0.04 implies that, 4% of the solutions obtained by MO-Jaya algorithm are dominated by the solutions obtained using MOQO-Jaya algorithm. Cov(*A*, *C*) = 0.04, Cov(*A*, *D*) = 0.06 and Cov(*A*, *E*) = 0.3 imply that, the solutions obtained by MOQO-Jaya algorithm dominate 4% of the solutions obtained by NSTLBO algorithm, 6% of the solutions obtained by MOTLBO algorithm and 30% of the solutions obtained by NSGA-II. It is observed that the MOQO-Jaya algorithm obtained a better value of spacing as compared to the other algorithms.

Now the performance of MOQO-Jaya, MO-Jaya, NSTLBO, MOTLBO and NSGA-II are compared on the basis of hypervolume. The values of hypervolume obtained by the respective algorithms are reported in Table 7.13.

It is observed that the MOQO-Jaya algorithm obtained a higher hypervolume as compared to NSGA-II. This is mainly because the Pareto-optimal solutions obtained using MOQO-Jaya and MO-Jaya algorithm are well distributed along the respective Pareto-fronts as compared to the Pareto-optimal solutions obtained using NSGA-II. The MOQO-Jaya algorithm obtained the same value of hypervolume as that of MO-Jaya, NSTLBO and MOTLBO algorithms within less number of function evaluations. Although for the purpose of fair comparison of results the maximum number of function evaluations for MOQO-Jaya, MO-Jaya, NSTLBO and MOTLBO algorithms is maintained as 100,000. However, the MOQO-Jaya, MO-Jaya, NSTLBO and MOTLBO algorithms required 3750, 3950, 4550 and 6050 function evaluations, respectively to converge at the non-dominated set of solutions. The computational time required by MO-Jaya, NSTLBO and MOTLBO algorithms to perform 100,000 function evaluations are 45.71, 63.06 and 255.47 s, respectively. However, the computational time required by NSGA-II to perform 100,000 function evaluations is not given in (Garg et al., 2012). Figure 7.5 shows the Pareto-fronts obtained by MOQO-Jaya, MO-Jaya, NSTLBO, MOTLBO and NSGA-II algorithms. The Pareto-optimal set of solutions obtained by MOQO-Jaya algorithm is reported in Table 7.14 for the purpose of demonstration.

Table 7.13 The values of hypervolume obtained using MOQO-Jaya, MO-Jaya, NSTLBO and MOTLBO algorithms for WEDM process (case study 2)

Algorithm	Hypervolume	No. of function evaluations required	Computational time (s)
MOQO-Jaya	**0.6599**	**3750**	50.81
MO-Jaya	**0.6599**	3950	**45.71**
NSTLBO	**0.6599**	4550	63.06
MOTLBO	**0.6599**	6050	255.47
NSGA-II	0.6373	100,000	NA

NA means not available in the literature
Bold values indicate better performance of the algorithm

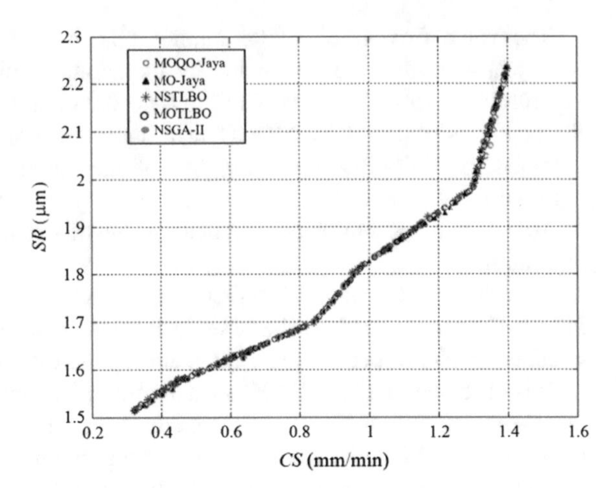

Fig. 7.5 The Pareto-fronts obtained using MOQO-Jaya, MO-Jaya, NSTLBO, MOTLBO and NSGA-II for WEDM (case study 2)

7.2.3.3 Case Study 3

This problem described in Sect. 7.1.3.3 was solved by Kuriachen et al. (2015) using PSO algorithm. In order to handle the multiple objectives Kuriachen et al. (2015) used a fuzzy logic based method and obtained a single optimal solution. However, the common control parameters and the algorithm-specific parameter used for PSO algorithm were not reported by Kuriachen et al. (2015). Now in order to obtain multiple Pareto-optimal solutions, in this work a posteriori approach is used to solve the multi-objective optimization problem of micro-WEDM process. The multi-objective optimization problem is solved using MOQO-Jaya, MO-Jaya, NSTLBO and MOTLBO algorithms and the results are compared.

A population size of 50 and a maximum number of generations equal to 40 (i.e. maximum number of function evaluations equal to 2000) are considered for the MOQO-Jaya and MO-Jaya algorithms. A population size of 50 and maximum number of generations equal 20 (i.e. maximum number of function evaluations = 2 × 50 × 20 = 2000) are considered for NSTLBO and MOTLBO algorithms.

The performances of MOQO-Jaya, MO-Jaya, NSTLBO and MOTLBO algorithms are compared on the basis of coverage and spacing. The MOQO-Jaya, MO-Jaya, NSTLBO and MOTLBO algorithms are executed 30 times independently and the best, mean and standard deviation of values of coverage and spacing obtained over 30 independent runs are reported in Table 7.15. Cov(A, B) = 0.02 implies that, 2% of the solutions obtained by MO-Jaya algorithm are dominated by the solutions obtained using MOQO-Jaya algorithm. The value Cov(A, C) = 0.44 implies that, the solutions obtained by MOQO-Jaya algorithm dominate 44% of the solutions obtained by NSTLBO algorithm. Cov(A, D) = 0.18 implies that, 18% of the solutions obtained by MOTLBO algorithm and dominated by the solutions obtained by MOQO-Jaya algorithm. It is observed that the MOQO-Jaya and MO-Jaya algorithms obtained a better value of spacing as compared to the other algorithms.

Table 7.14 The Pareto-optimal set of solutions for WEDM obtained MOQO-Jaya algorithm (case study 2)

S. No.	T_{on} (μs)	T_{off} (μs)	*IP* (A)	*SV* (V)	*WF* (m/min)	*WT* (g)	*CS* (mm/min)	*SR* (μm)
1	112	56	140.0084	55	10	8	0.3227	1.515
2	112	55.7456	140	55	10	8	0.3272	1.518
3	112	54.1329	140.0662	55	10	8	0.3605	1.5371
4	112	53.1498	140	55	10	8	0.3847	1.5485
5	112	52.3656	140	55	10	8	0.406	1.5577
6	112	52.5899	140	53.8734	10	8	0.4115	1.5601
7	112	51.2675	140	55	10	8	0.4391	1.5706
8	112	54.1704	140	45.2386	10	8	0.4495	1.5801
9	112	50.3989	140	54.802	10	8	0.4703	1.5817
10	112	49.7906	140.0138	53.3606	10	8	0.5101	1.5953
11	112	48	140	55	10	8	0.5593	1.609
12	112	49.4588	140.0129	48.3331	10	8	0.5876	1.6217
13	112	48.6735	140.0537	49.6516	10	8	0.6042	1.6251
14	112	48	140	49.6813	10	8	0.634	1.6327
15	112	48	140	48.0341	10	8	0.6571	1.6401
16	112.0003	48	140	45.8241	10	8	0.6882	1.65
17	112.0051	48.1251	140.0646	41.8557	10	8	0.7378	1.6668
18	112	48	140.0031	39.9919	10	8	0.7701	1.676
19	112	48	140	38.3046	10	8	0.7937	1.6835
20	112.0425	48	140.0926	36.8229	10	8	0.8182	1.6936
21	112	48	140.0029	35	10	8	0.8401	1.6983
22	112.083	48	140.2117	35	10	8	0.8475	1.7051
23	112.6033	48	140	35	10	8	0.8924	1.7425
24	112.7933	48	140	35	10	8	0.9088	1.7565
25	113.1242	48	140	35	10	8	0.9374	1.7807
26	113.8887	48	140	35	10	4.5849	0.9514	1.8038
27	113.5755	48.0004	140	35.0075	10	7.4396	0.9678	1.8097
28	115.0893	48	140	35	10	4	1.0464	1.8509
29	115.2332	48	140	35	10	4	1.0589	1.8573
30	115.6821	48	140	35	10	4	1.0977	1.877
31	116.0263	48	140	35	10	4	1.1275	1.8921
32	116.3251	48	140	35	10	4	1.1533	1.9052
33	116.4661	48.0039	140.0295	35.0206	9.9857	4.2675	1.169	1.9196
34	116.8284	48	140	35	10	4	1.1969	1.9273
35	117.4169	48	140	35	10	4	1.2478	1.9532
36	117.5523	48	140.0048	35.0009	9.9969	4.0652	1.2605	1.9617
37	117.7382	48	140	35	10	4	1.2756	1.9673
38	118	48	140.0288	35	9.9924	4.309	1.303	1.9915

(continued)

Table 7.14 (continued)

S. No.	T_{on} (μs)	T_{off} (μs)	*IP* (A)	*SV* (V)	*WF* (m/min)	*WT* (g)	*CS* (mm/min)	*SR* (μm)
39	118	48	140.0343	35	9.9935	4.7765	1.3102	2.0102
40	118	48	142.806	35	9.9348	4.8035	1.3124	2.0225
41	118	48	140.0287	35.0079	9.9959	5.5565	1.3219	2.0411
42	118	48	147.1012	35	9.9345	5.4216	1.3246	2.061
43	118	48	200	35	6	4	1.3375	2.0751
44	118	48	140.0363	35.0059	9.9953	6.9113	1.3426	2.0952
45	118	48	200	35	6	5.0274	1.3532	2.1161
46	118	48	200	35	6	5.9286	1.3669	2.152
47	118	48	200	35	6	6.3818	1.3739	2.1701
48	118	48	200	35	6	6.5716	1.3768	2.1777
49	118	48	200	35	6	7.5579	1.3918	2.217
50	118	48	200	35	6	7.9005	1.3971	2.2307

Table 7.15 The values of coverage and spacing obtained by MOQO-Jaya, MO-Jaya, NSTLBO and MOTLBO algorithms for micro-WEDM process

	Best	Mean	SD
Cov(*A*, *B*)	0.02	0.0118	0.0178
Cov(*B*, *A*)	0.02	0.02	0
Cov(*A*, *C*)	0.44	0.40	0.0116
Cov(*C*, *A*)	0.06	0.031	0.0256
Cov(*A*, *D*)	0.18	0.14	0.0315
Cov(*D*, *A*)	0.06	0.035	0.0288
S(*A*)	0.0106	0.0108	0.0120
S(*B*)	0.0106	0.0110	0.0152
S(*C*)	0.0306	0.0311	0.0133
S(*D*)	0.0100	0.0132	0.0135

A, *B*, *C* and *D* are the non-dominated set of solutions obtained using MOQO-Jaya, MO-Jaya, NSTLBO and MOTLBO algorithms
SD is the standard deviation

Now the performance of MOQO-Jaya, MO-Jaya, NSTLBO and MOTLBO algorithms compared on the basis of hypervolume. The values of hypervolume obtained by the respect algorithms are reported in Table 7.16. The MOQO-Jaya algorithm achieved a higher hypervolume as compared to MO-Jaya, NSTLBO and MOTLBO algorithms. The MOQO-Jaya, MO-Jaya, NSTLBO and MOTLBO algorithms required 1300, 1350, 1500 and 1850 function evaluations, respectively, to converge at the non-dominated set of solutions. The computational time required by MOQO-Jaya, MO-Jaya, NSTLBO and MOTLBO algorithms to perform 2000 function evaluations is 1.458, 1.876 and 5.19 s, respectively. The computational

Table 7.16 The values of hypervolume obtained by MOQO-Jaya, MO-Jaya, NSTLBO and MOTLBO algorithms for micro-WEDM process (case study 3)

Algorithm	Hypervolume	No. of function evaluations required	Computational time (s)
MOQO-Jaya	**0.1621**	**1300**	1.79
MO-Jaya	0.1594	1350	**1.458**
NSTLBO	0.1594	1500	1.876
MOTLBO	0.1594	1850	5.19

Values in bold indicate better performance of the algorithm

time required by PSO algorithm was not given in (Kuriachen et al., 2015). The MOQO-Jaya algorithm showed a faster convergence as compared to the other algorithms. However, the computational time required by MOQO-Jaya algorithm to complete one run of simulation is slightly higher than MO-Jaya algorithm.

The non-dominated set of solutions obtained using MOQO-Jaya algorithm is reported in Table 7.17. Figure 7.6 shows the Pareto-fronts for micro-WEDM obtained using MOQO-Jaya, MO-Jaya, NSTLBO and MOTLBO algorithms along with the unique solution obtained by PSO algorithm. It is observed that the unique solution obtained by PSO algorithm (Kuriachen et al. 2015) lies in the objective space which is dominated by the Pareto-fronts obtained by MOQO-Jaya and MO-Jaya algorithms.

The values of *MRR* and *SR* obtained by substituting the values of optimum process parameter combination obtained by PSO algorithm (Kuriachen et al., 2015) are 0.0230 mm^3/min and 1.3386 μm, respectively. Now, for a value of *MRR* higher than that provided by PSO algorithm, the corresponding value of *SR* provided by MOQO-Jaya algorithm is 1.1042 μm (refer to Table 7.17, solution no. 15), which is 17.51% less than the value of *SR* provided by PSO algorithm. On the other hand, for a value of *SR* lower than that provided by PSO algorithm, the corresponding value of *MRR* provided by MOQO-Jaya algorithm is 0.0261 mm^3/min (refer to Table 7.17, solution no. 33), which is 13.91% higher than the value of *MRR* obtained by PSO algorithm.

7.2.4 Electro-discharge Machining Process

7.2.4.1 Case Study 1

This optimization problem described in Sect. 7.1.4.1 is solved using a posteriori approach in order to obtain a set of multiple Pareto-optimal solutions. The multi-objective optimization problem is solved using MOQO-Jaya, MO-Jaya and NSTLBO algorithms and the results are compared. A population size of 50 and maximum number of function evaluations as 5000 are considered for MOQO-Jaya,

Table 7.17 The Pareto-optimal solutions for micro-WEDM obtained using MOQO-Jaya algorithm (case study 3)

S. No.	Gap voltage (V)	Capacitance (μF)	Feed rate (μm/s)	*MRR* (mm^3/min)	*SR* (μm)
1	125.7133	0.4	8.468	0.0116	1.0811
2	127.5512	0.01	6.683	0.0158	1.0801
3	127.1506	0.0149	6.809	0.0163	1.082
4	126.3742	0.0275	6.3936	0.0166	1.0869
5	127.5622	0.0417	6.51	0.0177	1.0883
6	127.5709	0.0616	6.8391	0.0192	1.0904
7	126.7396	0.0651	6.9987	0.0195	1.0918
8	126.6504	0.0755	7.0133	0.0201	1.0933
9	127.4166	0.0873	7.0936	0.0208	1.0949
10	126.8775	0.1041	7.0221	0.0212	1.0967
11	127.2741	0.1133	7.1204	0.0217	1.0977
12	128.5621	0.1179	7.2287	0.0221	1.1001
13	127.7182	0.1491	7.3145	0.0227	1.101
14	127.5904	0.1558	7.4536	0.0229	1.102
15	127.7345	0.1661	7.6185	0.0232	1.1042
16	127.4851	0.1869	7.989	0.0237	1.1111
17	128.4227	0.1793	8.1458	0.024	1.1192
18	128.2071	0.1816	8.3251	0.0243	1.1267
19	128.7842	0.1806	8.4176	0.0245	1.1332
20	128.2862	0.1793	8.564	0.0247	1.1413
21	127.5773	0.1886	8.7611	0.0249	1.1513
22	128.578	0.1772	8.8675	0.0252	1.165
23	129.8908	0.1917	9	0.0254	1.1756
24	129.0233	0.1614	9	0.0255	1.1839
25	131.8662	0.1649	9	0.0256	1.1997
26	133.3393	0.1657	8.9985	0.0257	1.2143
27	133.9554	0.162	9	0.0257	1.2234
28	134.9952	0.1681	8.9986	0.0258	1.2352
29	136.2422	0.1684	9	0.0259	1.2552
30	137.6607	0.1633	8.9934	0.0259	1.2829
31	138.8097	0.1631	8.9836	0.026	1.3062
32	138.9056	0.1656	8.9857	0.026	1.3075
33	139.6854	0.1653	9	0.0261	1.3269
34	140.5523	0.1629	8.9907	0.0261	1.3484
35	141.4885	0.1632	9	0.0262	1.3738
36	142.2547	0.1617	9	0.0262	1.3958
37	142.7115	0.1539	9	0.0262	1.4122
38	143.1763	0.1565	9	0.0263	1.4251
39	143.4845	0.1563	9	0.0263	1.4347

(continued)

Table 7.17 (continued)

S. No.	Gap voltage (V)	Capacitance (μF)	Feed rate (μm/s)	*MRR* (mm^3/min)	*SR* (μm)
40	144.8025	0.1588	9	0.0264	1.4764
41	145.2812	0.1522	9	0.0264	1.4953
42	145.4714	0.1557	9	0.0264	1.5005
43	146.1907	0.1539	9	0.0264	1.5268
44	146.8804	0.1618	9	0.0265	1.5492
45	147.0261	0.156	9	0.0265	1.557
46	147.5998	0.1534	9	0.0265	1.58
47	148.4937	0.1529	9	0.0266	1.6159
48	148.9593	0.1513	9	0.0266	1.6357
49	149.4353	0.1545	9	0.0266	1.6544
50	150	0.1522	9	0.0267	1.6796

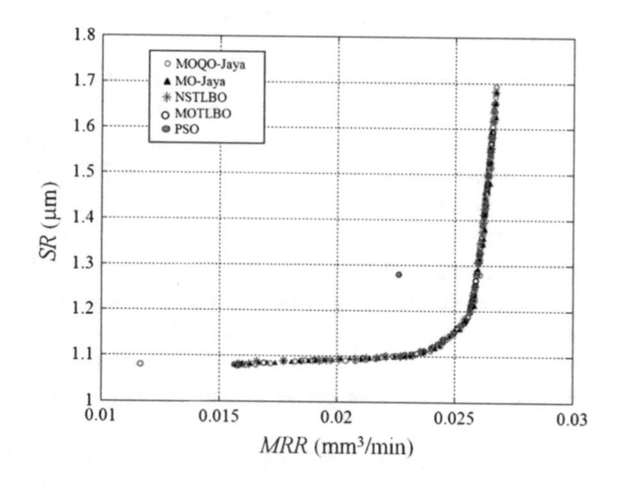

Fig. 7.6 The Pareto-fronts obtained using MOQO-Jaya, MO-Jaya, NSTLBO and MOTLBO algorithms and solution obtained by PSO algorithm for micro-WEDM (case study 3)

MO-Jaya and NSTLBO algorithms. The performances of MOQO-Jaya, MO-Jaya and NSTLBO algorithms are compared on the basis of coverage, spacing, and hypervolume performance measures.

The MOQO-Jaya, MO-Jaya and NSTLBO algorithms are executed 30 times independently and the best, mean and standard deviation of values of coverage and spacing obtained over 30 independent runs are reported in Table 7.18.

In Table 7.18, the value Cov(A, B) = 0.06 implies that, 6% of the solutions obtained by MO-Jaya algorithm are dominated by the solutions obtained by MOQO-Jaya algorithm. On the other hand, Cov(B, A) = 0.04 implies that, only 4% of the solutions obtained using MOQO-Jaya algorithm are dominated by the solutions obtained MO-Jaya algorithm. The value Cov(A, C) = 0.16 implies that, the solutions obtained using MOQO-Jaya algorithm dominate 16% of the solutions obtained using NSTLBO. On the other hand, Cov(C, A) = 0.04 implies that, only

Table 7.18 The values of coverage and spacing for the non-dominated solutions obtained by MOQO-Jaya, MO-Jaya and NSTLBO algorithms for EDM process (case study 1)

	Best	Mean	SD
Cov(*A*, *B*)	0.06	0.0577	0.0021
Cov(*B*, *A*)	0.04	0.0386	0.0013
Cov(*A*,*C*)	0.16	0.1534	0.0047
Cov(*C*, *A*)	0.04	0.0376	0.0014
S(*A*)	0.0501	0.0529	0.0010
S(*B*)	0.064	0.0669	0.0017
S(*C*)	0.0668	0.0702	0.0020

A, *B* and *C* are the non-dominated set of solutions obtained using MOQO-Jaya, MO-Jaya and NSTLBO algorithms
SD is the standard deviation

4% of the solutions obtained using MOQO-Jaya algorithm are dominated by the solutions obtained using NSTLBO algorithm. It is observed that the MOQO-Jaya obtained a better value of spacing as compared to MO-Jaya and NSTLBO algorithms.

Now the performances of MOQO-Jaya, MO-Jaya and NSTLBO algorithms are compared on the basis of hypervolume. The values of hypervolume obtained by the respective algorithms are reported in Table 7.19. The hypervolume obtained by MOQO-Jaya and MO-Jaya algorithms is 16,782 which is higher than the hypervolume obtained by NSTLBO algorithm which is 15,392. The MOQO-Jaya algorithm required 4600 function evaluations, MO-Jaya algorithm required 4750 function evaluations and NSTLBO algorithm required 5000 function evaluations to converge at the Pareto-optimal set of solutions. The computational time required by MOQO-Jaya, MO-Jaya and NSTLBO algorithms to complete 5000 function evaluations is 9.59, 7.77 and 13.51 s, respectively. The MOQO-Jaya algorithm showed a faster convergence as compared to MO-Jaya algorithm and NSTLBO algorithm.

Figure 7.7 shows the Pareto-fronts obtained by MOQO-Jaya, MO-Jaya and NSTLBO algorithms. The results of MOQO-Jaya algorithm shows that the optimum value of *MRR* lies in the range of 0.12453–3.10207 (mg/s). The best compromised value for *TWR* achieved by MOQO-Jaya algorithm lies in the range of 0.00965–25.64 mg/s. The taper angle increases with increase in pulse on time because the workpiece debris result in abrasive action on the walls of the workpiece

Table 7.19 The values of hypervolume obtained by MOQO-Jaya, MO-Jaya and NSTLBO algorithms for EDM process (case study 1)

Algorithm	Hypervolume	No. of function evaluations required	Computational time (s)
MOQO-Jaya	**0.1621**	**4600**	9.59
MO-Jaya	0.1594	4750	**7.77**
NSTLBO	0.1594	5000	13.51

Values in bold indicate better performance of the algorithm

during flushing. The best compromised values for taper angle suggested by MOQO-Jaya algorithm lies in the range of 0.0811°–3.8046°. The best compromised values for delamination factor suggested by MOQO-Jaya algorithm lies in the range of 1.0749–1.2849.

Table 7.20 shows the Pareto-optimal set of solutions obtained by MOQO-Jaya algorithm for the purpose of demonstration.

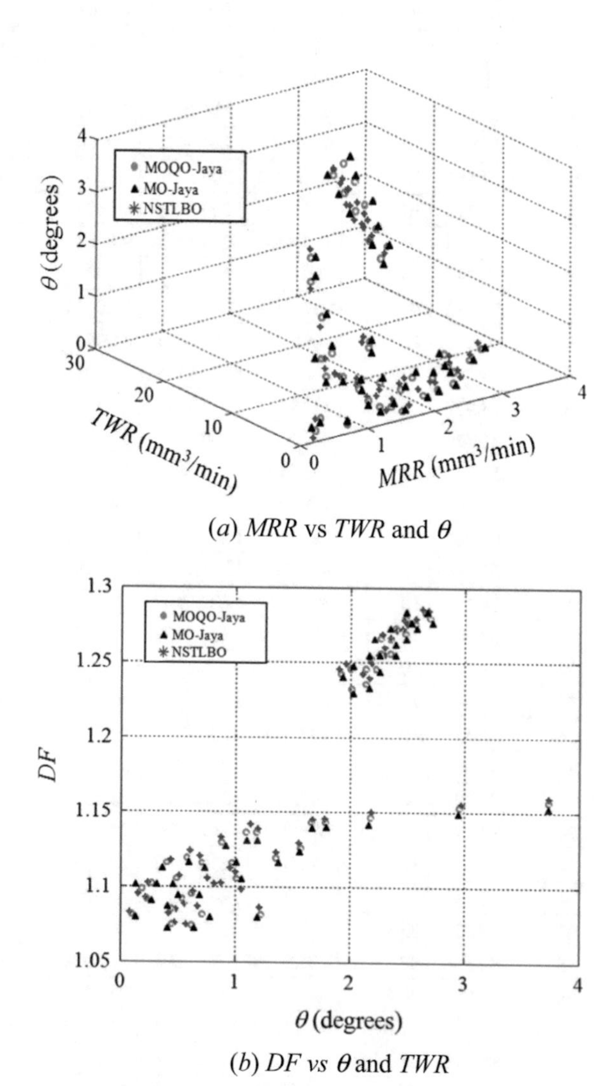

Fig. 7.7 The Pareto-optimal solution obtained by MOQO-Jaya, MO-Jaya and NSTLBO algorithms for EDM process

Table 7.20 The Pareto optimal solution set provided by MOQO-Jaya algorithm EDM process (case study 1)

S. No.	V_g (V)	I_p (A)	T_{on} (μs)	*N* (rpm)	*MRR* (0.1 mg/s)	*TWR* (0.1 mg/s)	Θ (°)	*DF*
1	25	10	1913.724	200	1.2453	0.0965	3.3476	1.1574
2	25.049	10	1844.116	200	1.2865	0.0986	3.0562	1.1558
3	25	10	1757.623	200	1.3199	0.0996	2.7192	1.1536
4	26.268	10	2000	200	1.4191	0.1162	3.7259	1.1603
5	25	10	300	200	1.4245	0.2215	0.0811	1.079
6	31.700	10	2000	200	2.5179	0.2405	3.8046	1.1629
7	28.5	10	932.73	212.1907	3.0999	0.2672	0.6472	1.1259
8	33.883	10	980.8407	214.6995	5.0426	0.4827	0.7417	1.13
9	39.456	10	1366.835	200	5.5058	0.5499	1.7016	1.1488
10	39.512	10	893.006	200	6.1636	0.6041	0.6878	1.1325
11	43.100	10	785.4233	214.3395	9.0452	1.0027	0.5488	1.1238
12	60.542	10	300	200	9.4074	1.871	0.159	1.0949
13	50.252	10	951.2899	200	10.0145	1.1314	0.943	1.1347
14	50.462	10	1094.945	209.8391	11.047	1.3154	1.1924	1.1359
15	95	10	300	370.8176	11.201	1.547	0.5758	1.0749
16	53.920	10	1193.568	203.7774	11.3644	1.3979	1.5711	1.1404
17	61.759	10	417.817	200	11.776	1.8677	0.269	1.1054
18	52.378	10	997.8612	216.286	12.7649	1.5735	0.9942	1.1302
19	59.060	10	1199.503	212.3769	14.2355	1.9259	1.6368	1.1357
20	62.064	10	782.3541	212.8126	16.3417	2.1735	0.7642	1.1214
21	57.646	10	899.7264	241.1095	17.1198	2.3377	0.8244	1.1187
22	78.169	10	300	303.9107	18.6777	3.184	0.3371	1.0817
23	63.966	10	721.4555	233.2439	19.5525	2.7339	0.6496	1.1129
24	81.445	10.4816	300	263.1196	20.3185	4.6091	0.3181	1.0889
25	82.047	10.242	300	276.3105	20.4793	4.1102	0.3279	1.0849
26	81.535	10	407.3847	289.2469	22.0527	3.4706	0.4306	1.0863
27	93.309	10	460.6347	290.8624	23.1194	3.5922	0.5781	1.0837
28	77.298	10	847.0946	243.5197	23.7081	3.6683	1.0182	1.1097
29	84.271	11.0556	628.0503	247.9736	24.8563	5.7193	0.7768	1.1108
30	95	10	680.5518	230.7705	25.9749	4.2886	1.0235	1.101
31	95	10	726.217	247.0427	26.8204	4.4199	1.0638	1.0983
32	63.175	35.7338	815.4502	250.6803	26.8784	141.1848	1.9473	1.2461
33	46.866	45	704.2118	262.2001	26.8928	168.1049	2.1411	1.2377
34	66.097	36.3491	644.0377	251.924	27.032	153.657	1.8928	1.2522
35	63.469	37.0986	865.4543	259.8581	27.2314	155.9124	2.0329	1.2489
36	65.779	37.3432	876.1797	259.7034	27.5357	164.2039	2.1005	1.2531
37	48.778	45	750.8355	259.6498	27.6576	176.157	2.1703	1.2431
38	53.815	45	571.8286	249.4247	27.9858	200.6127	2.1753	1.2581

(continued)

Table 7.20 (continued)

S. No.	V_g (V)	I_p (A)	T_{on} (μs)	N (rpm)	*MRR* (0.1 mg/s)	*TWR* (0.1 mg/s)	θ (°)	*DF*
39	55.327	45	591.4365	277.3743	28.3591	208.3648	2.2865	1.2568
40	52.183	45	875.1416	246.9805	28.454	185.9188	2.3039	1.254
41	55.671	45	867.8184	251.2169	29.5533	205.7847	2.3407	1.2608
42	56.978	45	895.9178	245.2088	29.6352	208.5368	2.4141	1.265
43	59.199	45	664.6343	264.8876	29.8885	224.7545	2.3068	1.2662
44	58.104	45	835.4176	253.878	30.121	218.1782	2.3559	1.2653
45	60.387	45	738.0855	255.0349	30.42	228.327	2.3521	1.27
46	62.181	45	846.2731	248.4064	30.6445	233.4057	2.473	1.2745
47	64.936	45	937.1024	251.8109	30.7501	246.4366	2.577	1.278
48	64.786	45	770.7681	249.6904	30.8257	243.3128	2.4831	1.2792
49	69.817	45	810.0259	259.0902	30.8293	260.9632	2.5826	1.2849
50	68.095	45	836.1816	252.7234	31.0207	256.4056	2.5864	1.2836

7.2.4.2 Case Study 2

This optimization problem described in Sect. 7.1.4.2 is solved using a posteriori approach in order to obtain a set of Pareto-optimal solutions. The multiobjective optimization problem is solved using MOQO-Jaya, MO-Jaya and NSTLBO algorithms and the results are compared. A population size of 50 and maximum number of function evaluations as 5000 are considered for MOQO-Jaya, MO-Jaya and NSTLBO algorithms. The performances of MOQO-Jaya, MO-Jaya and NSTLBO algorithms are compared on the basis of coverage, spacing, and hypervolume performance measures.

The MOQO-Jaya, MO-Jaya and NSTLBO algorithms are executed 30 times independently and the best, mean and standard deviation of values of coverage and spacing obtained over 30 independent runs are reported in Table 7.21. The value Cov(A, B) = 0.12 implies that, 12% of the solutions obtained by MO-Jaya algorithm are dominated by the solutions obtained by MOQO-Jaya algorithm. On the other hand, the value Cov(B, A) = 0.08 implies that, 8% of the solutions obtained by MOQO-Jaya algorithm are dominated by the solutions obtained by MO-Jaya algorithm. Cov(A, C) = 0.16 implies that, solutions obtained using MOQO-Jaya algorithm dominate 16% of the solutions obtained using NSTLBO. It is observed that the MOQO-Jaya obtained a better value of spacing as compared to MO-Jaya and NSTLBO algorithms.

Now the performance of MOQO-Jaya, MO-Jaya and NSTLBO algorithms is compared in terms of hypervolume. The value of hypervolume obtained by the respective algorithms is reported Table 7.22. The hypervolume obtained by MOQO-Jaya and MO-Jaya algorithm is 100.575 and 100.568 which is higher than the hypervolume obtained by NSTLBO algorithm which is 98.948. The

Table 7.21 The values of coverage and spacing for the non-dominated solutions obtained by MOQO-Jaya, MO-Jaya and NSTLBO for micro-EDM process (case study 2)

	Best	Mean	SD
Cov(*A*, *B*)	0.12	0.1131	0.0036
Cov(*B*, *A*)	0.08	0.0758	0.0028
Cov(*A*,*C*)	0.16	0.1498	0.0039
Cov(*C*, *A*)	0.04	0.0380	0.0013
S(*A*)	0.0206	0.0216	0.0006
S(*B*)	0.0215	0.0228	0.0004
S(*C*)	0.0321	0.0333	0.0010

A, *B* and *C* are the non-dominated set of solutions obtained using MOQO-Jaya, MO-Jaya and NSTLBO algorithms
SD is the standard deviation

Table 7.22 The values of hypervolume obtained by MOQO-Jaya, MO-Jaya and NSTLBO algorithms for micro-EDM process (case study 2)

Algorithm	Hypervolume	No. of function evaluations required	Computational time (s)
MOQO-Jaya	**100.575**	**4850**	8.56
MO-Jaya	100.568	4950	**6.08**
NSTLBO	98.948	5000	10.98

Values in bold indicate better performance of the algorithm

MOQO-Jaya, MO-Jaya and NSTLBO algorithms required 4850, 4950 and 5000 function evaluations, respectively, for convergence. The computational time required by MOQO-Jaya, MO-Jaya and NSTLBO algorithms are 8.56, 6.08 and 10.98 s. Thus the MOQO-Jaya algorithm has obtained a better Pareto-front as compared to MO-Jaya and NSTLBO algorithms with a faster convergence rate.

Table 7.23 shows the Pareto-optimal set of solutions obtained by MOQO-Jaya algorithm. The results of MOQO-Jaya algorithm have revealed that, in order to achieve a trade-off between *MRR* and *TWR* the optimal setting for pulse energy is 2000 μJ and any deviation from this value may result in non-optimal values of *MRR* and *TWR*. With aspect ratio fixed at 0.5 an increase in *MRR* is observed with increase in feed rate and speed. However, at extreme values of feed rate and speed the *MRR* increases with increase in aspect ratio.

A low value of feed rate, speed and aspect ratio results in minimum tool wear ($TWR = 0.3307 \times 10^{-3}$ mm^3/min, refer to solution 1, Table 7.23). On the other hand, a high value of feed rate, speed and aspect ratio gives a high *MRR* (32.1458×10^{-3} mm^3/min, refer to solution 50, Table 7.23) but at the expense of significant increase in *TWR* (7.040×10^{-3} mm^3/min). Figure 7.8 shows the Pareto fronts obtained by MOQO-Jaya, MO-Jaya and NSTLBO algorithms.

Table 7.23 The Pareto-optimal set of solutions obtained by MOQO-Jaya algorithm for micro-EDM process (case study 2)

S. No.	*E* (μJ)	*F* (μm/s)	*S* (rpm)	*A*	*MRR* (10^{-3} mm^3/min)	*TWR* (10^{-3} mm^3/min)
1	2000	10	100	0.5	2.6219	0.3307
2	2000	12.9182	100.0819	0.5	2.9267	0.3709
3	2000	19.5363	100.2894	0.5	3.561	0.4689
4	2000	22.5177	100	0.6364	4.4926	0.6837
5	2000	15.471	100	0.8852	4.6108	0.7671
6	2000	11.1055	525.5919	0.5	5.6157	0.8431
7	2000	14.9048	494.173	0.5018	6.2136	0.9265
8	2000	17.4029	520.7207	0.5028	7.0729	1.0511
9	2000	12.1938	660.3727	0.5	7.2438	1.0534
10	2000	16.9822	648.2183	0.5023	8.5232	1.2308
11	2000	15.3951	800	0.5	9.9806	1.389
12	1999.999	16.3754	800	0.5	10.3308	1.4354
13	2000	18.2665	800	0.5	10.9949	1.5265
14	2000	23.1415	729.9279	0.5	11.4719	1.6347
15	2000	22.9902	800	0.5	12.6004	1.7614
16	2000	25.9902	781.586	0.5	13.2492	1.8773
17	1999.999	26.7446	800	0.5	13.8351	1.9551
18	2000	29.8817	793.8799	0.5	14.7207	2.1072
19	2000	31.589	800	0.5022	15.4229	2.2233
20	2000	33.7982	800	0.5	16.0848	2.3343
21	2000	37.0695	800	0.5	17.1046	2.5167
22	2000	37.6695	800	0.5	17.2904	2.5507
23	2000	40.8566	800	0.5033	18.3284	2.751
24	2000	42.1834	800	0.5	18.6761	2.8103
25	2000	45.1411	800	0.5	19.5745	2.9847
26	2000	48.3518	800	0.5	20.5423	3.1778
27	2000	50.9093	779.8629	0.5	20.7107	3.2622
28	2000	50.6162	800	0.5	21.2207	3.3163
29	2000	52.5044	800	0.5	21.7841	3.4334
30	2000	54.6281	800	0.5	22.4155	3.5667
31	2000	56.967	800	0.5	23.1082	3.7154
32	2000	60	800	0.5	24.0027	3.9115
33	2000	60	800	0.5253	24.5151	4.0939
34	2000	60	800	0.5524	25.0289	4.2814
35	2000	60	800	0.5779	25.4824	4.4508
36	2000	60	794.4076	0.6082	25.7814	4.6148
37	2000	60	800	0.6292	26.3089	4.7687
38	2000	60	800	0.652	26.6441	4.9011

(continued)

Table 7.23 (continued)

S. No.	*E* (μJ)	*F* (μm/s)	*S* (rpm)	*A*	*MRR* (10^{-3} mm^3/min)	*TWR* (10^{-3} mm^3/min)
39	2000	60	800	0.6732	26.9399	5.0197
40	2000	60	800	0.686	27.1113	5.0891
41	1999.999	60	800	0.7105	27.4242	5.2172
42	1999.999	60	800	0.7441	27.8258	5.3843
43	2000	60	800	0.8279	28.6992	5.758
44	2000	60	800	0.8464	28.8696	5.8326
45	1999.999	60	800	0.891	29.2525	6.0023
46	2000	60	800	0.9016	29.3378	6.0404
47	2000	60	800	0.9574	29.7536	6.2287
48	2000	60	800	1.0504	30.3372	6.499
49	2000	60	800	1.1582	30.8707	6.7526
50	2000	60	800	1.906	32.1458	7.404

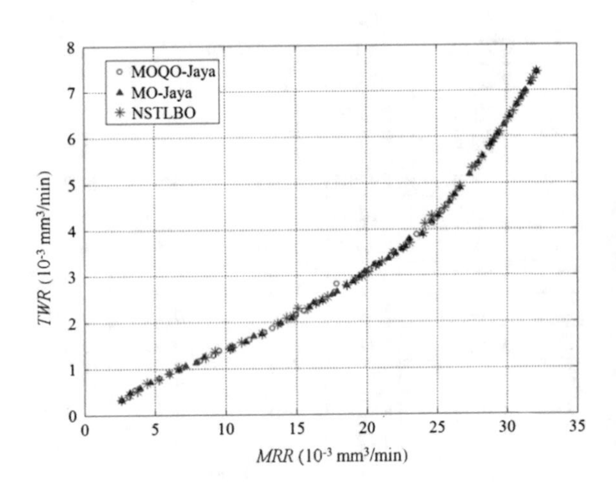

Fig. 7.8 The Pareto-front obtained by MOQO-Jaya, MO-Jaya and NSTLBO algorithms for micro-EDM process

7.2.5 *Electro-chemical Machining Process*

7.2.5.1 Case Study 1

This problem described in Sect. 7.1.5.1 was solved by Mukherjee and Chakraborty (2013) BBO algorithm. In order to provide multiple choices to the process planner Mukherjee and Chakraborty (2013) obtained three trade-off solutions for the multi-objective optimization problem of ECM process using the BBO algorithm. The algorithm–specific parameters for BBO algorithm are as follows: habitat modification probability equal to 1; mutation probability equal to 0.005; maximum species count equal to 500; maximum immigration rate equal to 1; maximum emigration rate equal to 1; maximum mutation rate equal to 0; elitism parameter equal to 2; generation count limit equal to 50; number of genes in each population member equal to 20 (Mukherjee and Chakraborty 2013). However, the number of habitats required by the BBO algorithm to solve the multi-objective optimization problem of ECM process is unknown. Thus, the number of function evaluations required by BBO algorithm cannot be determined.

Now in order to obtain multiple trade-off solutions for the multi-objective optimization problem of ECM process is solved using a posteriori approach. The multi-objective optimization problem is now solved using MOQO-Jaya, MO-Jaya and NSTLBO algorithms. For this purpose a population size of 50 and maximum number of iterations equal to 50 (i.e. maximum number of function evaluations = 50 × 50 = 2500) are selected for the MOQO-Jaya and MO-Jaya algorithms and a population size of 50 and maximum number of iterations equal to 25 is selected for NSTLBO algorithm (i.e. maximum number of function evaluations = 2 × 50 × 25 = 2500). In addition the same problem is solved using SQP and MC simulation and the results are compared on the basis of coverage, spacing and hypervolume performance measure.

The MOQO-Jaya, MO-Jaya and NSTLBO algorithms are executed 30 times independently and the best, mean and standard deviation of the values of coverage and spacing obtained over 30 independent runs is reported in Table 7.24. It is observed that 6% of the solutions obtained using NSTLBO are dominated by the solutions obtained by MOQO-Jaya algorithm. On the other hand, 4% of the solutions obtained using MOQO-Jaya algorithm are dominated by the non-dominated solutions obtained using NSTLBO. It is observed that, considering best values, 74% of the solutions obtained using SQP are dominated by the obtained using MOQO-Jaya algorithm. On the other hand, the none of the solutions obtained using MOQO-Jaya algorithm are inferior to the solutions obtained using SQP.

Table 7.24 The values of coverage and spacing obtained by MOQO-Jaya, MO-Jaya and NSTLBO algorithms for ECM process (case study 1)

	Best	Mean	SD
Cov(*A*, *B*)	0.04	0.023	0.0016
Cov(*B*, *A*)	0.04	0.041	0.0024
Cov(*A*,*C*)	0.06	0.033	0.019
Cov(*C*, *A*)	0.04	0.012	0.012
Cov(*A*, *D*)	0.74	0.708	0.012
Cov(*D*, *A*)	0	0	0
S(*A*)	0.016	0.034	0.058
S(*B*)	0.016	0.021	0.026
S(*C*)	0.018	0.021	0.037

A, *B*, *C* and *D* are the non-dominated set of solutions obtained by MOQO-Jaya, MO-Jaya algorithm, NSTLBO and SQP, respectively

SD is the standard deviation

Table 7.25 The values of hypervolume obtained by MOQO-Jaya, MO-Jaya, NSTLBO, SQP and BBO algorithms for ECM process (case study 1)

Algorithm	Hypervolume	No. of FEs required for convergence	Computational time (s)
MOQO-Jaya	**0.2341**	**2350**	3.97
MO-Jaya	**0.2341**	2400	**2.65**
NSTLBO	0.2331	2500	3.46
SQP	0.1957	7750	4.25
BBO	0.1271	NA	NA

NA means not available in the literature
Values in bold indicate better performance of the algorithm

Now the performance of MOQO-Jaya, MO-Jaya, NSTLBO, SQP and BBO are compared on the basis of hypervolume. The values of hypervolume obtained by the respective algorithms are reported in Table 7.25. It is observed that the MOQO-Jaya and MO-Jaya algorithms obtained a better hypervolume as compared to NSTLBO, SQP and BBO algorithms. Furthermore, the MOQO-Jaya algorithm showed a faster convergence speed as compared to the other algorithms. However, the computational time required by MO-Jaya algorithm to perform one complete simulation run is the lowest as compared to the other algorithms. The number of function evaluations required by BBO algorithm for convergence and the computational time required by BBO algorithm to perform one complete simulation run is not reported by Mukherjee and Chakraborty (2013). The Pareto-fronts obtained by MOQO-Jaya, MO-Jaya, NSTLBO, BBO, SQP and MC simulation are shown in Fig. 5.9. Some of the solutions obtained using SQP seem to coincide with the Pareto-front obtained using MOQO-Jaya algorithm. Although the solutions obtained using SQP are competitive, the MOQO-Jaya algorithm has provided more number of Pareto-optimal solutions as compared to SQP. It is observed that all the solutions obtained using BBO and MC simulation are inferior to the solutions obtained using MOQO-Jaya algorithm (Fig. 7.9).

The Pareto-optimal set of solutions obtained using MOQO-Jaya algorithm is reported in Table 7.26 for the purpose of demonstration.

7.2.5.2 Case Study 2

This problem described in Sect. 7.1.5.2 was solved by Acharya et al. (1986) using goal programming (GP) approach, Choobineh and Jain (1990) fuzzy set theory, Jain and Jain (2007) using GA and Rao et al. (2008) using PSO algorithm. However, the results obtained by Jain and Jain (2007) using GA is infeasible as it violates the passivity constraint. In order to solve the multi-objective optimization problem using PSO algorithm considering a population size of 75 and maximum number of generations equal to 50 (i.e. maximum number of function evaluations $= 75 \times 50 = 3750$) were considered by Rao et al. (2008). The results obtained using GP, fuzzy sets, and PSO algorithm are shown in Fig. 7.10.

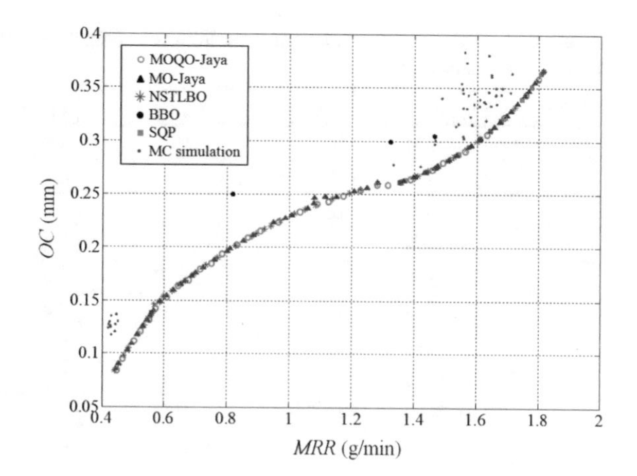

Fig. 7.9 The Pareto-fronts obtained by MOQO-Jaya, MO-Jaya, NSTLBO, BBO, SQP and MC simulation for ECM process (case study 1)

Now in order to satisfy the three objective simultaneously and to obtain a set of multiple trade-off solutions the same problem is solved using a posteriori approach. For this purpose the MOQO-Jaya, MO-Jaya and NSTLBO algorithms are applied to solve the multi-objective optimization problem of ECM process. A population size of 50 and maximum number of generations equal to 75 are considered for MOQO-Jaya and MO-Jaya algorithms (i.e. maximum number of function evaluations = 50 × 75 = 3750, which is same as PSO algorithm). A population size of 50 and maximum number of function evaluations equal to 3750 is considered for NSTLBO algorithm. In addition the same problem is solved using SQP and MC simulation and the results are compared.

Now the performance of MOQO-Jaya, MO-Jaya and NSTLBO algorithms are compared on the basis of coverage and spacing. For this purpose, MOQO-Jaya, MO-Jaya and NSTLBO algorithms are executed 30 times independently and the best, mean and standard deviation of the values of coverage and spacing are reported in Table 7.27.

The value Cov(*A*, *B*) = 0.02 implies that, the solutions obtained using MOQO-Jaya algorithm dominate 2% of the solutions obtained by MO-Jaya algorithm. The value Cov(*A*, C) = 0.06 implies that, 6% of the solutions obtained using NSTLBO are dominated by the solutions obtained using MOQO-Jaya algorithm. On the other hand, Cov(*C*, *A*) = 0 implies that, none of the solutions obtained using MOQO-Jaya algorithm are inferior to solutions obtained using NSTLBO. Cov(*A*, *D*) = 1 and Cov(*D*, *A*) = 0 imply that all the solutions obtained using SQP are inferior to the solutions obtained using MOQO-Jaya algorithm.

Now the performance of MOQO-Jaya, MO-Jaya, NSTLBO and SQP are compared on the basis of hypervolume. The values of hypervolume obtained by the respective algorithms are reported in Table 5.25. The MOQO-Jaya and MO-Jaya algorithm obtained a higher hypervolume as compared to NSTLBO algorithm and SQP. This indicates that the Pareto-front obtained by MOQO-Jaya and MO-Jaya algorithms are better than NSTLBO algorithm and SQP (Table 7.28).

Table 7.26 The Pareto-optimal set of solutions obtained by MOQO-Jaya algorithm for ECM process (case study 1)

S. No.	x_1 (g/l)	x_2 (l/min)	x_3 (V)	x_4 (mm)	*MRR*	*OC*
1	15.0754	10	10	0.4	0.4413	0.084
2	16.3339	10	10	0.4	0.4537	0.0898
3	18.0159	10	10	0.4	0.4703	0.0976
4	19.5263	10	10	0.4	0.4852	0.1045
5	22.6566	10	10.05	0.4	0.516	0.1188
6	24.4525	10	10	0.4	0.5341	0.1265
7	25.918	10	10	0.4	0.5487	0.1328
8	27.1363	10	10	0.4	0.5608	0.138
9	28.2178	10.0533	10	0.4	0.5685	0.1456
10	19.163	14	10	0.4	0.5838	0.1483
11	30.7621	10	10	0.4	0.5972	0.1533
12	22.6917	14	10	0.4	0.6259	0.1586
13	24.7859	14	10	0.4	0.6511	0.1645
14	26.1559	14	10	0.4	0.6675	0.1683
15	28.4139	14	10	0.4	0.6947	0.1744
16	31.6673	14	10	0.4	0.7341	0.1828
17	34.5679	14	10	0.4	0.7693	0.1899
18	37.5067	14	10	0.4	0.805	0.1969
19	39.4034	14	10	0.4	0.8282	0.2012
20	43.6015	14	10	0.4	0.8796	0.2103
21	44.7639	14	10	0.4	0.8939	0.2127
22	46.9081	14	10	0.4015	0.9189	0.217
23	48.5128	14	10	0.4	0.9401	0.2201
24	50.9989	14	10	0.4	0.9708	0.2248
25	54.7242	14	10	0.4	1.0171	0.2313
26	57.3731	14	10	0.4007	1.0493	0.2356
27	59.9367	14	10	0.4	1.0821	0.2396
28	64.3497	14	10	0.4	1.1375	0.2459
29	68.7189	14	10	0.4	1.1925	0.2513
30	71.6128	14	10	0.4017	1.2273	0.2545
31	75	14	10.7564	0.4013	1.2812	0.2604
32	75	14	30	1.1953	1.3545	0.2615
33	75	14	30	1.1908	1.3575	0.2618
34	75	14	30	1.1194	1.4043	0.2668
35	75	14	30	1.1094	1.4108	0.2676
36	75	14	30	1.0586	1.4433	0.2718
37	75	14	30	1.004	1.4777	0.2768
38	75	14	30	0.9455	1.5139	0.2827
39	75	14	30	0.9223	1.528	0.2852

(continued)

Table 7.26 (continued)

S. No.	x_1 (g/l)	x_2 (l/min)	x_3 (V)	x_4 (mm)	*MRR*	*OC*
40	75	14	30	0.8583	1.5665	0.2927
41	75	14	30	0.8557	1.568	0.293
42	75	14	30	0.8017	1.5998	0.2998
43	75	14	30	0.7822	1.6111	0.3024
44	75	14	30	0.7324	1.6397	0.3092
45	75	14	30	0.6429	1.6898	0.3227
46	75	14	30	0.5862	1.7206	0.3319
47	75	14	30	0.517	1.7574	0.3439
48	75	14	30	0.4441	1.7951	0.3574
49	75	14	30	0.4029	1.8159	0.3654
50	75	14	30	0.4	1.8174	0.366

The MOQO-Jaya algorithm obtained same hypervolume as that of MO-Jaya algorithm with a faster convergence rate as compared to MO-Jaya algorithm. The Pareto-optimal solutions obtained using MOQO-Jaya algorithm are reported in Table 7.29 for the purpose of demonstration. The MOQO-Jaya algorithm obtained 50 new solutions, in a single simulation run, which are completely different than the single solutions provided by PSO algorithm, GP and fuzzy set theory, providing more choices to the process planner. None of the solutions obtained using MOQO-Jaya algorithm is inferior to the solution obtained using PSO algorithm, GP and fuzzy set theory. The computational time required by GP, Fuzzy sets, GA and PSO algorithm are not available in the literature. The solutions obtained by MOQO-Jaya, MO-Jaya, NSTLBO, SQP, MC simulation, PSO algorithm, GP and fuzzy sets are shown in Fig. 7.10

7.2.6 *Focused Ion Beam Micro-milling Process*

This problem described in Sect. 7.1.6 was solved by Bhavsar et al. (2015) using NSGA-II considering a population size of 60 and a maximum number of generations equal to 1000 (i.e. maximum number of function evaluations equal to 60,000). The values of algorithm-specific parameters required by NSGA-II are as follows: Pareto fraction equal to 0.7, cross-over probability equal to 0.9 (Bhavsar et al. 2015). The value of the mutation probability required by NSGA-II was not reported by Bhavsar et al. (2015). Bhavsar et al. (2015) used a posteriori approach to solve the multi-objective optimization problem. Therefore, for the purpose of fair comparison, in this work a posteriori approach is used to solve the multi-objective optimization problem of FIB micro-milling process. For this purpose the same problem is solved using MOQO-Jaya, MO-Jaya and NSTLBO algorithms in order to see whether any improvement in the results can be achieved.

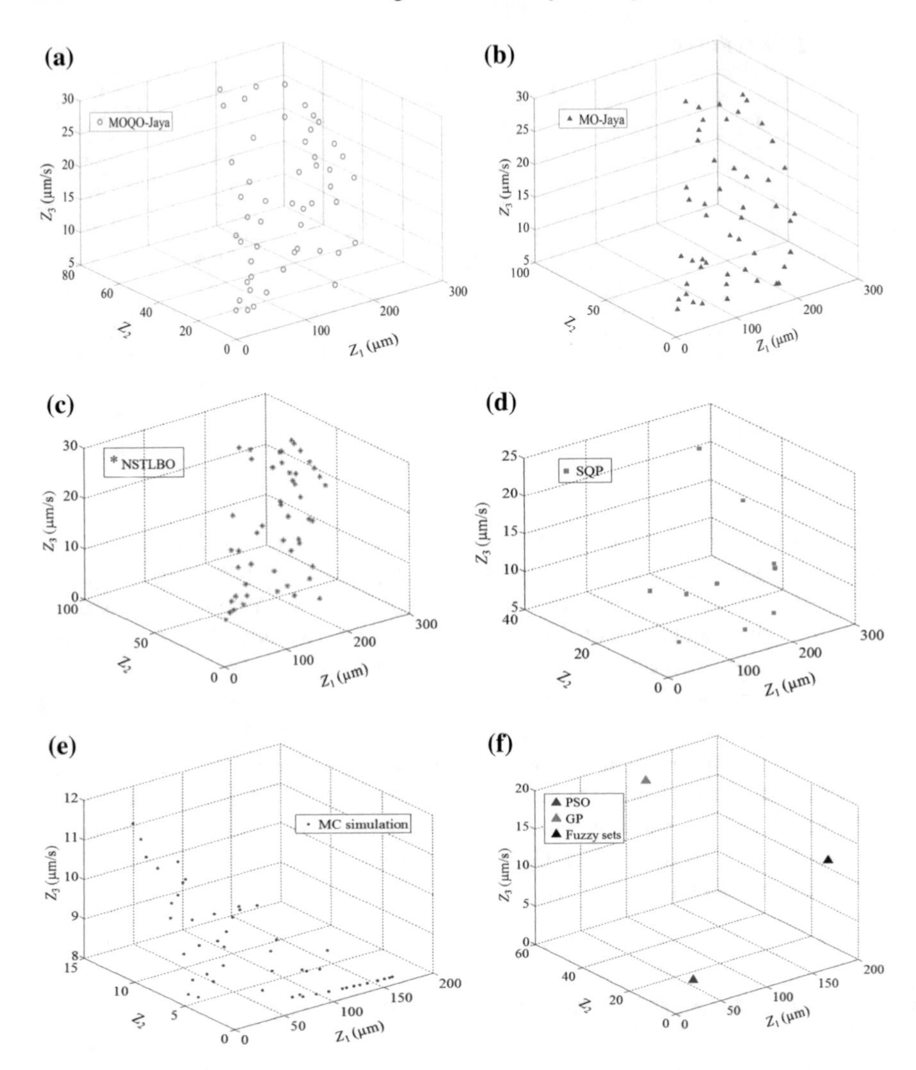

Fig. 7.10 Solutions obtained by **a** MOQO-Jaya, **b** MO-Jaya, **c** NSTLBO, **d** SQP, **e** MC simulation, and **f** GP, Fuzzy sets and PSO for ECM process (case study 2)

For the purpose of fair comparison of results, the maximum number of function evaluations considered by MOQO-Jaya, MO-Jaya and NSTLBO algorithms are is maintained as 60,000. A population size of 50 and maximum number of iterations equal to 1200 (i.e. maximum number of function evaluations = 50 × 1200 = 60,000) is considered for the MOQO-Jaya and MO-Jaya algorithms. A population size of 50 and maximum number of iterations equal 600 (i.e. maximum number of function evaluations = 2 × 50 × 600 = 60,000) in considered for NSTLBO algorithm. In addition, the same problem is solved using SQP and MC simulation and results obtained are compared.

Table 7.27 The values of coverage and spacing obtained by MOQO-Jaya, MO-Jaya and NSTLBO for ECM process (case study 2)

	Best	Mean	SD
Cov(*A*, *B*)	0.02	0.031	0.0014
Cov(*B*, *A*)	0.02	0.036	0.0023
Cov(*A*, *C*)	0.06	0.0565	0.0018
Cov(*C*, *A*)	0	0	0
Cov(*A*, *D*)	1	1	0
Cov(*D*, *A*)	0	0	0
S(*A*)	0.0494	0.0642	0.0070
S(*B*)	0.0494	0.0589	0.0063
S(*C*)	0.0573	0.0649	0.0071
S(*D*)	0.2414	0.2524	0.0082

A, *B*, *C* and *D* are the non-dominated set of solutions obtained by MOQO-Jaya, MO-Jaya algorithm, NSTLBO and SQP, respectively
SD is the standard deviation

Table 7.28 The values of hypervolume obtained using MOQO-Jaya, MO-Jaya, NSTLBO and SQP algorithms for ECM process (case study 2)

Algorithm	Hypervolume	No. of FEs required for convergence	Computational time (s)
MOQO-Jaya	**57,305**	**750**	3.56
MO-Jaya	**57,305**	900	**1.94**
NSTLBO	55,289	1150	2.38
SQP	44,296	841	5.85

NA means not available in the literature. Values in bold indicate better performance of the algorithm

Now the performance of MOQO-Jaya, MO-Jaya, NSTLBO and NSGA-II algorithms are compared on the basis of coverage and spacing measure. For this purpose, the MOQO-Jaya, MO-Jaya and NSTLBO algorithms are executed 30 times independently and the best, mean and standard deviation of coverage and spacing obtained over 3 independent runs are reported in Table 7.30.

The value Cov(*A*, *B*) = 0.02 implies that, 2% of the solutions obtained using MO-Jaya algorithm are inferior to the solutions obtained using MOQO-Jaya algorithm. Cov(*A*, *C*) = 0.06 implies that, 6% of the solutions obtained using NSTLBO are dominated by the solutions obtained using MOQO-Jaya algorithm. On the other hand, Cov(*C*, *A*) = 0.04 implies that, 4% of the solutions obtained using MOQO-Jaya algorithm are dominated by the solutions obtained using NSTLBO. Cov(*A*, *D*) = 1 and Cov(*D*, *A*) = 0 imply that, all the solutions obtained using SQP are dominated by the solutions obtained using MOQO-Jaya algorithm. Cov(*A*, *E*) = 1 and Cov(*E*, *A*) = 0 imply that, all the solutions obtained by NSGA-II are inferior to the solution obtained using MOQO-Jaya algorithm. All the results

Table 7.29 The Pareto-optimal set of solutions obtained by MOQO-Jaya algorithm for ECM process (case study 2)

S. No.	f (μm/s)	U (cm/s)	V (volts)	Z_1 (μm)	Z_2	Z_3 (μm/s)
1	8	300	10.1768	17.4445	6.5539	8
2	9.0954	300	11.0404	23.6876	8.3929	9.0954
3	9.627	300	12.337	34.3629	7.7501	9.627
4	8.6197	300	13.0213	39.0633	4.5794	8.6197
5	10.7132	300	12.8934	41.1364	10.1111	10.7132
6	8.2352	300	14.2126	50.6042	3.1259	8.2352
7	13.4621	300	13.663	53.8859	19.5552	13.4621
8	13.0112	300	14.3593	62.2236	15.2969	13.0112
9	12.3107	300	14.8564	67.8325	11.5483	12.3107
10	13.0748	300	15.2212	74.9278	13.4346	13.0748
11	12.6819	300	15.4231	77.207	11.6687	12.6819
12	8	300	17.0401	88.7228	1.7857	8
13	9.8903	300	16.8247	92.4075	3.8973	9.8903
14	11.0067	300	16.9397	98.3431	5.5871	11.0067
15	18.4577	300	16.4137	108.4109	37.4908	18.4577
16	17.107	300	16.7805	112.9235	27.1186	17.107
17	19.419	300	16.7029	116.7896	42.9134	19.419
18	18.0541	300	17.0295	120.7478	31.6011	18.0541
19	15.0196	300	17.5779	124.4132	15.2404	15.0196
20	11.0073	300	18.2914	125.3006	4.6044	11.0073
21	14.4935	300	18.037	133.1369	12.5925	14.4935
22	9.2147	300	19.304	138.7948	2.1466	9.2147
23	12.6301	300	18.8737	145.7666	6.9115	12.6301
24	19.2338	300	18.1307	150.7405	33.7324	19.2338
25	16.5329	300	18.5132	151.9832	18.7631	16.5329
26	17.4524	300	18.7717	162.0896	21.9305	17.4524
27	8	300	20.87	168.2179	1.0707	8
28	8	300	21	171.5455	1.0541	8
29	21.5278	300	18.6463	171.9101	46.7718	21.5278
30	22.9972	300	18.5758	174.1964	59.6091	22.9972
31	9.9366	300	20.9168	183.9988	2.2878	9.9366
32	20.8112	300	19.184	185.6341	38.6332	20.8112
33	23.753	300	18.9022	186.3199	63.9413	23.753
34	20.0266	300	19.4987	192.5727	32.3781	20.0266
35	11.6323	300	21	197.848	3.9492	11.6323
36	24.4505	300	19.2886	200.8078	67.2917	24.4505
37	25.696	300	19.2424	203.1034	80.6711	25.696
38	17.0148	300	20.2481	203.8474	16.5653	17.0148
39	25.3123	300	19.4773	209.8227	74.1953	25.3123

(continued)

Table 7.29 (continued)

S. No.	f (μm/s)	U (cm/s)	V (volts)	Z_1 (μm)	Z_2	Z_3 (μm/s)
40	15.1901	300	20.7927	212.2743	10.3822	15.1901
41	19.4944	300	20.2596	215.078	26.7328	19.4944
42	15.8757	300	21	222.7415	11.8324	15.8757
43	24.4776	300	20.0245	226.0861	61.4664	24.4776
44	25.5155	300	20.1534	234.3895	70.0231	25.5155
45	20.3198	300	21	244.7071	28.2659	20.3198
46	24.961	300	20.5252	246.2355	61.8796	24.961
47	22.6241	300	21	254.932	41.2924	22.6241
48	24.0104	300	21	260.7751	50.9326	24.0104
49	25.7064	300	21	267.6465	64.801	25.7064
50	26.1138	300	21	269.2554	68.4986	26.1138

Table 7.30 The values of coverage and spacing obtained by MOQO-Jaya, MO-Jaya and NSTLBO for FIB micro-milling process

	Best	Mean	SD
Cov(A, B)	0.02	0.0196	0.0023
Cov(B, A)	0.02	0.0106	0.019
Cov(A, C)	0.06	0.0213	0.022
Cov(C, A)	0.04	0.0044	0.029
Cov(A, D)	1	1	0
Cov(D, A)	0	0	0
Cov(A, E)	1	0.8	0.015
Cov(D, E)	0	0	0
$S(A)$	0.0106	0.0112	0.0019
$S(B)$	0.0106	0.0133	0.002
$S(C)$	0.0117	0.0140	0.0016
$S(D)$	0.0525		

A, B, C, D and E are the non-dominated set of solutions obtained by MOQO-Jaya, MO-Jaya, NSTLBO, SQP and NSGA-II, respectively
SD is the standard deviation

obtained using MC simulation are inferior to the results obtained using MOQO-Jaya algorithm. The MOQO-Jaya algorithm obtained a better value of spacing as compared to MO-Jaya, NSTLBO, SQP and NSGA-II.

The performance of MOQO-Jaya, MO-Jaya, NSTLBO, SQP and NSGA-II algorithms are now compared on the basis of hypervolume. The values of hypervolume obtained by the respective algorithms are reported in Table 7.31.

It is observed that the MOQO-Jaya and MO-Jaya algorithms obtained a higher hypervolume as compared to NSTLBO, NSGA-II and SQP. The number of function evaluations required by MOQO-Jaya, MO-Jaya, NSTLBO and NSGA-II for convergence are 900, 1000, 7740 and 60,000, respectively. It is observed that the

Table 7.31 The values of hypervolume obtained using MOQO-Jaya, MO-Jaya, NSTLBO, SQP and NSGA-II algorithms for FIB micro-milling process

Algorithm	Hypervolume	No. of FEs required for convergence	Computational time (s)
MOQO-Jaya	**28.0263**	**900**	52.01
MO-Jaya	**28.0263**	1000	**50.93**
NSTLBO	27.7893	7740	54.46
SQP	23.0530	1800	4.8
NSGA-II	16.8477	60,000	NA

NA means not available in the literature
Values in bold indicate better performance of the algorithm

MOQO-Jaya algorithm obtained the same value of hypervolume in less number of function evaluations as compared to NSGA-II. The computational time required by MOQO-Jaya, MO-Jaya and NSTLBO are 52.01, 50.93 and 54.46 s, respectively. However, the computational time required by NSGA-II was not reported by Bhavsar et al. (2015).

The non-dominated set of solutions obtained by MOQO-Jaya algorithm is shown in Table 7.32 for the purpose of demonstration. Figure 7.11 shows the Pareto-fronts obtained by MOQO-Jaya, MO-Jaya, NSTLBO SQP and NSGA-II algorithms. The minimum value of R_a observed in the Pareto front obtained by NSGA-II is 13.97 nm and the maximum value of *MRR* is 0.5314 μm^3/s. The minimum value of R_a observed in the Pareto front obtained by MO-Jaya is 6.0187 nm which is 56.91% better than the minimum value of R_a obtained using NSGA-II and the maximum value of *MRR* obtained by MO-Jaya is 0.6301 μm^3/s which is 15.66% higher than the maximum value of *MRR* obtained by NSGA-II.

7.2.7 Laser Cutting Process

This problem described in Sect. 7.1.7 was solved by Pandey and Dubey (2012) using GA. The algorithm-specific parameters and common control parameters for GA were tuned by Pandey and Dubey (2012) as follows: crossover probability = 0.8, mutation probability = 0.07, population size = 200 and maximum no. of generations = 800. Therefore, the maximum number of function evaluations required by GA is 200 × 800 = 160,000.

Pandey and Dubey (2012) used a posteriori approach to obtain a set of Pareto-optimal solutions. The same problem was solved by Kovacevic et al. (2014) using iterative search method and a Pareto-optimal set of solutions was reported. Therefore, for a fair comparison of results in the present work a posteriori approach is used to solve the multi-objective optimization problem of laser cutting process. The multi-objective optimization problem is now solved using MOQO-Jaya, MO-Jaya and NSTLBO algorithms. For fair comparison of results the allowable number of

Table 7.32 The Pareto-optimal set of solutions obtained by MOQO-Jaya algorithm for FIB micro-milling process

S. No.	x_1 (kV)	x_2 (°)	x_3 (nA)	x_4 (μs)	x_5 (%)	*MRR* (μm^3/s)	R_a (nm)
1	30	10	0.0618	1	30	0.0261	6.0187
2	30	10	0.2328	1	30	0.0436	10.4116
3	30	10	0.3432	1	30	0.055	13.2461
4	30	10	0.5342	1	30	0.0746	18.1549
5	30	10	0.608	1	30	0.0822	20.0495
6	30	10	0.6352	1	30	0.085	20.7495
7	30	10	0.724	1	30	0.0941	23.0287
8	30	10	0.7843	1	30	0.1003	24.5776
9	30	10	0.8935	1	30	0.1115	27.3845
10	30	10	0.9842	1	30	0.1209	29.7153
11	30	10	1.0369	1	30	0.1263	31.0691
12	30	10	1.1218	1	30	0.135	33.2501
13	30	10	1.2955	1	30	0.1528	37.7105
14	30	10	1.3739	1	30	0.1609	39.7265
15	30	10	1.4474	1	30	0.1685	41.6141
16	30	10	1.5316	1	30	0.1771	43.7768
17	30	10	1.6594	1	30	0.1902	47.0609
18	30	10	1.7219	1	30	0.1967	48.6653
19	30	10	1.8151	1	30	0.2062	51.0591
20	30	12.4057	1.905	1	30	0.2164	53.5506
21	30	70	1.0475	1	75	0.2294	54.2325
22	30	70	1.1377	1	75	0.2442	55.6306
23	30	70	1.257	1	75	0.2637	57.4792
24	30	70	1.3668	1	75	0.2816	59.1796
25	30	70	1.441	1	75	0.2937	60.3285
26	30	70	1.5147	1	75	0.3057	61.4695
27	30	70	1.6038	1	75	0.3203	62.8505
28	30	70	1.6699	1	75	0.3311	63.8747
29	30	70	1.7144	1	75	0.3384	64.5641
30	30	70	1.8456	1	75	0.3598	66.5959
31	30	70	1.9103	1	75	0.3704	67.5974
32	30	70	1.9318	1	75	0.3739	67.9306
33	30	70	2.0261	1	75	0.3893	69.3924
34	30	70	2.1936	1	75	0.4167	71.9857
35	30	70	2.2881	1	75	0.4321	73.4503
36	30	70	2.362	1	75	0.4442	74.5956
37	30	70	2.404	1	75	0.4511	75.245
38	30	70	2.5022	1	75	0.4671	76.7672
39	30	70	2.5467	1	75	0.4744	77.4564

(continued)

Table 7.32 (continued)

S. No.	x_1 (kV)	x_2 (°)	x_3 (nA)	x_4 (μs)	x_5 (%)	*MRR* (μm^3/s)	R_a (nm)
40	30	70	2.6352	1	75	0.4888	78.8261
41	30	70	2.7231	1	75	0.5032	80.1878
42	30	70	2.837	1	75	0.5218	81.9533
43	30	70	2.8809	1	75	0.529	82.6331
44	30	70	2.9862	1	75	0.5462	84.2633
45	30	70	3.0427	1	75	0.5554	85.1392
46	30	70	3.1238	1	75	0.5687	86.3957
47	30	70	3.221	1	75	0.5846	87.9006
48	30	70	3.3153	1	75	0.6	89.3617
49	30	70	3.407	1	75	0.615	90.7822
50	30	70	3.5	1	75	0.6302	92.2225

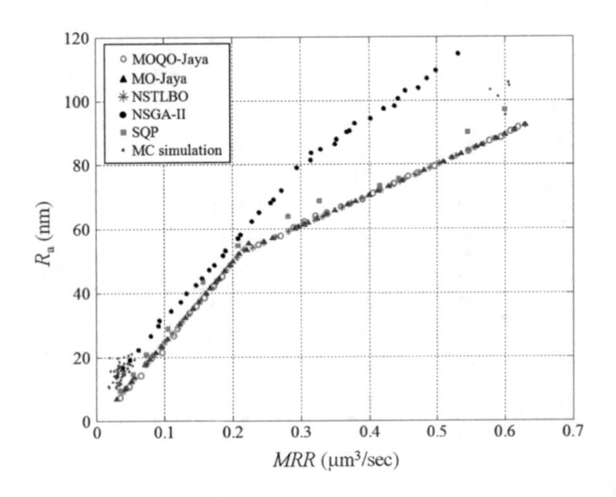

Fig. 7.11 The Pareto-fronts obtained MOQO-Jaya, MO-Jaya, NSTLBO, NSGA-II, SQP and MC simulation for FIB micro-milling process

function evaluations for MOQO-Jaya, MO-Jaya and NSTLBO algorithms is maintained as 160,000 (which is same as that of GA). Therefore, a population size of 50 and maximum number of generations equal to 3200 (i.e. maximum number of function evaluations = 50 × 3200 = 160,000) is considered for MOQO-Jaya and MO-Jaya algorithms. A population size of 50 and maximum number of iterations equal to 1600 (i.e. maximum number of function evaluations = 2 × 50 × 1600 = 160,000) is considered for NSTLBO algorithm. In addition the same problem is solved using SQP and MC simulation and the results are compared.

Now the performances of MOQO-Jaya, MO-Jaya and NSTLBO are compared on the basis of coverage and spacing measure. For this purpose, the MOQO-Jaya, MO-Jaya and NSTLBO are executed 30 times independently and the best, mean and standard deviation of the values of coverage and spacing obtained over 30 independent runs are reported in Table 7.33.

Table 7.33 The values of coverage and spacing obtained by MOQO-Jaya, MO-Jaya, NSTLBO, SQP, iterative search method and GA for laser cutting

	Best	Mean	SD
Cov(*A*, *B*)	0.04	0.036	0.0019
Cov(*B*, *A*)	0.02	0.019	0.0014
Cov(*A*, *C*)	0.12	0.113	0.075
Cov(*C*, *A*)	0	0.002	0.0019
Cov(*A*,*D*)	0.84	0.785	0.056
Cov(*D*, *A*)	0	0	0
Cov(*A*, *E*)	0	0.005	0.0021
Cov(*E*, *A*)	0	0	0
Cov(*A*, *F*)	1	1	0
Cov(*F*, *A*)	0	0	0
S(*A*)	0.0141	0.0162	0.014
S(*B*)	0.0141	0.0154	0.010
S(*C*)	0.0153	0.0166	0.013
S(*D*)	0.0630	0.0669	0.0022

A, *B*, *C*, *D*, *E* and *F* represent the non-dominated set of solutions obtained using MOQO-Jaya algorithm, MO-Jaya algorithm, NSTLBO algorithm, SQP, iterative search method (Kovacevic et al. 2014) and GA (Pandey and Dubey 2012)
SD is the standard deviation

The value Cov(*A*, *B*) = 0.04 implies that, 4% of the solutions obtained using MO-Jaya algorithm are dominated by the solutions obtained using MOQO-Jaya algorithm. Cov(*A*, *C*) = 0.12 implies that, 12% of the solutions obtained using NSTLBO are dominated by the solutions obtained using MOQO-Jaya algorithm. Cov(*A*, *D*) = 0.84 implies that, 84% of the solutions obtained using SQP are inferior to the solutions obtained by MOQO-Jaya algorithm. Cov(*E*, *A*) = 0 implies that none of the solutions obtained using MOQO-Jaya algorithm are inferior to the solutions obtained using iterative search method. Therefore, the non-dominated solutions obtained using MOQO-Jaya algorithm are equally good as compared to the non-dominated solutions obtained using iterative search method. Cov(*A*, *F*) = 1 implies that all the solutions obtained using GA are inferior to the solutions obtained using MOQO-Jaya algorithm. It is observed that all the solutions obtained using MC simulation are inferior to the solutions obtained by MOQO-Jaya algorithm. The value of spacing measure obtained by iterative search method and GA are *S*(*E*) = 0.0167 and *S*(*F*) = 0.0138, respectively. The MOQO-Jaya algorithm obtained a better value of spacing as compared to MO-Jaya, NSTLBO, GA, SQP and iterative search methods.

Now the performances of MOQO-Jaya, MO-Jaya, NSTLBO, GA, SQP and iterative search methods are compared on the basis of hypervolume. The values of hypervolume are reported in Table 7.34. It is observed that MOQO-Jaya and MO-Jaya algorithms have obtained a better value of hypervolume as compared to other algorithms The number of function evaluations required by MOQO-Jaya, MO-Jaya and NSTLBO algorithms for convergence are 3200, 3450 and 6950,

Table 7.34 The values of hypervolume achieved by MOQO-Jaya, MO-Jaya algorithm, NSTLBO algorithm, iterative search method, GA and SQP for laser cutting process

Algorithm	Hypervolume	No. of function evaluations required	Computational time (s)
MOQO-Jaya	**15.580**	**3200**	103.61
MO-Jaya	**15.580**	3450	**98.56**
NSTLBO	15.370	6950	102.94
GA	13.552	160,000	NA
Iterative search method	14.460	NA	NA
SQP	13.783	2500	2.87

NA means not available in the literature
Values in bold indicate better performance of the algorithm

respectively. The computational time required by MOQO-Jaya, MO-Jaya and NSTLBO algorithms are 103.61, 98.56 and 102.94 s, respectively.

The number function evaluations and computational time required by the iterative search method for solving the multi-objective optimization problem of laser cutting process are not reported by Kovacevic et al. (2014). The computational time required by GA is not reported by Pandey and Dubey (2012). The Pareto-optimal set of solutions obtained by MOQO-Jaya algorithm is reported in Table 7.35 for the purpose of demonstration. Figure 7.12 shows the Pareto-fronts obtained by MOQO-Jaya, MO-Jaya, NSTLBO, SQP, GA and iterative search method.

7.2.8 Abrasive Waterjet Machining Process

This problem described in Sect. 7.1.8 was solved by Shukla and Singh (2016) optimized using PSO, SA, FA, BH, CSA, BBO and NSGA algorithms. Now Jaya and QO-Jaya algorithms are applied to solve the same problem. Shukla and Singh (2016) optimized Ktw and taper angle individually using PSO, SA, FA, BH, CSA and BBO algorithms. However, Shukla and Singh (2016) did not provide the size of initial population used for these algorithms, therefore, the maximum number of function evaluations required by these algorithms cannot be determined. For this purpose a population size of 25 and maximum number of iterations equal to 20 (i.e. maximum number of function evaluations = 25 × 20 = 500) are selected for Jaya algorithm and QO-Jaya algorithms.

The results of Jaya and QO-Jaya algorithms are compared with the results of other algorithms used by Shukla and Singh (2016) and are reported in Table 7.36. It can be observed that the Jaya and QO-Jaya algorithms could achieve the best value of *Ktw* in only 50 function evaluations (i.e. 2 iterations) and 100 function evaluations (i.e. 4 iterations), respectively, which are very low as compared to the number of iterations required by PSO, SA, FA, BH, CSA and BBO algorithms. The

Table 7.35 The Pareto-optimal set of solutions for laser cutting obtained using MOQO-Jaya algorithm for laser cutting process

S. No.	x_1 (kg/cm^2)	x_2 (ms)	x_3 (Hz)	x_4 (mm/min)	K_t (°)	R_a (μm)
1	5.9298	1.4	14	15	0.3822	11.9839
2	5.7709	1.4	14	15	0.3875	11.7467
3	5.7119	1.4	14	15	0.3921	11.655
4	5.4499	1.4	14	15	0.4304	11.224
5	5.3084	1.4	14	15	0.463	10.9752
6	5.2023	1.4	14	15	0.493	10.7814
7	5.1048	1.4	14	15	0.5246	10.5977
8	5.0488	1.4	14	15	0.5446	10.4896
9	5	1.4	14	15.3798	0.6206	10.3306
10	5	1.4	14	15.8594	0.6931	10.2482
11	5	1.4	14	16.3194	0.7626	10.1668
12	5	1.4	14	16.4867	0.7879	10.1366
13	5	1.4	14	16.937	0.856	10.0539
14	5	1.4	14	17.0702	0.8761	10.0289
15	5	1.4	14	17.563	0.9506	9.9351
16	5	1.4	14	18.0241	1.0204	9.8448
17	5	1.4	14	18.1223	1.0352	9.8253
18	5	1.4	14	18.5354	1.0977	9.742
19	5	1.4	14	18.6638	1.1171	9.7158
20	5	1.4	14	19.1591	1.192	9.6127
21	5	1.4	14	19.7952	1.2882	9.4764
22	6.3525	1.4	6	16.3638	1.3263	9.4353
23	5	1.4	14	20.4346	1.3848	9.3349
24	6.5309	1.4	6	18.8029	1.4152	9.1705
25	6.2758	1.4	6	17.886	1.4587	9.054
26	6.5752	1.4	6	20.3265	1.4989	8.8913
27	6.6942	1.4	6	22.3554	1.5939	8.5401
28	6.5582	1.4	6	22.2012	1.626	8.4342
29	7.2098	1.4	6	25	1.6757	8.3015
30	7.0769	1.4	6	25	1.6881	8.1998
31	6.4669	1.4	6	23.9453	1.7717	7.8893
32	6.57	1.4	6	25	1.8034	7.7205
33	6.478	1.4	6	25	1.8359	7.6181
34	6.3617	1.4	6	25	1.882	7.4817
35	6.2488	1.4	6	25	1.9322	7.3421
36	6.2097	1.4	6	25	1.9508	7.292
37	6.0326	1.4	6	25	2.0432	7.0545
38	5.9994	1.4	6	25	2.062	7.0081
39	5.8668	1.4	6	25	2.1415	6.8163

(continued)

Table 7.35 (continued)

S. No.	x_1 (kg/cm^2)	x_2 (ms)	x_3 (Hz)	x_4 (mm/min)	K_t (°)	R_a (μm)
40	5.8503	1.4	6	25	2.152	6.7918
41	5.6983	1.4	6	25	2.2533	6.5584
42	5.6113	1.4	6	25	2.3157	6.4189
43	5.4194	1.4	6	25	2.4644	6.0962
44	5.3476	1.4	6	25	2.524	5.9702
45	5.2938	1.4	6	25	2.5701	5.8738
46	5.2574	1.4	6	25	2.6019	5.8077
47	5.162	1.4	6	25	2.6881	5.6308
48	5.1244	1.4	6	25	2.7231	5.5599
49	5.061	1.4	6	25	2.7835	5.4381
50	5.0062	1.4	6	25	2.837	5.3312

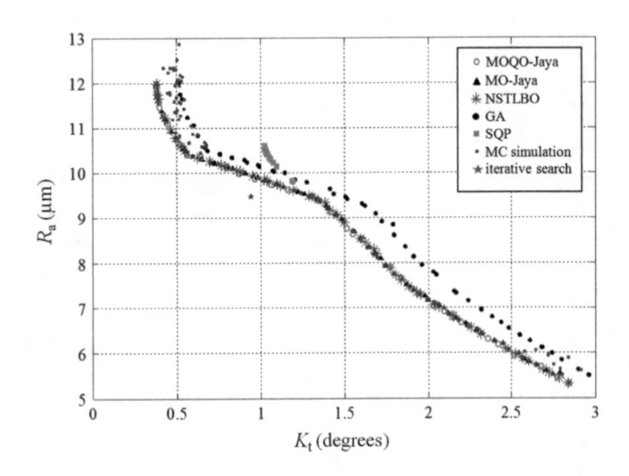

Fig. 7.12 The Pareto-fronts obtained using MOQO-Jaya, MO-Jaya, NSTLBO, GA, SQP, iterative search method and MC simulation for laser cutting process

Jaya and QO-Jaya algorithms could achieve a value of *taper angle* which is better than the best value of *taper angle* obtained by Shukla and Singh (2016). The Jaya and QO-Jaya algorithms required only 6 iterations to obtain the minimum value of *taper angle* which is very low as compared to the number of iterations required by PSO, SA, FA, BH, CSA and BBO algorithms. Furthermore, the Jaya and QO-Jaya algorithm are executed 10 times independently and the mean, standard deviation (SD) and CPU time required by Jaya algorithm for 10 independent runs is recorded and the same is compared with mean, SD and CPU time required by BBO algorithm in Table 7.37. It is observed that the Jaya and QO-Jaya algorithms achieved the same value of *Ktw* and *taper angle* for all 10 independent runs. Therefore, Jaya and QO-Jaya algorithms have shown more consistency and robustness and required less CPU time as compared to BBO algorithm. Figures 7.13 and 7.14 give the convergence graphs of Jaya and QO-Jaya algorithms for *Ktw* and *taper angle*, respectively.

Table 7.36 The optimal values of Kerf top width and taper angle obtained by Jaya and QO-Jaya algorithms for AWJM process (case study 1) (Rao et al. 2017b)

	Algorithm							
	PSO	FA	SA	CSA	BH	BBO	Jaya	QO-Jaya
Kerf top width	2.8840	2.9127	2.9136	2.9184	2.8698	**2.9187**	**2.9187**	**2.9187**
No. of iterations	20	180	680	5	150	10	**2**	4
Taper angle	1.0916	1.0672	1.0720	1.0692	1.0794	1.0631	**1.0539**	**1.0539**
No. of iterations	220	370	270	760	620	15	**6**	**6**

The results of PSO, FA, SA, CSA, BH and BBO are obtained from Shukla and Singh (2016)

In order to achieve multiple trade-off solutions for *Ktw* and *taper angle* in AWJM process Shukla and Singh (2016) used NSGA algorithm. A population size

Table 7.37 The mean, standard deviation and CPU time required by BBO, Jaya and QO-Jaya algorithms

	Kerf top width		Taper angle		
	BBO	Jaya	BBO	Jaya	QO-Jaya
Mean	2.9135	**2.9187**	1.1019	**1.0539**	**1.0539**
SD	0.0045	**0**	0.0035	**0**	**0**
CPU time	0.5112	**0.017**	1.1220	**0.0166**	**0.0166**

The values of mean, SD and CPU time for BBO algorithm are obtained from (Shukla and Singh 2016)

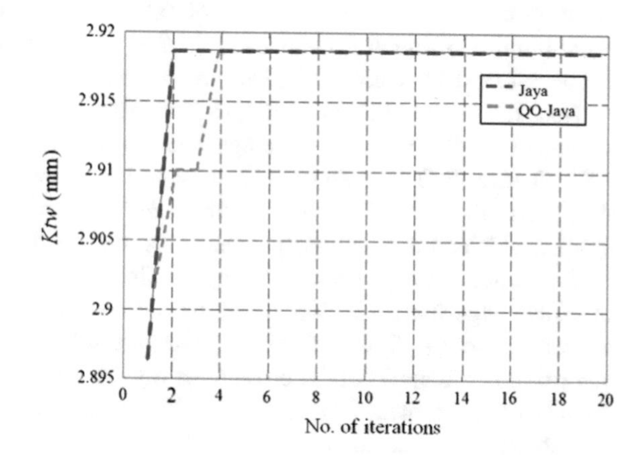

Fig. 7.13 The convergence graph for Jaya and QO-Jaya algorithms for maximization of *Ktw* in AWJM process (case study 1)

of 50 and maximum number of iterations as 1000 (i.e. maximum number of function evaluations as 50,000) was considered for NSGA. The algorithm-specific parameters required by NSGA were chosen by Shukla and Singh (2016) as follows: sharing fitness 1.2, dummy fitness 50. However, the values of crossover and mutation parameters required by NSGA were not reported by Shukla and Singh (2016).

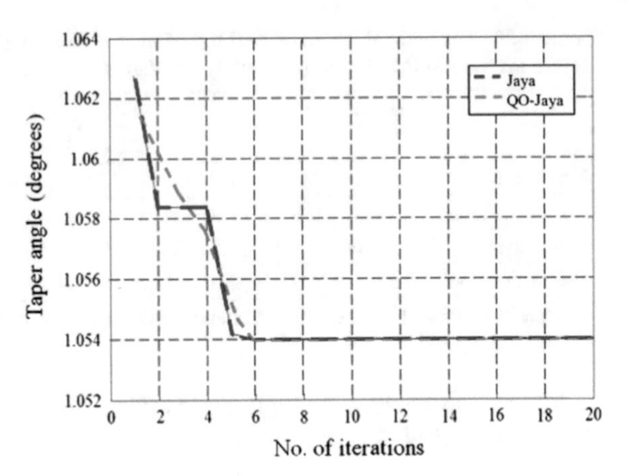

Fig. 7.14 The convergence graph for Jaya and QO-Jaya algorithms for minimization of taper angle for AWJM process (case study 1)

Shukla and Singh (2016) used a posteriori approach to solve the multiobjective optimization problem. Therefore, for the purpose of fair comparison, in the present work the same problem is solved using a posteriori approach. The multiobjective optimization problem is now solved using MOQO-Jaya, MO-Jaya and NSTLBO algorithms and the results are compared with the results of NSGA. For fair comparison of results the maximum number of function evaluations for MOQO-Jaya, MO-Jaya and NSTLBO algorithms are maintained same as that of NSGA. For this purpose a population size of 100 and maximum number of iterations equal to 500 are chosen for MOQO-Jaya and MO-Jaya algorithms (i.e. maximum number of function evaluations = 100 × 500 = 50,000). A population size of 50 and maximum number of iterations equal to 250 (i.e. maximum number of function evaluations = 2 × 100 × 250 = 50,000) are chosen for NSTLBO algorithm.

Now the performances of MOQO-Jaya, MO-Jaya and NSTLBO algorithms are evaluated on the basis of coverage and spacing measure. For this purpose the MOQO-Jaya, MO-Jaya and NSTLBO algorithms are executed 30 times independently and the best, mean and standard deviation of the values of coverage and spacing obtained over 30 independent runs are reported in Table 7.38.

The value Cov(A, B) = 0.05 implies that, 5% of the solutions obtained by MO-Jaya algorithm are dominated by the solutions obtained by MOQO-Jaya algorithm. Cov(A, C) = 0.07 implies that, the solutions obtained by MOQO-Jaya algorithm dominate 7% of the solutions obtained by NSTLBO algorithm. Cov(A, D) = 0.6 implies that, 60% of the solutions obtained by NSGA are dominated by the solutions obtained by MOQO-Jaya algorithm. It is observed that the MOQO-Jaya algorithm obtained a better value of spacing as compared to the other algorithms.

Now the performances of MOQO-Jaya, MO-Jaya and NSTLBO algorithms are compared on the basis of hypervolume. The values of hypervolume achieved by MOQO-Jaya, MO-Jaya, NSTLBO and NSGA are reported in Table 7.39. It is observed that the MOQO-Jaya and MO-Jaya algorithms achieved a higher value of hypervolume as compared to NSTLBO and NSGA algorithms. Only for fair

Table 7.38 The values of coverage and spacing obtained by MOQO-Jaya, MO-Jaya and NSTLBO algorithms for AWJM process (case study 1)

	Best	Mean	SD
Cov(*A*, *B*)	0.05	0.0481	0.0017
Cov(*B*, *A*)	0.09	0.0869	0.0030
Cov(*A*, *C*)	0.07	0.0671	0.0021
Cov(*C*, *A*)	0.10	0.0940	0.0036
Cov(*A*, *D*)	0.60	0.5737	0.0121
Cov(*D*, *A*)	0.06	0.0567	0.0016
S(*A*)	0.0063	0.0060	0.00021
S(*B*)	0.0069	0.0066	0.00025
S(*C*)	0.0066	0.0062	0.00019
S(*D*)	0.0090	0.0084	0.00021

A, *B*, *C* and *D* are the non-dominated set of solutions obtained using MOQO-Jaya, MO-Jaya, NSTLBO and NSGA algorithms
SD is the standard deviation

comparison of results the maximum number of function evaluations for MOQO-Jaya, MO-Jaya and NSTLBO algorithms are maintained as 50,000. However, The number of functions evaluations required by MOQO-Jaya, MO-Jaya and NSTLBO algorithms for convergence are 650, 700 and 1600, respectively. The MOQO-Jaya algorithm obtained the same value of hypervolume within less number of function evaluations as compared to MO-Jaya algorithm. The computational time required by MOQO-Jaya, MO-Jaya and NSTLBO algorithms to perform one complete simulation run is 80.01, 77.086 and 80.67 s, respectively. However, the CPU time required NSGA is not reported by Shukla and Singh (2016).

The Pareto-optimal solutions obtained by MOQO-Jaya algorithm are reported in Table 7.40 for the purpose of demonstration. The maximum value of kerf top width achieved by NSGA is 2.8768 mm. However, the maximum value of kerf top width achieved by MOQO-Jaya algorithm is 2.9168 mm which is 1.39% higher than NSGA. The minimum value of taper angle achieved by NSGA is 1.0884. The minimum value of taper angle achieved by MOQO-Jaya algorithm is 1.054 which is 3.16% lower than NSGA.

Table 7.39 The values of hypervolume achieved by MOQO-Jaya, MO-Jaya, NSTLBO and NSGA

Algorithm	Hypervolume	No. of function evaluations required	Computational time (s)
MOQO-Jaya	**2.6028**	**650**	80.01
MO-Jaya	**2.6028**	700	**77.086**
NSTLBO	2.5995	1600	80.67
NSGA	2.5516	50,000	NA

NA means not available in the literature
Value in bold indicates better performance

Table 7.40 The Pareto-optimal set of solutions obtained using MOQO-Jaya algorithm for AWJM process (case study 1) (Rao et al. 2017b)

S. No.	x_1	x_2	x_3	Taper angle	*Ktw* (mm)
1	192.8341	1	2.2969	1.054	1.4364
2	190.3569	1	2.3256	1.0575	1.4692
3	189.0174	1	2.3893	1.0646	1.4941
4	184.8933	1	2.4494	1.0963	1.5571
5	183.2294	1	2.4553	1.1135	1.5817
6	181.7402	1	2.3581	1.126	1.5926
7	183.1224	1	2.6216	1.14	1.6088
8	179.2901	1	2.3349	1.1615	1.6297
9	178.3591	1	2.4828	1.1835	1.6612
10	177.8894	1	2.536	1.1976	1.6759
11	176.6961	1	2.5357	1.2189	1.6958
12	177.2182	1	2.6921	1.2393	1.7109
13	175.1161	1	2.5117	1.2469	1.7202
14	175.9205	1	2.7512	1.2787	1.7427
15	174.0249	1	2.6695	1.2961	1.7614
16	171.902	1	2.5579	1.3245	1.7846
17	170.3026	1	2.3523	1.3534	1.7963
18	173.3885	1	2.8888	1.376	1.8122
19	169.3082	1	2.5461	1.39	1.8335
20	170.1231	1	2.7713	1.4141	1.8491
21	167.8296	1	2.5368	1.4308	1.8625
22	168.6643	1	2.7213	1.4397	1.8692
23	167.3267	1	2.6468	1.4618	1.8856
24	165.781	1	2.4384	1.4866	1.897
25	165.6612	1	2.548	1.4981	1.9091
26	164.5781	1	2.4693	1.527	1.9256
27	164.9521	1	2.6523	1.5363	1.9358
28	163.6484	1	2.5488	1.5646	1.953
29	161.9246	1	2.4196	1.6184	1.9825
30	163.1274	1	2.7874	1.6282	1.9934
31	160.903	1	2.5471	1.6628	2.0153
32	160	1	2.4136	1.6911	2.0277
33	160	1	2.7105	1.7223	2.053
34	160	1	2.8204	1.7502	2.0677
35	160.3105	1.1206	2.6007	1.7757	2.0684
36	160.0922	1	3.001	1.8118	2.0961
37	160	1	3.0497	1.8369	2.1077
38	160	1	3.111	1.8665	2.1206
39	160	1	3.153	1.8884	2.1299
40	160.4974	1	3.2392	1.9194	2.1398

(continued)

Table 7.40 (continued)

S. No.	x_1	x_2	x_3	Taper angle	*Ktw* (mm)
41	160.2468	1	3.2538	1.9373	2.1487
42	160	1.1717	3.0692	1.9652	2.1606
43	160	1	3.3132	1.9839	2.1694
44	160.9715	1.2477	3.1271	2.0104	2.1735
45	160.3978	1	3.4275	2.0493	2.193
46	160	1	3.4407	2.0733	2.2052
47	160.1419	1.2956	3.1936	2.1078	2.2194
48	160	1.441	3.0533	2.1343	2.2305
49	160	1.1261	3.428	2.1519	2.2375
50	160	1.2261	3.3659	2.1754	2.2477
51	160	1.5655	3.0326	2.202	2.2588
52	160	1.4557	3.2306	2.2374	2.2742
53	160	1.4891	3.2233	2.2539	2.2811
54	160	1.2508	3.5	2.2913	2.2939
55	160	1.6316	3.2604	2.3633	2.327
56	160	1.7969	3.1443	2.3931	2.3403
57	160	1.9638	3.0159	2.4234	2.3532
58	160	1.5122	3.5	2.4592	2.3638
59	160	1.8912	3.2031	2.4785	2.3766
60	160	1.8239	3.2803	2.4874	2.3796
61	160	1.6459	3.4822	2.5264	2.3926
62	160	2.117	3.1381	2.5628	2.4139
63	160	2.2039	3.1078	2.591	2.4265
64	160	1.7834	3.5	2.6206	2.4322
65	160	2.0761	3.3131	2.6451	2.4474
66	160	2.6073	2.8463	2.6702	2.4622
67	160	2.7077	2.7865	2.6946	2.4729
68	160	2.5349	3.0574	2.7214	2.4856
69	160	2.2855	3.3213	2.7553	2.4959
70	160	2.9505	2.7807	2.7868	2.5168
71	160	3.0759	2.672	2.8067	2.5249
72	160	2.1918	3.5	2.839	2.5271
73	173.4034	5	2.6611	2.8604	2.5525
74	160	3.3135	2.6889	2.8864	2.5655
75	160	2.5849	3.3567	2.9157	2.5674
76	160	3.2944	2.8314	2.9162	2.5814
77	169.481	5	2.6634	2.9194	2.6011
78	167.0056	5	2.4094	2.9389	2.609
79	168.88	5	2.763	2.954	2.624
80	166.0315	5	2.667	2.9868	2.6491

(continued)

Table 7.40 (continued)

S. No.	x_1	x_2	x_3	Taper angle	*Ktw* (mm)
81	163.9872	5	2.4796	3.0097	2.6606
82	165.0937	5	2.7381	3.0232	2.6724
83	163.3434	5	2.6442	3.0445	2.6867
84	162.0704	5	2.4139	3.0569	2.6876
85	163.0084	5	2.787	3.0851	2.7107
86	160	5	2.4514	3.1155	2.7254
87	160	5	2.7439	3.1538	2.7527
88	160.4855	5	2.8104	3.1577	2.7539
89	160	4.9389	2.9001	3.1995	2.7721
90	160	4.9319	2.9085	3.2024	2.7731
91	160	5	3.0448	3.2586	2.8024
92	161.5897	5	3.1958	3.2944	2.8118
93	160	5	3.1407	3.3059	2.8227
94	160	5	3.1851	3.3301	2.8329
95	162.227	4.8823	3.3429	3.369	2.8335
96	160	5	3.2683	3.3792	2.8533
97	160	4.9694	3.2964	3.3969	2.8587
98	160	5	3.3313	3.4198	2.8698
99	160	5	3.3725	3.448	2.8811
100	160	5	3.4936	3.5378	2.9168

7.2.9 Plasma Arc Machining Process

This problem described in Sect. 7.1.9 solved using a posteriori approach in order to optimize the *MRR* and *DFR* and to obtain a set of Pareto-optimal solutions. The MOQO-Jaya, MO-Jaya and NSTLBO algorithms are applied to solve the multi-objective optimization problem. A population size of 50 and maximum number of iterations 100 (i.e. maximum number of function evaluations = 50 × 100 = 5000) is chosen for MOQO-Jaya algorithm and MO-Jaya algorithm. A population size of 50 and maximum number of iterations equal to 50 (i.e. maximum number of function evaluations = 2 × 50 × 50 = 5000) is chosen for NSTLBO algorithm.

Now the performances of MOQO-Jaya, MO-Jaya and NSTLBO algorithms are compared on the basis of coverage and spacing measure. For this purpose, the MOQO-Jaya, MO-Jaya and NSTLBO algorithms are executed 30 times independently and the best, mean and standard deviation of the values of coverage and spacing obtained over 30 independent runs is reported in Table 7.41.

Table 7.41 The values of coverage and spacing obtained by MOQO-Jaya, MO-Jaya and NSTLBO algorithms for PAM process

	Best	Mean	SD
Cov(*A*, *B*)	0.06	0.0564	0.0012
Cov(*B*, *A*)	0.04	0.0378	0.0012
Cov(*A*, *C*)	0.12	0.1150	0.0029
Cov(*C*, *A*)	0	0.0096	0.00028
S(*A*)	0.016	0.0154	0.00040
S(*B*)	0.016	0.0151	0.00050
S(*C*)	0.018	0.0169	0.00061

A, *B* and *C* are the non-dominated set of solutions obtained using MOQO-Jaya, MO-Jaya and NSTLBO algorithms
SD is the standard deviation

In Table 7.41, the value Cov(*A*, *B*) = 0.06 implies that, 6% of the solutions obtained by MO-Jaya algorithm are dominated by the solutions obtained by MOQO-Jaya algorithm. Cov(*A*, *C*) = 0.12 implies that, the solutions obtained by MOQO-Jaya algorithm dominate 12% of the solutions obtained by NSTLBO algorithm. It is observed that the MOQO-Jaya and MO-Jaya algorithm obtained a better value of spacing as NSTLBO algorithms.

Now the performances of MOQO-Jaya, MO-Jaya and NSTLBO algorithms are compared on the basis of hypervolume. The values of hypervolume obtained by the respective algorithms are reported in Table 7.42. It is observed that the value of hypervolume obtained by MOQO-Jaya and MO-Jaya algorithms are better than that of NSTLBO algorithm. The MOQO-Jaya algorithm obtained the same value of hypervolume within less number of function evaluations as compared to MO-Jaya algorithm. The number of function evaluations required by MOQO-Jaya, MO-Jaya and NSTLBO algorithms to converge at the Pareto-optimal solutions are 4450, 4600 and 5000 respectively. The computational time required by MOQO-Jaya, MO-Jaya and NSTLBO algorithms to perform one complete simulation run are 10.56, 7.2 and 13.98 s, respectively.

The Pareto-optimal solutions obtained using MOQO-Jaya algorithm are reported in Table 7.43 for the purpose of demonstration. The results of MOQO-Jaya algorithm have revealed that, the optimal value for current and speed are 45 (A) and 800 (mm/min) to achieve a trade-off between *MRR* and *DFR*. *MRR* increases

Table 7.42 The values of hypervolume achieved by MOQO-Jaya, MO-Jaya and NSTLBO algorithms for PAM process

Algorithm	Hypervolume	No. of function evaluations required	Computational time (s)
MOQO-Jaya	**2.064**	**4450**	10.56
MO-Jaya	**2.064**	4600	**7.2**
NSTLBO	1.890	5000	13.98

Values in bold values indicate better performance

Table 7.43 The Pareto-optimal set of solutions obtained by MOQO-Jaya algorithm for PAM process

S. No.	x_1 (mm)	x_2 (A)	x_3 (V)	x_4 (mm/min)	*MRR* (g/s)	*DFR* (g/s)
1	2.5	45	128.2032	800	0.2342	0.0004
2	2.5	45	130.0663	800	0.2573	0.0004
3	2.5	45	134.3141	800	0.3127	0.0004
4	2.5	45	137.139	800	0.3508	0.0005
5	2.5	45	140.1553	800	0.392	0.0005
6	2.5	45	141.3426	800	0.4082	0.0005
7	2.5	45	142.5303	800	0.4243	0.0005
8	2.4048	45	142.5775	800	0.4589	0.0006
9	2.4115	45	144.2314	800	0.4804	0.0006
10	2.3928	45	144.823	800	0.4961	0.0006
11	2.3038	45	143.0715	800	0.5033	0.0006
12	2.4004	45	147.4697	800	0.5308	0.0007
13	2.3995	45	148.7112	800	0.5483	0.0007
14	2.3583	45	148.559	800	0.564	0.0007
15	2.2949	45	148.3905	800	0.5889	0.0008
16	2.3668	45	150.6993	800	0.5898	0.0008
17	2.3144	45	151.253	800	0.6215	0.0008
18	2.268	45	150.9664	800	0.6388	0.0008
19	2.0579	45	146.8665	800	0.6613	0.0009
20	2.1508	45	150.1631	800	0.6794	0.0009
21	2.1876	45	152.6231	800	0.7005	0.001
22	2.1182	45	151.5029	800	0.7152	0.001
23	2.0995	45	152.0345	800	0.7319	0.001
24	2.0861	45	153.623	800	0.7628	0.0011
25	2.0661	45	153.5103	800	0.7701	0.0011
26	2.0207	45	153.3698	800	0.788	0.0011
27	1.985	45	153.4289	800	0.804	0.0012
28	1.9782	45	154.6089	800	0.8259	0.0012
29	1.8797	45	153.3452	800	0.8429	0.0013
30	1.9448	45	155.8998	800	0.8601	0.0013
31	2.0101	45	158.4338	800	0.8681	0.0014
32	1.8571	45	155.9489	800	0.8943	0.0014
33	1.8422	45	156.0147	800	0.9005	0.0015
34	1.7788	45	155.7694	800	0.9156	0.0015
35	1.8493	45	157.9314	800	0.928	0.0016
36	1.8344	45	158.1038	800	0.9358	0.0016
37	1.901	45	160.6495	800	0.9464	0.0017
38	1.8744	45	160.3258	800	0.9528	0.0017
39	1.8186	45	159.9524	800	0.9681	0.0017

(continued)

Table 7.43 (continued)

S. No.	x_1 (mm)	x_2 (A)	x_3 (V)	x_4 (mm/min)	*MRR* (g/s)	*DFR* (g/s)
40	1.8681	45	161.9012	800	0.9757	0.0018
41	1.8435	45	162.0781	800	0.9874	0.0018
42	1.6871	45	160.0925	800	1.004	0.0019
43	1.6935	45	160.9864	800	1.0155	0.002
44	1.7365	45	162.4991	800	1.0263	0.002
45	1.7111	45	163.0979	800	1.0397	0.0021
46	1.7012	45	163.4105	800	1.0457	0.0022
47	1.7513	45	165	800	1.0518	0.0022
48	1.7091	45	165	800	1.0624	0.0023
49	1.6118	45	165	800	1.076	0.0025
50	1.5829	45	165	800	1.0769	0.0026

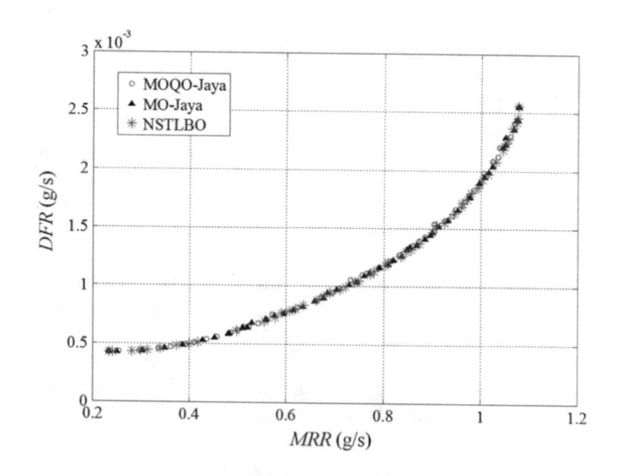

Fig. 7.15 The Pareto-fronts obtained by MOQO-Jaya, MO-Jaya and NSTLBO algorithms for PAM process

continuously from a minimum value of 0.2342–1.0769 (g/s) as the arc gap voltage increases from 128.2032 to 165 (V). However, the increase in *MRR* is achieved on the expense of increase in *DFR*. Therefore, the best compromised values for *DFR* lie in the range of 0.0004–0.0026 (g/s). The *DFR* shows an inverse trend with respect to thickness of workpiece. However, as the arc gap voltage increases the *DFR* also increases. Figure 7.15 shows the Pareto-fronts obtained by MOQO-Jaya, MO-Jaya and NSTLBO algorithms.

References

Acharya, B. G., Jain, V. K., & Batra, J. L. (1986). Multiobjective optimization of ECM process. *Precision Engineering, 8,* 88–96.

Baskar, N., Saravanan, R., Asokan, P., & Prabhaharan, G. (2004). Ants colony algorithm approach for multi-objective optimization of surface grinding operations. *International Journal of Advanced Manufacturing Technology, 23,* 311–317.

Bhattacharyya, B., & Sorkhel, S. K. (1999). Investigation for controlled electrochemical machining through response surface methodology-based approach. *Journal of Materials Processing Technology, 86,* 200–207.

Bhavsar, S. N., Aravindan, S., & Rao, P. V. (2015). Investigating material removal rate and surface roughness using multi-objective optimization for focused ion beam (FIB) micro-milling of cemented carbide. *Precision Engineering, 40,* 131–138.

Choobineh, F., Jain, V. K., (1990) Selection of ECM parameters: A fuzzy sets approach. In: *Proceedings of the 1990 International Conference on Systems, Man and Cybernetics*, IEEE, Los Angeles, CA, USA, pp. 430–435.

Garg, M. P., Jain, A., & Bhushan, G. (2012). Modelling and multi-objective optimization of process parameters of wire electrical discharge machining using non-dominated sorting genetic algorithm-II. *Proceedings of the Institution of Mechanical Engineers, Part B: Journal of Engineering Manufacture, 226*(12), 1986–2001.

Jain, N. K., & Jain, V. K. (2007). Optimization of electrochemical machining process parameters using genetic algorithm. *Machining Science and Technology, 11,* 235–258.

Kovacevic, M., Madic, M., Radovanovic, M., & Rancic, D. (2014). Software prototype for solving multi-objective machining optimization problems: Application in non-conventional machining processes. *Expert Systems with Applications, 41,* 5657–5668.

Kuriachen, B., Somashekhar, K. P., & Mathew, J. (2015). Multiresponse optimization of micro-wire electrical discharge machining process. *The International Journal of Advanced Manufacturing Technology, 76*(1–4), 91–104.

Kuriakose, S., & Shunmugam, M. S. (2005). Multi-objective optimization of wire-electro discharge machining process by non-dominated sorting genetic algorithm. *Journal of Materials Processing Technology, 170,* 133–141.

Mukherjee, R., & Chakraborty, S. (2013). Selection of the optimal electrochemical machining process parameters using biogeography-based optimization algorithm. *International Journal of Advanced Manufacturing Technology, 64,* 781–791.

Palanikumar, K., Latha, B., Senthilkumar, V. S., & Karthikeyan, R. (2009). Multiple performance optimization in machining of GFRP composites by a PCD tool using non-dominated sorting genetic algorithm (NSGA-II). *Metals and Materials International, 15*(2), 249–258.

Pandey, A. K., & Dubey, A. K. (2012). Simultaneous optimization of multiple quality characteristics in laser cutting of titanium alloy sheet. *Optics & Laser Technology, 44,* 1858–1865.

Pawar, P. J., & Rao, R. V. (2013). Parameter optimization of machining processes using teaching—learning-based optimization algorithm. *International Journal of Advanced Manufacturing Technology, 67,* 995–1006.

Pawar, P. J., Rao, R. V., & Davim, J. P. (2010). Multiobjective optimization of grinding process parameters using particle swarm optimization algorithm. *Materials and Manufacturing Processes, 25,* 424–431.

Rao, R. V. (2010). *Advanced modelling and optimization of manufacturing processes: international research and development*. London: Springer Verlag.

Rao, R. V., Pawar, P. J., & Shankar, R. (2008). Multi-objective optimization of electrochemical machining process parameters using a particle swarm optimization algorithm. *Journal of Engineering Manufacture, 222,* 949–958.

Rao, R. V., Rai, D. P., & Balic, J. (2017a). A multi-objective algorithm for optimization of modern machining processes. *Engineering Applications of Artificial Intelligence, 61,* 103–125.

Rao, R. V., Rai, D. P., Ramkumar, J., & Balic, J. (2016a). A new multiobjective Jaya algorithm for optimization of modern machining processes. *Advances in Production Engineering and Management, 11*(4), 271–286.

Rao, R. V., Rai, D. P., Balic, J. (2016b) Multi-objective optimization of machining and micro-machining processes using non-dominated sorting teaching–learning-based optimization algorithm. *Journal of Intelligent Manufacturing*. https://doi.org/10.1007/s10845-016-1210-5.

Rao, R. V., Rai, D. P., Balic, J., Cus, F. (2017b) Optimization of abrasive waterjet machining process using multiobjective Jaya algorithm. *Materials Today: Proceedings*.

Saravanan, R., Asokan, P., & Sachidanandam, M. (2002). A multiobjective genetic algorithm approach for optimization of surface grinding operations. *International Journal of Machine Tools and Manufacture, 42,* 1327–1334.

Shukla, R., & Singh, D. (2016). Experimentation investigation of abrasive water jet machining parameters using Taguchi and evolutionary optimization techniques. *Swarm and Evolutionary Computation, 32,* 167–183.

Wen, X. M., Tay, A. A. O., & Nee, A. Y. C. (1992). Microcomputer based optimization of the surface grinding process. *Journal of Materials Processing Technology, 29,* 75–90.

Zou, F., Wang, L., Hei, X., Chen, D., & Wang, B. (2014). Multi-objective optimization using teaching–learning-based optimization algorithm. *Engineering Applications of Artificial Intelligence, 26,* 1291–1300.

Chapter 8
Single- and Multi-objective Optimization of Nano-finishing Processes Using Jaya Algorithm and Its Variants

Abstract This chapter describes the formulation of process parameters optimization models for nano-finishing processes of rotational magnetorheological abrasive flow finishing, magnetic abrasive finishing, magnetorheological fluid based finishing, and abrasive flow machining. The application of TLBO and NSTLBO algorithms, Jaya algorithm and its variants such as Quasi-oppositional (QO) Jaya and multi-objective (MO) Jaya is made to solve the single and multi-objective optimization problems of the selected nano-finishing processes. The results of Jaya algorithm and its variants are found better as compared to those given by the other approaches.

8.1 Process Parameters Optimization Models for Nano-finishing Processes

8.1.1 Rotational Magnetorheological Abrasive Flow Finishing Process

8.1.1.1 Case Study 1

The out-of-roundness of the internal surfaces of stainless steel tubes finished by the rotational–magnetorheological abrasive flow finishing process was reported by Das et al. (2010, 2011). The optimization problem formulated in this work is based on the empirical models developed by Das et al. (2010, 2011). The objective considered in this work is maximization of improvement in out-of-roundness at the internal surfaces of the steel tubes and improvement in out-of-roundness per finishing cycle $\Delta OOR'$ (nm/cycle) of internal surfaces of stainless steel tubes. The process input parameters considered are: hydraulic extrusion pressure '*P*' (bar), number of finishing cycles '*N*', rotational speed of magnet '*S*' (RPM), and mesh size of abrasive '*M*'.

R. Venkata Rao, *Jaya: An Advanced Optimization Algorithm and its Engineering Applications*, https://doi.org/10.1007/978-3-319-78922-4_8

Objective function

The objective functions in terms of coded values of the process input parameters are expressed by Eqs. (8.1) and (8.2). The process input parameters are coded between −2 and 2 and the bounds on the process input parameters in terms of actual values are expressed by Eqs. (8.3)–(8.6).

$$\begin{aligned} \textit{maximize}\ \Delta OOR = {} & 1.24 + 0.18P + 0.18N - 0.083S - 0.11M + 0.068PN \\ & - 0.083PS - 0.16PM - 3.125E - 3NS - 0.26NM \\ & - 0.063SM - 0.11P^2 - 0.13N^2 - 0.20S^2 - 0.13M^2 \end{aligned} \tag{8.1}$$

$$\begin{aligned} \textit{maximize}\ \overline{\Delta OOR} = {} & 1.24 + 0.17P + 0.041N - 0.086S - 0.074M + 0.013PN \\ & - 0.08PS - 0.13PM + 0.022NS - 0.26NM - 0.064SM \\ & - 0.11P^2 - 0.14N^2 - 0.20S^2 - 0.13M^2 \end{aligned} \tag{8.2}$$

Process input parameter bounds

The bounds on the process input parameters are expressed as follows

$$32.5 < P < 42.5 \tag{8.3}$$

$$600 < N < 1400 \tag{8.4}$$

$$50 < S < 250 \tag{8.5}$$

$$90 < M < 210 \tag{8.6}$$

8.1.1.2 Case Study 2

The rotational–magnetorheological abrasive flow finishing process on flat workpieces was reported by Das et al. (2012). The optimization problem formulated in this work is based on the empirical models developed by Das et al. (2012). The objective considered is maximization of percentage improvement in surface roughness 'ΔR_a' (%). The process input parameters considered are: hydraulic extrusion pressure '*P*' (bar), number of finishing cycles '*N*', rotational speed of magnet '*S*' (RPM) and volume ratio of CIP/SiC '*R*'.

Objective functions

The objective functions are expressed by Eqs. (8.7) and (8.8).

$$\begin{aligned} \textit{maximize}\ \%\Delta Ra_{SS} = & -403 + 17.66P + 0.2N + 0.83S + 10.38R \\ & + 1.89E - 4PN - 1.67E - 3PS - 3.56E - 3PR \\ & - 6.53E - 5NS - 7E - 3NR + 8.46E - 3SR - 0.23P^2 \\ & - 1.39E - 4N^2 - 5.56E - 3S^2 - 1.35R^2 \end{aligned} \tag{8.7}$$

$$\begin{aligned} \textit{maximize}\ \%\Delta Ra_{BR} = & -912.47 + 39.27P + 0.41N + 1.67S + 28.49R \\ & - 4.02E - 3PN - 0.01PS - 0.08PR + 3.93E - 4NS \\ & - 2.76E - 3NR - 0.10SR - 0.46P^2 - 2.07E - 4N^2 \\ & - 9.26E - 3S^2 - 3.83R^2 \end{aligned} \tag{8.8}$$

Process input parameter bounds

The bounds on the process input parameters are expressed as follows:

$$32.5 \leq P \leq 42.5 \tag{8.9}$$

$$400 \leq N \leq 800 \tag{8.10}$$

$$20 \leq S \leq 100 \tag{8.11}$$

$$0.34 \leq R \leq 4 \tag{8.12}$$

8.1.2 Magnetic Abrasive Finishing Process

The optimization problem formulated in this work is based on the empirical model percentage improvement in surface roughness ΔRa (%) developed by Mulik and Pandey (2011) for the MAF process. The objective considered in this work is maximization of percentage improvement in surface roughness. The process input parameters considered are: voltage 'X_1'(volts), mesh number 'X_2', natural logarithm of rotation per minute of electromagnet 'X_3' (RPM), and % wt of abrasives 'X_4'.

Objective function

The objective function is expressed by Eq. (8.13).

$$\begin{aligned} \textit{maximize}\ \Delta Ra = & -674 + 5.48X_1 + 0.0628X_2 + 115X_3 - 15.8X_4 - 0.000018X_2^2 \\ & - 0.976X_1X_3 - 0.00203X_2X_4 - 2.41X_3X_4 \end{aligned} \tag{8.13}$$

Process input parameter bounds

The bounds on the process input parameters are expressed as follows

$$50 \le X_1 \le 90 \tag{8.14}$$

$$400 \le X_2 \le 1200 \tag{8.15}$$

$$5.193 \le X_3 \le 6.1092 \tag{8.16}$$

$$15 \le X_4 \le 35 \tag{8.17}$$

8.1.3 Magnetorheological Fluid Based Finishing Process

The optimization problem formulated in this work is based on the empirical models developed by Sidpara and Jain (2012, 2013) for MRFF process. The objective considered in this work are: maximization of axial force 'F_a' (N), maximization of normal force 'F_n' (N) and maximization of tangential force 'F_t' (N) acting on the workpiece during MRFF process. The process input parameters considered are: angle of curvature 'θ' (°), tool rotational speed 'S' (RPM), feed rate 'F' (mm/min).

Objective functions

The objective functions are expressed by Eqs. (8.18)–(8.20).

$$\begin{aligned} \mathit{maximize}\ F_n = & -15.027 - 0.266\theta + 0.042S + 0.757F - 0.0002\theta S \\ & - 0.0056\theta F + 0.0001SF + 0.0085\theta^2 - 0.00002S^2 \\ & - 0.115F^2 \end{aligned} \tag{8.18}$$

$$\begin{aligned} \mathit{maximize}\ F_t = & -7.257 - 0.191\theta + 0.021S + 0.526F - 0.00008\theta S \\ & - 0.008\theta F + 0.0001SF + 0.006\theta^2 - 0.00001S^2 \\ & - 0.077F^2 \end{aligned} \tag{8.19}$$

$$\begin{aligned} \mathit{maximize}\ F_a = & -4.372 - 0.081\theta + 0.012S + 0.247F - 0.00004\theta S \\ & - 0.0021\theta F + 0.00009SF + 0.002\theta^2 - 0.000006S^2 \\ & - 0.044F^2 \end{aligned} \tag{8.20}$$

Process input parameter bounds

The bounds on the process input parameters are expressed as follows

$$5 \leq \theta \leq 25 \tag{8.21}$$

$$700 \leq S \leq 1100 \tag{8.22}$$

$$1 \leq F \leq 5 \tag{8.23}$$

8.1.4 Abrasive Flow Machining Process

The optimization problem formulated in this work is based on the empirical models developed by Jain and Jain (2000) for AFM process. The objective considered is maximization of material removal rate '*MRR*' (mg/min). The process input parameters considered are: media flow '*v*' (cm/min); percentage concentration of abrasives '*c*'; abrasive mesh size '*d*' and number of cycles '*n*'. The constraint is on maximum allowable value of surface roughness '*Ra*' (μm).

Objective functions

The objective function is expressed by Eq. (8.24).

$$maximize\ MRR = 5.285E - 7v^{1.6469}c^{3.0776}d^{-0.9371}n^{-0.1893} \tag{8.24}$$

Constraint

$$Ra \leq Ra_{\max} \tag{8.25}$$

where Ra is the surface roughness in μm and $Ra_{\max}$ is the maximum allowable value of surface roughness.

$$Ra = 282751.0v^{-1.8221}c^{-1.3222}d^{0.1368}n^{-0.2258} \tag{8.26}$$

Process input parameter bounds

The bounds on the process input parameters are expressed as follows

$$40 \leq v \leq 85 \tag{8.27}$$

$$33 \leq c \leq 45 \tag{8.28}$$

$$100 \leq d \leq 240 \tag{8.29}$$

$$20 \leq n \leq 120 \tag{8.30}$$

8.2 Results of Application of Optimization Algorithms on Nano-finishing Processes

Rao and Rai (2016) described the results of optimization of advanced finishing processes using TLBO algorithm. Now, the results of application of Jaya algorithm and its variants and the comparison of results with those of TLBO and other approaches are presented.

8.2.1 Results of Rotational Magnetorheological Abrasive Flow Finishing Process

8.2.1.1 Case Study 1

This problem described in Sect. 8.1.1.1 was solved by Das et al. (2011) using desirability approach. Das et al. (2011) had optimized percentage change in out-of-roundness and percentage change in out-of roundness per cycle by considering each objective separately. Now the same problem is solved using TLBO, Jaya and QO-Jaya algorithms to see whether any improvement in results can be achieved. A population size of 25 and maximum number of function evaluation equal to 1000 are considered for TLBO, Jaya and QO-Jaya algorithms. The optimal combination of process input parameters obtained by TLBO, Jaya and QO-Jaya algorithms is reported in Table 8.1.

Figures 8.1 and 8.2 show the convergence graph for TLBO, Jaya and QO-Jaya algorithms. It is observed that the TLBO, Jaya and QO-Jaya algorithms converged at the same maximum value of ΔOOR. The TLBO, Jaya and QO-Jaya algorithms has achieved improvements of 46.57% in ΔOOR As compared to the values obtained by Das et al. (2011) using desirability function approach. However, the

Table 8.1 The optimal combination of process parameters for ΔOOR and $\Delta OOR'$ obtained by the TLBO, Jaya and QO-Jaya algorithms in R-MRFF process

S. No.	P (bar)	N	S (RPM)	M	ΔOOR (μm)	$\Delta OOR'$ (nm/cycle)	% improvement
1	42.5	1400	133.8	99	2.7	–	46.57
2	42.5	1400	140.7	90	–	1.9	21.02

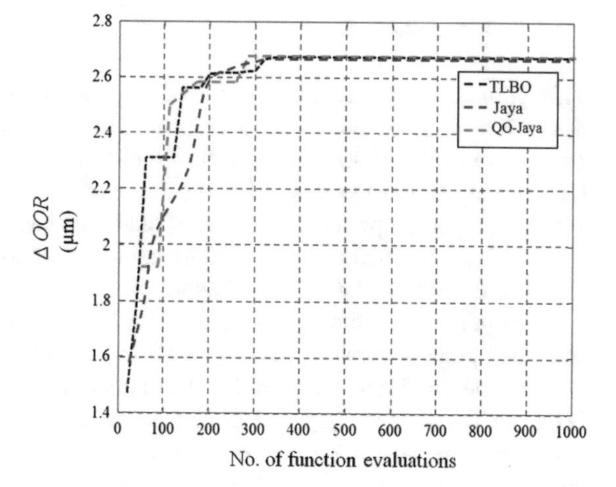

Fig. 8.1 The convergence graphs for TLBO, Jaya and QO-Jaya algorithms for $\Delta\ OOR$

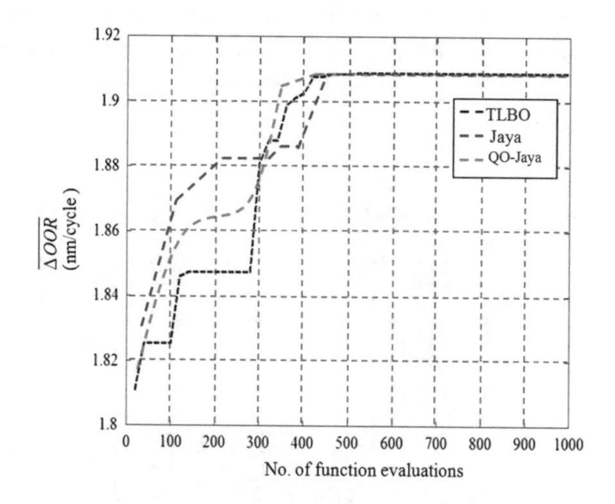

Fig. 8.2 The convergence graphs for TLBO, Jaya and QO-Jaya algorithms for $\Delta OOR'$

number of function evaluations required by TLBO, Jaya and QO-Jaya algorithms to converge at the optimal solutions is 360, 365 and 280, respectively.

Now the TLBO, Jaya and QO-Jaya algorithms are applied to maximize $\Delta OOR'$. It is observed that the TLBO, Jaya and QO-Jaya algorithm converged at the maximum value of $\Delta OOR'$ in 460, 450 and 440 function evaluations, respectively. The TLBO, Jaya and QO-Jaya algorithms obtained a 21.02% improvement in $\Delta OOR'$ as compared to the values obtained by Das et al. (2011) using desirability approach.

8.2.1.2 Case Study 2

This problem described in Sect. 8.1.1.2 was solved by Das et al. (2012) using desirability function approach to determine the optimum combination of process parameters for R-MRAFF process. Now the same problem is solved using Jaya and QO-Jaya algorithms to see whether any improvement in the results can be achieved. For this purpose, a population size of 10 and maximum number of iterations equal to 100 (i.e. maximum number of function evaluations = 10 × 100 = 1000) is considered for Jaya and QO-Jaya algorithms.

The optimum value of % ΔR_{aSS} and % ΔR_{aBR} obtained by Jaya algorithm along with the optimum combination of process input parameters of R-MAFF process is also reported in Tables 8.2 and 8.3. The Jaya and QO-Jaya algorithms achieved a better value of % ΔR_{aSS} and % ΔR_{aBR} as compared to desirability function approach. The Jaya algorithm required are 310 and 210 function evaluations to converge at the best values of % ΔR_{aSS} and % ΔR_{aBR}. The QO-Jaya algorithm required 290 and 200 function evaluations to converge at the best values of % ΔR_{aSS} and % ΔR_{aBR}. Figures 8.3 and 8.4 show the convergence graphs for Jaya and QO-Jaya algorithms for the best values of % ΔR_{aSS} and % ΔR_{aBR}, respectively.

Table 8.2 The optimal combination of process parameters for % ΔR_{aSS} obtained using Jaya and QO-Jaya algorithms

S. No.	Algorithm	*P* (bar)	*N*	*S* (RPM)	*R*	% ΔR_{aSS}	Computational time (s)
1	desirability approach (Das et al. 2012)	38.92	660	6	2.3	45.27 (42.73*)	–
2	Jaya	38.40	673.0	66.637	2.257	**42.822**	**0.415**
3	QO-Jaya	38.40	673.0	66.637	2.257	**42.822**	0.550

*Corrected value

Table 8.3 The optimal combination of process parameters for % ΔR_{aBR} obtained using Jaya algorithms

S. No.	Algorithm	*P* (bar)	*N*	*S* (RPM)	*R*	% ΔR_{aBR}	Computational time (s)
1	desirability approach (Das et al. 2012)	38.03	690	73	2.22	70.95 (76.319*)	–
2	Jaya	38.80	667.5	71.856	2.13	**76.6582**	**0.280**
3	QO-Jaya	38.80	667.5	71.856	2.13	**76.6582**	0.921

*Corrected value

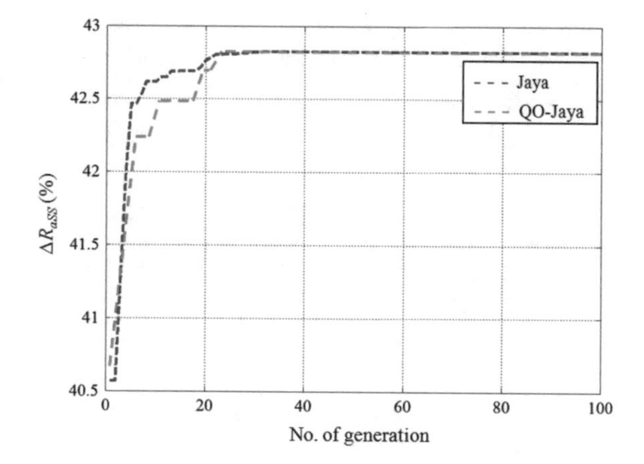

Fig. 8.3 The convergence graphs for Jaya and QO-Jaya algorithms for % ΔR_{aSS} in R-MRAFF process

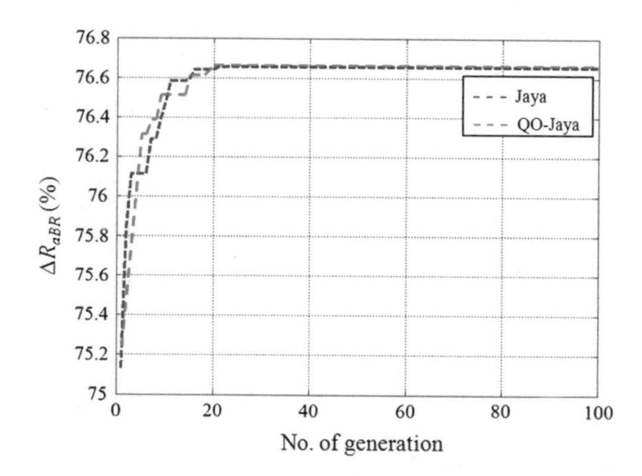

Fig. 8.4 The convergence graphs for Jaya and QO-Jaya algorithms for % ΔR_{aBR} in R-MRAFF process

8.2.2 Results of Magnetic Abrasive Finishing Process

This optimization problem described in Sect. 8.1.2 is now solved using the TLBO, Jaya and QO-Jaya algorithms. A population size of 10 is considered for TLBO algorithm with a maximum number of iterations equal to 100 (i.e. maximum number of function evaluations = 2 × 10 × 100 = 2000). A population size of 10 is considered for Jaya and QO-Jaya algorithms with maximum number of iterations equal to 200 (i.e. maximum number of function evaluations = 10 × 200 = 2000). The results obtained by TLBO, Jaya (Rao and Rai 2017) and QO-Jaya algorithms are reported in Table 8.4 along with the optimum combination of process input parameters suggested by Mulik and Pandey (2011) based on experimentation.

The TLBO, Jaya and QO-Jaya algorithms converged at the same optimum value of percentage improvement in surface roughness. The TLBO, Jaya and QO-Jaya algorithms achieved a better value of percentage improvement in surface roughness

Table 8.4 The optimal combination of process input parameters for maximization of ΔR_a obtained using the TLBO, Jaya and QO-Jaya algorithms

S. No.	X_1 (V)	X_2	$\ln(X_3)$	X_4	ΔR_a (%)	Computational time (s)	Methodology
1	70	800	5.6348	15	20.2944	–	Based on experimentation (Mulik and Pandey 2011)
2	90	400	5.193	35	**69.0348**	0.078737	TLBO
3	90	400	5.193	35	**69.0348**	0.058331	Jaya
4	90	400	5.193	35	**69.0348**	0.069754	QO-Jaya

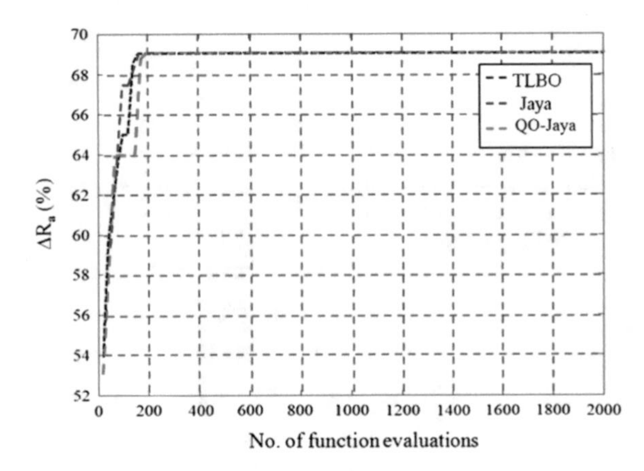

Fig. 8.5 The convergence graph for the TLBO, Jaya and QO-Jaya algorithms for ΔR_a in MAF process

as compared to that found experimentally by Mulik and Pandey (2011). Figure 8.5 shows the convergence graph of the TLBO, Jaya and QO-Jaya algorithms for ΔR_a (%). It is observed that the number of function evaluations required by the TLBO, Jaya and QO-Jaya algorithms to obtain the maximum value of ΔR_a (%) are 260, 245 and 245 respectively.

It is observed from Table 8.4 that the TLBO, Jaya and QO-Jaya algorithms have achieved a better value of ΔR_a (i.e. 69.0348) as compared to the ΔR_a value obtained by Mulik and Pandey (2011) which is 20.2944.

8.2.3 Results of Magnetorheological Fluid Based Finishing Process

This problem described in Sect. 8.1.3 is now solved using a posteriori approach in order to optimize F_n, F_t and F_a, simultaneously, and obtain a set of Pareto-optimal solutions. The MOQO-Jaya, MO-Jaya and NSTLBO algorithms are applied to solve the multi-objective optimization problem and the results are compared.

A population size of 50 and maximum number of iterations equal to 40 (i.e. maximum number of function evaluations = 50 × 40 = 2000) are considered for MOQO-Jaya and MO-Jaya algorithms. A population size of 50 and maximum number iterations equal to 20 (i.e. maximum number of function evaluations = 2 × 50 × 20 = 2000) is considered for NSTLBO algorithm.

Now the performance of MOQO-Jaya, MO-Jaya and NSTLBO algorithms are compared on the basis of coverage and spacing measure. For this purpose the MOQO-Jaya, MO-Jaya and NSTLBO algorithms are executed 30 times independently and the best, mean and standard deviation of the values of coverage and spacing obtained over 30 independent runs are reported in Table 8.5.

In Table 8.5 the value *Cov(A, B)* = 0.08 implies that, 8% of the solutions obtained by MO-Jaya algorithm are dominated by the solutions obtained by MOQO-Jaya algorithm. On the other hand, the value *Cov(B, A)* = 0.08 implies that, 8% of the solutions obtained by MOQO-Jaya algorithm are dominated by the solutions obtained using MO-Jaya algorithm. *Cov(A, C)* = 0.28 implies that, the solutions obtained using MOQO-Jaya algorithm dominate 28% of the solutions obtained using NSTLBO. *Cov(C, A)* = 0.06 implies that, only 6% of the solutions obtained using MOQO-Jaya algorithm are dominated by the solutions obtained by MOQO-Jaya algorithm. It is observed that the MOQO-Jaya obtained a better value of spacing as compared to MO-Jaya and NSTLBO algorithms.

Now the performance of MOQO-Jaya, MO-Jaya and NSTLBO algorithms are compared on the basis of hypervolume. The values of hypervolume obtained by the respective algorithms are reported in Table 8.6.

It is observed that the MOQO-Jaya and MO-Jaya algorithms obtained a higher hypervolume as compared to NSTLBO algorithm within less number of function evaluations. The MOQO-Jaya algorithm achieved the same value of hypervolume as that of MO-Jaya algorithm within less number of function evaluations as compared to MO-Jaya algorithm. The computational time required by MOQO-Jaya, MO-Jaya and NSTLBO algorithms is 10.56, 7.2 and 13.98 s, respectively, to complete one simulation run.

Table 8.5 The values of coverage and spacing for the non-dominated solutions obtained by MOQO-Jaya, MO-Jaya and NSTLBO algorithms for MRFF process

	Best	Mean	SD
Cov(A, B)	0.08	0.077	0.0028
Cov(B, A)	0.08	0.0773	0.0026
Cov(A, C)	0.28	0.2684	0.0082
Cov(C, A)	0.06	0.0564	0.0022
S(A)	0.0406	0.0429	0.00082
S(B)	0.0461	0.0482	0.0012
S(C)	0.0566	0.0595	0.0017

A, *B* and *C* are the non-dominated set of solutions obtained using MOQO-Jaya, MO-Jaya and NSTLBO algorithms, respectively; SD is the standard deviation

Table 8.6 The values of hypervolume obtained by MOQO-Jaya, MO-Jaya and NSTLBO algorithms for MRFF process

Algorithm	Hypervolume	No. of function evaluations required	Computational time (s)
MOQO-Jaya	**0.0407**	**4450**	10.56
MO-Jaya	**0.0407**	4600	**7.2**
NSTLBO	0.0248	5000	13.98

Values in bold indicate better performance of the algorithm

The Pareto-optimal set of solutions obtained by MOQO-Jaya algorithm is shown in Table 8.7 for the purpose of demonstration. The Pareto-fronts obtained by MOQO-Jaya, MO-Jaya and NSTLBO algorithms are shown in Fig. 8.6.

Table 8.7 The Pareto-optimal set of solutions obtained by MOQO-Jaya algorithm for MRFF process

S. No.	θ (°)	S (RPM)	F	F_n (N)	F_t (N)	F_a (N)
1	5	1011.2484	3.7188	6.3609	3.662	1.6794
2	5	1011.59	3.7151	6.3613	3.6622	1.6794
3	5	1012.0802	3.7228	6.3615	3.6627	1.6794
4	5	1012.3268	3.7152	6.362	3.6628	1.6794
5	5	1013.9747	3.7488	6.3624	3.6643	1.6794
6	5	1013.1812	3.7152	6.3627	3.6634	1.6794
7	5	1047.8376	3.8495	6.3631	3.677	1.6711
8	5	1014.6201	3.7394	6.3632	3.6647	1.6794
9	5	1049.2008	3.8269	6.3634	3.677	1.6707
10	5	1014.0137	3.7095	6.3635	3.6639	1.6794
11	5	1015.9952	3.7483	6.364	3.6657	1.6793
12	5	1015.5636	3.727	6.3643	3.6652	1.6793
13	5	1015.7605	3.731	6.3643	3.6654	1.6793
14	5	1048.0612	3.8147	6.3646	3.677	1.6712
15	5	1045.6634	3.8398	6.3646	3.6769	1.6721
16	5	1017.4313	3.759	6.3647	3.6667	1.6792
17	5	1017.1226	3.7491	6.3648	3.6664	1.6792
18	5	1017.5541	3.7501	6.365	3.6667	1.6792
19	5	1043.1844	3.847	6.3652	3.6767	1.673
20	5	1046.744	3.8074	6.3656	3.6769	1.6718
21	5	1044.412	3.8225	6.366	3.6768	1.6727
22	5	1019.9951	3.7663	6.366	3.6683	1.6789
23	5	1045.1149	3.812	6.3662	3.6768	1.6725
24	5	1020.9165	3.7717	6.3663	3.6689	1.6788

(continued)

Table 8.7 (continued)

S. No.	θ (°)	S (RPM)	F	F_n (N)	F_t (N)	F_a (N)
25	5	1019.5586	3.7416	6.3666	3.6678	1.679
26	5	1019.1767	3.7219	6.3669	3.6674	1.6791
27	5	1018.1628	3.6831	6.367	3.6661	1.6791
28	5	1037.447	3.8325	6.3673	3.6757	1.675
29	5	1024.1356	3.7829	6.3674	3.6707	1.6784
30	5	1042.3921	3.8015	6.3676	3.6765	1.6736
31	5	1028.4713	3.7926	6.3685	3.6727	1.6776
32	5	1034.6946	3.8053	6.3687	3.6749	1.676
33	5	1025.0412	3.7562	6.3688	3.6709	1.6783
34	5	1039.8154	3.7796	6.3691	3.676	1.6746
35	5	1025.8243	3.7526	6.3692	3.6712	1.6782
36	5	1042.1433	3.7608	6.3692	3.6761	1.6738
37	5	1040.7126	3.7675	6.3693	3.676	1.6743
38	5	1032.2731	3.7827	6.3695	3.674	1.6767
39	5	1033.7131	3.7799	6.3697	3.6745	1.6764
40	5	1039.0695	3.7592	6.37	3.6756	1.6748
41	5	1027.7906	3.7429	6.37	3.672	1.6778
42	5	1038.7944	3.7433	6.3705	3.6754	1.6749
43	5	1030.3408	3.7392	6.3707	3.6729	1.6773
44	5	1039.5436	3.7223	6.3709	3.6752	1.6746
45	5	1036.8261	3.7306	6.3711	3.6748	1.6755
46	5	1026.3003	3.615	6.3715	3.6685	1.6774
47	5	1034.7668	3.7088	6.3717	3.6739	1.6761
48	5	1028.0661	3.6122	6.3719	3.6692	1.6771
49	5	1029.7373	3.6064	6.3723	3.6696	1.6766
50	5	1035.6173	3.6296	6.3726	3.6722	1.6753

8.2.4 Abrasive Flow Machining (AFM) Process

This problem described in Sect. 8.1.4 was solved by Jain and Jain (2000) using GA considering a population size equal to 50, maximum number of generations equal to 200 (i.e. maximum number of function evaluations equal to 10,000), The other algorithm specific parameters for GA considered by Jain and Jain (2000) are total string length equal to 40, crossover probability equal to 0.8, and mutation probability equal to 0.01. Now the same problem is solved using the Jaya and QO-Jaya algorithms in order to see whether in improvement in the results can be achieved. For the purpose of fair comparison of results, the maximum number of function evaluations considered by Jaya and QO-Jaya algorithms is maintained as 10,000. For this purpose a population size of 10 and number of iterations equal to 1000 (i.e. maximum number of function evaluations = 10 × 1000 = 10,000) are chosen for

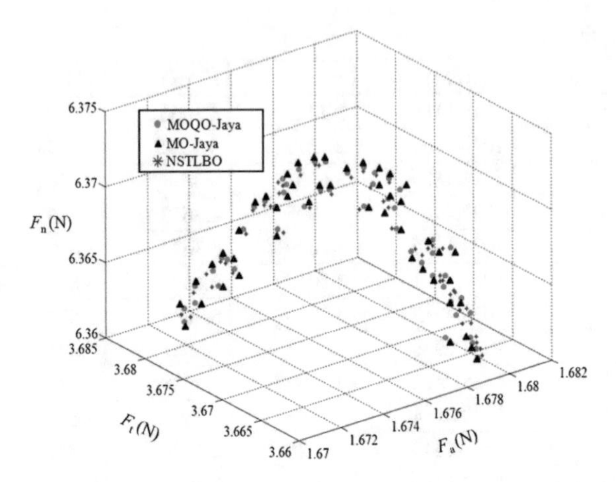

Fig. 8.6 The Pareto-fronts obtained by MOQO-Jaya, MO-Jaya and NSTLBO algorithms for MRFF process

the Jaya algorithm and QO-Jaya algorithms. The results obtained by Jaya and QO-Jaya algorithms for different values of maximum allowable surface roughness (i.e. Ra_{max} = 0.7, 0.6, 0.5 and 0.4) are reported in Table 8.8.

The comparison of results obtained using Jaya algorithm, QO-Jaya algorithm and GA is shown in Table 8.9. The values of *MRR* provided by Jaya algorithm 6.41, 6.27 and 5.68% are higher than the values of *MRR* provided by GA for Ra_{max} = 0.7, 0.6 and 0.5, respectively. For the purpose of demonstration the convergence graphs of Jaya and QO-Jaya algorithms for Ra_{max} = 0.4 is shown in Fig. 8.7.

It can be seen from the results of application of the advanced optimization algorithms on the nano-finishing processes that the Jaya algorithm and its variants have shown better performance and convergence compared to the other approaches suggested by the previous researchers.

Table 8.8 The optimal combination of process input parameters for AFM process obtained by Jaya and QO-Jaya algorithms

S. No.	Ra_{max} (μm)	*v*	*c*	*d*	*n*	*Ra* (μm)	*MRR* (mg/min)	Computational time (s)		No. of function evaluations required	
								Jaya	QO-Jaya	Jaya	QO-Jaya
1	0.7	85	45	100	20	0.53	0.738	0.160	0.56	20	20
2	0.6	85	45	100	20	0.53	0.738	0.158	0.68	60	60
3	0.5	85	45	100	27.37	0.5	0.6954	0.241	0.78	190	170
4	0.4	85	45	100	73.54	0.4	0.5768	0.206	0.59	230	200

Table 8.9 The results obtained using Jaya and QO-Jaya algorithms and GA for AFM process

S. No.	Ra_{max}	GA (Jain and Jain 2000)		Jaya and QO-Jaya algorithm		% increase in *MRR*
		Ra (μm)	*MRR* (mg/min)	*Ra* (μm)	*MRR* (mg/min)	
1	0.7	0.5433	0.6935	0.5367	**0.738**	6.41
2	0.6	0.5113	0.6944	0.5367	**0.738**	6.27
3	0.5	0.4812	0.6580	0.5	**0.6954**	5.68
4	0.4	0.4171*	0.5803	0.4	**0.5768**	Constraint is violated

*Constraint violated by GA

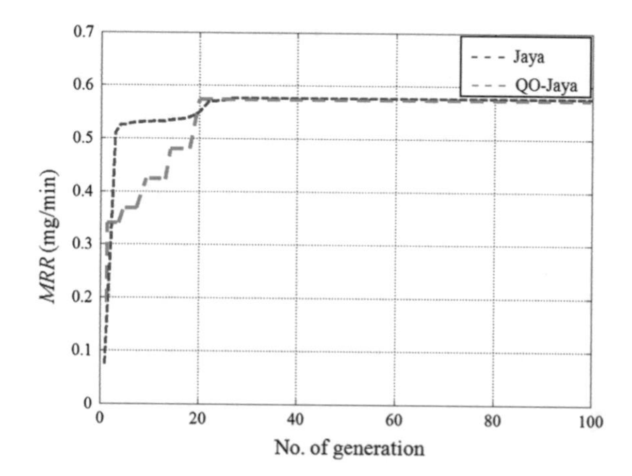

Fig. 8.7 The convergence graphs for Jaya and QO-Jaya algorithms for AFM process

References

Das, M., Jain, V. K., & Ghoshdastidar, P. S. (2010). Nano-finishing of stainless-steel tubes using rotational magnetorheological abrasive flow finishing process. *Machining Science and Technology, 14,* 365–389.

Das, M., Jain, V. K., & Ghoshdastidar, P. S. (2011). The out-of-roundness of the internal surfaces of stainless steel tubes finished by the rotational–magnetorheological abrasive flow finishing process. *Materials and Manufacturing Processes, 26,* 1073–1084.

Das, M., Jain, V. K., & Ghoshdastidar, P. S. (2012). Nano-finishing of flat workpieces using rotational–magnetorheological abrasive flow finishing (R-MRAFF) process. *International Journal of Advanced Manufacturing Technology, 62,* 405–420.

Jain, R. K., & Jain, V. K. (2000). Optimum selection of machining conditions in abrasive flow machining using neural network. *Journal of Materials Processing Technology, 108,* 62–67.

Mulik, R. S., & Pandey, P. M. (2011). Magnetic abrasive finishing of hardened AISI 52100 steel. *International Journal of Advanced Manufacturing Technology, 55,* 501–515.

Rao, R. V., & Rai, D. P. (2016). Optimization of advanced finishing processes using teaching-learning-based optimization algorithm. In V. K. Jain (Ed.), *Nanofinishing science and technology: Basic and advanced finishing processes*. Florida, USA: CRC Press.

Rao, R. V., & Rai, D. P. (2017). A new algorithm for parameter optimization of nano-finishing processes. *Scientia Iranica: Transactions on Industrial Engineering, 24*(2), 868–875.

Sidpara, A., & Jain, V. K. (2012). Theoretical analysis of forces in magnetorheological fluid based finishing process. *International Journal of Mechanical Sciences, 56,* 50–59.

Sidpara, A., & Jain, V. K. (2013). Analysis of forces on the freeform surface in magnetorheological fluid based finishing process. *International Journal of Machine Tools and Manufacture, 69,* 1–10.

Chapter 9
Single- and Multi-objective Optimization of Casting Processes Using Jaya Algorithm and Its Variants

Abstract In the case of casting processes, the effectiveness of Jaya and QO-Jaya algorithms is tested on optimization problems of squeeze casting process, continuous casting process, pressure die casting process and green sand casting process. The results of Jaya and QO-Jaya algorithms are compared with the results of GA, PSO, SA, TLBO algorithms and Taguchi method used by the previous researchers on the basis of objective function value, convergence speed and computational time. The results of Jaya and QO-Jaya algorithm are found better.

9.1 Optimization of Casting Processes

Manufacturing industries widely use casting processes for production of components with complex geometry that are difficult to manufacture by machining process. Casting processes such as sand casting, die casting, continuous casting, squeeze casting, investment casting, etc. are being used on a large scale. A number of input parameters govern the performance of these casting processes and hence selection of input parameters, meticulously, is important for success of these processes.

In order to achieve the ideal values of output parameters the researchers had used various optimization techniques for selection of input process parameters of casting processes. Kulkarni and Babu (2005) used simulated annealing (SA) algorithm for optimizing quality characteristics in continuous casting process. Guharaja et al. (2006) used Taguchi method for optimization of quality characteristics in green sand casting process. Vijian and Arunachalam (2006) used Taguchi method for optimization of squeeze casting process parameters. Chiang et al. (2008) casting density, warpage and flowmark in die casting process using grey-fuzzy based algorithm. Zheng et al. (2009) used artificial neural networks (ANN) to optimization of high pressure die casting process.

Kumar et al. (2011) used Taguchi method for optimization of quality characteristics in green sand casting process. Park and Yang (2011) used linear programming for maximizing average efficiency of pressure die casting process.

R. Venkata Rao, *Jaya: An Advanced Optimization Algorithm and its Engineering Applications*, https://doi.org/10.1007/978-3-319-78922-4_9

Senthil and Amirthagadeswaran (2012) used Taguchi method for optimization of squeeze casting parameters. Zhang and Wang (2013) developed a GA based intelligent system for low pressure die casting process parameters optimization. Kumaravadivel and Natarajan (2013) used response surface methodology (RSM) and Taguchi technique to minimize the percentage of defects in sand casting process.

Kittur et al. (2014) used RSM for optimization of pressure die casting process. Patel et al. (2016) used RSM to map the relationship between input and output parameters of squeeze casting process. Senthil and Amirthagadeswaran (2014) used genetic algorithm (GA) to improve the yield strength of components in squeeze casting process. Rao et al. (2014) applied teaching-learning-based optimization (TLBO) algorithm for process parameters optimization of squeeze casting process, continuous casting process and pressure die casting process. Surekha et al. (2012) used GA and PSO algorithm for multi-objective optimization of green sand mould system. Zhang and Hu (2016) used RSM and artificial fish swarm (AFS) algorithm for optimization of process parameters of vacuum casting process.

Patel et al. (2016) developed an intelligent system to predict the performance of squeeze casting process using GA and neural networks. Wang et al. (2016) used PSO algorithm for optimization of secondary cooling process in slab continuous casting. Li et al. (2017) used numerical simulation for minimization of shrinkage cavity and shrinkage defects in squeeze casting process. Rao and Rai (2017) used Jaya algorithm for the process parameters optimization of selected casting processes.

9.2 Process Parameters Optimization Models and Techniques for Casting Processes

9.2.1 Process Parameters Optimization Model for Squeeze Casting Process

The optimization problem formulated in this work is based on the empirical models developed by Senthil and Amirthagadeswaran (2014) for squeeze casting process. The objectives considered are: maximization of average hardness 'H' (BHN) and maximization of tensile strength 'TS' (MPa). The process input parameters considered are: squeeze pressure 'P' (MPa), melting temperature 'T' (C), die preheating temperature 'T_p' (C) and compression holding time 't_c' (s).

Objective functions

The objective functions are expressed by Eqs. (9.1) and (9.2)

$$\begin{aligned} \textit{Maximize}\ H = & -3.82542 + 0.8787 \times P + 0.46587 \times T_p + 0.30411 \times t_c \\ & - 0.00393 \times P^2 - 0.00116 \times T_p^2 + 0.00097 \times t_c^2 \\ & + 0.00051 \times P \times T_p - 0.00333 \times P \times t_c - 0.00018 \times T_p \times t_c \end{aligned} \quad (9.1)$$

$$\begin{aligned} \textit{Maximize}\ TS = & -11.2606 + 2.5778 \times P + 1.3316 \times T_p + 0.7552 \times t_c - 0.0116 \times P^2 \\ & - 0.0034 \times T_p^2 + 0.0031 \times t_c^2 + 0.0015 \times P \times T_p - 0.0097 \times P \times t_c \\ & - 0.001 \times T_p \times t_c \end{aligned} \quad (9.2)$$

Process input parameter bounds

The bounds on the process input parameters are expressed as follows

$$50 \leq P \leq 125 \quad (9.3)$$

$$675 \leq T \leq 750 \quad (9.4)$$

$$150 \leq T_p \leq 300 \quad (9.5)$$

$$15 \leq t_c \leq 60 \quad (9.10)$$

This problem was solved by Senthil and Amirthagadeswaran (2014) using Taguchi's method. Rao et al. (2014) solved the same problem using TLBO algorithm considering a total number of function evaluations equal to 400. The previous researchers had optimized the average hardness and average tensile strength individually. Therefore, for fair comparison of results the Jaya algorithm and QO-Jaya algorithm are applied to optimize average hardness and average tensile strength individual to see whether any improvement in result is achieved. The maximum function evaluations for Jaya algorithm and QO-Jaya algorithm is considered as 400 for fair comparison of results. Therefore, a population size equal to 25 and maximum number of iterations equal to 16 (i.e. maximum number of function evaluation = 25 × 16 = 400) is chosen for Jaya and QO-Jaya the algorithms. Table 9.1 shows the results of Taguchi method, TLBO algorithm, Jaya algorithm and QO-Jaya algorithm for average hardness.

Figure 9.1 shows the convergence graph of Jaya algorithm and QO-Jaya algorithm. It is observed that the Jaya algorithm and QO-Jaya algorithm could achieve a better hardness value as compared to Taguchi method and TLBO algorithm. The TLBO algorithm required 140 function evaluations to converge at a solution.

Table 9.1 The optimal values of average hardness obtained by TLBO, Jaya and QO-Jaya algorithms

Parameters	Taguchi method (Senthil and Amirthagadeswaran 2014)	TLBO (Rao et al. 2014)	Jaya	QO-Jaya
P	100	119	120.2278	120.2278
T	725	686	717.5017	717.5017
T_p	200	225	226.2574	226.2574
t_c	45	15	15.0000	15.0000
Average hardness	100.76	103.068	**103.719**	**103.719**

Values in bold indicate better performance of the algorithm

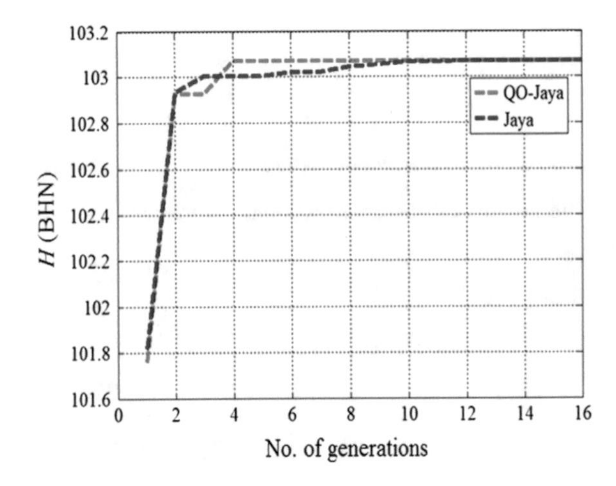

Fig. 9.1 Convergence graphs for Jaya and QO-Jaya algorithms for average hardness for squeeze casting process

On the other hand, the Jaya algorithm and QO-Jaya algorithm required on 100 and 40 function evaluations, respectively, to converge at the optimal solution.

Table 9.2 shows the best, mean and standard deviation (*SD*) of the values of average hardness obtained by TLBO algorithm, Jaya algorithm and QO-Jaya algorithm, obtained over 50 independent runs. The mean and *SD* obtained by Jaya algorithm and QO-Jaya algorithm is better than TLBO algorithm. Thus it can be said that the Jaya algorithm and QO-Jaya algorithm are more consistent as

Table 9.2 The best, mean and SD of average hardness values obtained by TLBO algorithm, Jaya algorithm and QO-Jaya algorithm over 50 independent runs

	TLBO (Rao et al. 2014)	Jaya	QO-Jaya
Best	103.068	**103.0719**	**103.719**
Mean	102.738	103.0695	**103.68**
SD	0.438	0.01487	**0.0085**

Values in bold indicate better performance

compared to TLBO and possess complete robustness. For a single independent run the Jaya algorithm and QO-Jaya algorithm required a computational time of 0.0922 and 0.134 s, respectively.

Table 9.3 shows results of Taguchi method, TLBO algorithm, Jaya algorithm and QO-Jaya algorithm for average tensile strength. The Jaya algorithm and QO-Jaya algorithm could achieve a better tensile strength as compared to Taguchi method and TLBO algorithm. The TLBO algorithm required 80 function evaluations to converge. On the other hand, the Jaya algorithm and QO-Jaya algorithm required 60 and 50 function evaluations, respectively to converge at the optimal solution. The convergence graph of Jaya algorithm and QO-Jaya algorithm for average tensile strength is shown in Fig. 9.2.

Table 9.4 shows the best, mean and *SD* of the values of average hardness obtained by TLBO algorithm, Jaya algorithm and QO-Jaya algorithm, over 50 independent runs. The mean objective function value and *SD* obtained over 50 independent run show that the Jaya algorithm and QO-Jaya algorithm are more

Table 9.3 Theoptimal values of average tensile strength obtained by TLBO algorithm, Jaya algorithm and QO-Jaya algorithm

Parameters	Taguchi method (Senthil and Amirthagadeswaran 2014)	TLBO (Rao et al. 2014)	Jaya	QO-Jaya
P	100	119	118.9875	118.9875
T	725	675	749.8668	749.8668
T_p	200	220	219.3900	219.3900
t_c	45	15	15.0000	15.0000
TS	278.45	290.21	**290.303**	**290.303**

Values in bold indicate better performance

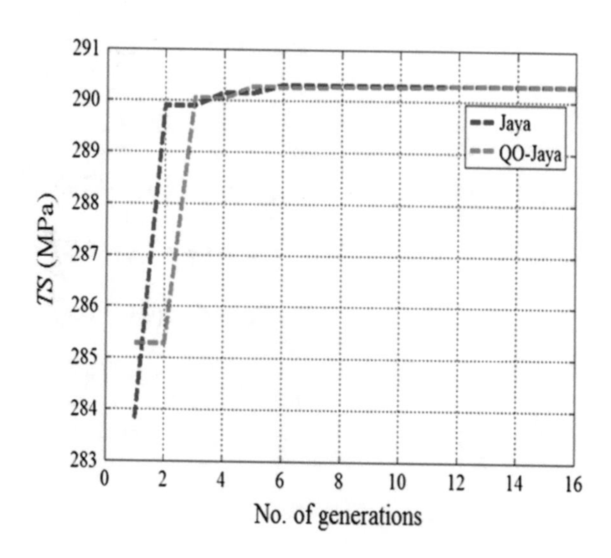

Fig. 9.2 Convergence graphs for Jaya and QO-Jaya algorithms for average tensile strength for squeeze casting process

Table 9.4 The best, mean and standard deviation of average hardness values obtained by TLBO algorithm, Jaya algorithm and QO-Jaya algorithm

	TLBO (Rao et al. 2014)	Jaya	QO-Jaya
Best	290.21	**290.303**	**290.303**
Mean	289.22	290.301	**290.302**
SD	1.59	0.008173	**0.001339**

Values in bold indicate better performance

consistent and robust as compared to TLBO algorithm. For a single simulation run the Jaya algorithm and QO-Jaya algorithm required a computational time of 0.0911 and 0.129 s, respectively.

9.2.2 *Process Parameters Optimization Model for Continuous Casting Process*

The optimization problem formulated in this work is based on the analysis given by Kulkarni and Babu (2005) for continuous casting process. The objective considered is minimization of the total loss function 'Z'. The process input parameters considered are: viscosity 'η' (poise), frequency 'f' (cpm), stroke 's' (mm), flux solidus temperature 'T_{sol}' (°C), drain rate 'R_p' (mm/min), flux density 'ρ' (kg/m^3) and mold length 'L_m' (mm).

Objective function

The objective function is expressed as follows:

$$Minimize\ Z = \sum_{i=1}^{9} L_i + L_{13} \tag{9.11}$$

$$\mathrm{L}_1 = (\mathrm{Q}_1 - 2)^2 \tag{9.12}$$

$$\mathrm{L}_2 = 0.0016(\mathrm{Q}_2 - 1050)^2 \tag{9.13}$$

$$\mathrm{L}_3 = 44.5(\mathrm{Q}_3 - 0.3)^2 \tag{9.14}$$

$$\mathrm{L}_4 = 44.5(\mathrm{Q}_4 - 0.3)^2 \tag{9.15}$$

$$\mathrm{L}_5 = 0.0016(\mathrm{Q}_5)^2 \tag{9.16}$$

$$\mathrm{L}_6 = 6.25 \times 10^{-6}(\mathrm{Q}_6)^2 \tag{9.17}$$

$$L_7 = 1.44(1/Q_7) \quad (9.18)$$

$$L_8 = 44.5(Q_8 - 0.3)^2 \quad (9.19)$$

$$L_9 = 44.5(Q_8 - 0.3)^2 \quad (9.20)$$

$$L_{13} = (1/Q_{13})^2 \quad (9.21)$$

Process input parameter bounds

The bounds on the process input parameters are expressed as follows:

$$1 \leq \eta \leq 2 \quad (9.22)$$

$$100 \leq f \leq 200 \quad (9.23)$$

$$6 \leq s \leq 12 \quad (9.24)$$

$$1000 \leq T_{sol} \leq 1100 \quad (9.25)$$

$$1 \leq R_p \leq 3 \quad (9.26)$$

$$2000 \leq \rho \leq 3000 \quad (9.27)$$

$$600 \leq L_m \leq 800 \quad (9.28)$$

Constraints

The constraints are expressed by Eqs. (9.29)–(9.38)

$$1 \leq Q_1 \leq 3 \quad (9.29)$$

$$1025 \leq Q_2 \leq 1075 \quad (9.30)$$

$$0.15 \leq Q_3 \leq 0.45 \quad (9.31)$$

$$0.15 \leq Q_4 \leq 0.45 \quad (9.32)$$

$$5 \leq Q_5 \leq 25 \quad (9.33)$$

$$Q_6 < 400 \quad (9.34)$$

$$Q_7 > 1.2 \quad (9.35)$$

$$0.15 \le Q_8 \le 0.45 \quad (9.36)$$

$$0.15 \le Q_9 \le 0.45 \quad (9.37)$$

$$Q_{13} \ge 1 \quad (9.38)$$

This problem was solved by Kulkarni and Babu (2005) using simulated annealing (SA) algorithm. Rao et al. (2014) solved the same problem using TLBO algorithm considering a maximum number of function evaluations equal to 10,000. Now, Jaya algorithm and QO-Jaya algorithm are applied to solve the same problem. For fair comparison of results, the maximum no. of function evaluations for Jaya algorithm and QO-Jaya algorithm is maintained as 10,000. Thus a population size equal to 25 and maximum number of iterations equal to 400 (i.e. maximum number of function evaluations = 25 × 400 = 10,000) is considered for Jaya algorithm and QO-Jaya algorithm. The results of SA, TLBO, Jaya and QO-Jaya algorithms are given in Table 9.5.

The Jaya algorithm and QO-Jaya algorithm could obtain a value of total loss function which is lower than SA and TLBO algorithms, while satisfying all the constraints. Table 9.6 shows the comparison of best value, mean value and *SD* of total loss function achieved by TLBO algorithm, Jaya algorithm and QO-Jaya algorithm over 50 independent runs. It is observed that, the best and average values obtained by Jaya algorithm and QO-Jaya algorithm is much lower than those obtained by TLBO algorithm with an acceptable value of standard deviation. This shows that the Jaya algorithm and QO-Jaya algorithm are more consistent and robust as compared to TLBO algorithm.

Table 9.5 The optimal values of total loss function obtained by TLBO algorithm, Jaya algorithm and QO-Jaya algorithm

Loss function	SA	TLBO (Rao et al. 2014)	Jaya result	QO-Jaya result
L1	0.24	0.23	0.1225	0.1225
L2	0.80	0.002	0.0145	0.0145
L3	0.09	0.01	0.000783	0.000783
L4	0.12	0.05	0.0250	0.0250
L5	0.19	0.14	0.1379	0.1379
L6	0.15	0.13	0.1367	0.1367
L7	0.30	0.30	0.3025	0.3025
L8	0.67	0.0004	6.0684E−6	6.0684E−6
L9	0.80	0.77	0.2665	0.2665
L13	0.87	0.91	1.2634	1.2634
Z	4.23	2.54	**2.27**	**2.27**

Values in bold indicate better performance of the algorithm

Table 9.6 The best, mean and standard deviation of total loss function values obtained by TLBO, Jaya and QO-Jaya algorithms

	TLBO (Rao et al. 2014)	Jaya	QO-Jaya
Best	2.54	**2.27**	**2.27**
Mean	2.57	**2.27**	**2.27**
SD	0.057	**0**	**0**

Values in bold indicate better performance of the algorithm

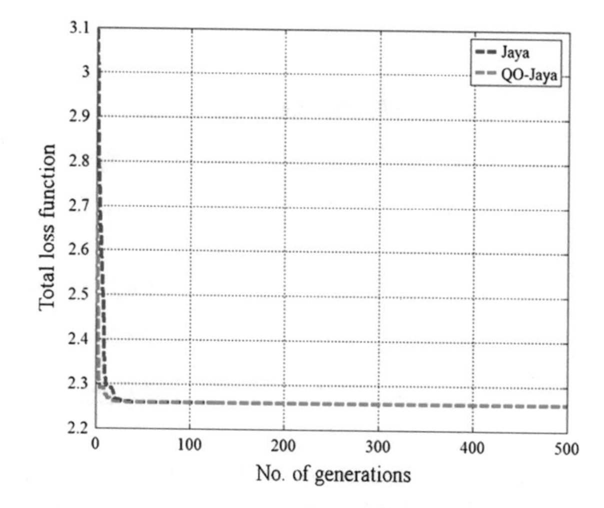

Fig. 9.3 Convergence graph for Jaya algorithm and QO-Jaya algorithm for total loss function

Figure 9.3 shows the convergence graphs for Jaya algorithm and QO-Jaya algorithm. It is seen that the Jaya algorithm and QO-Jaya algorithm could achieve convergence within 43 generations (1075 function evaluations) and 31 generations (775 function evaluations), respectively, which is much lower than the no. of function evaluations required by TLBO algorithm which is 6800 function evaluations. For a single independent run the Jaya algorithm and QO-Jaya algorithm required a computational time of 2.089 and 3.118 s, respectively.

9.2.3 Process Parameters Optimization Model for Pressure Die Casting Process

The optimization problem formulated in this work is based on the analysis given by Tsoukalas (2008) for pressure die casting process. The objective is to minimize the porosity level of an aluminium alloy component produced by pressure die casting process '*P*'. The process input parameters considered are: holding furnace temperature '*F*' (°C), die temperature '*D*' (°C), plunger velocity in the first stage '*S*' (m/s), plunger velocity in the second stage '*H*' (m/s) and multiplied pressure '*M*'.

Objective function

The objective function is expressed by Eq. (9.39).

$$\begin{aligned} Minimize\ P = 1.623 - 0.766 \times 10^{-3}F - 1.301 \times 10^{-3}D - 0.136S + 0.029H \\ - 1.636 \times 10^{-3}M \end{aligned} \tag{9.39}$$

Process input parameter bounds

$$610 \le F \le 730 \tag{9.40}$$

$$190 \le D \le 210 \tag{9.41}$$

$$0.02 \le S \le 0.34 \tag{9.42}$$

$$1.2 \le H \le 3.8 \tag{9.43}$$

$$120 \le M \le 280 \tag{9.44}$$

This problem was solved by Tsoukalas (2008) using GA considering maximum number of generations as 1000. However, the population size, cross-over rate, mutation rate required by GA was not reported by Tsoukalas (2008). The same problem was solved by Rao et al. (2014) using the TLBO algorithm using a population size of 10 and maximum number of iterations equal to 10 (i.e. maximum no. of function evaluations = 2 × 10 × 10 = 200). Now the same problem is solved using Jaya and QO-Jaya algorithms. For a fair comparison of results population size of and maximum number of iterations equal to 20 (i.e. maximum no. of function evaluations = 10 × 20 = 200, which is same as TLBO algorithm) is selected for Jaya algorithm and QO-Jaya algorithms. Table 9.7 shows the results of Jaya algorithm and QO-Jaya algorithm and the same are compared with the results of GA and TLBO algorithm.

Table 9.7 The optimal values of porosity obtained by GA, TLBO, Jaya and QO-Jaya algorithms

Parameters	GA (Tsoukalas 2008)	TLBO (Rao et al. 2014)	Jaya	QO-Jaya
F	729.4	730	730	730
D	269.9	270	270	270
S	0.336	0.34	0.34	0.34
H	1.2	1.2	1.2	1.2
M	275.7	280	280	280
Porosity	0.251	**0.243**	**0.243**	**0.243**

Values in bold indicate better performance

Figure 9.4 gives the convergence graphs for Jaya and QO-Jaya algorithms. It is observed that the Jaya algorithm and QO-Jaya algorithm could achieve the same solution as compared to TLBO algorithm. However, the number of function evaluations required by Jaya algorithm and QO-Jaya algorithm is 50 and 40, respectively, which is very less as compared to TLBO algorithm which is 80. For a single independent run the Jaya algorithm and QO-Jaya algorithm required a computational time of 0.0856 and 0.234 s, respectively.

Table 9.8 gives the best value, mean value and standard deviation of values of porosity obtained by TLBO algorithm, Jaya algorithm and QO-Jaya algorithm over 50 independent runs. The mean value of objective function obtained by Jaya algorithm and QO-Jaya algorithm is better than that of TLBO algorithm. Similarly, the standard deviation obtained by Jaya algorithm and QO-Jaya algorithm is zero. Therefore, the Jaya algorithm and QO-Jaya algorithm have shown more robustness and consistency as compared to TLBO algorithm.

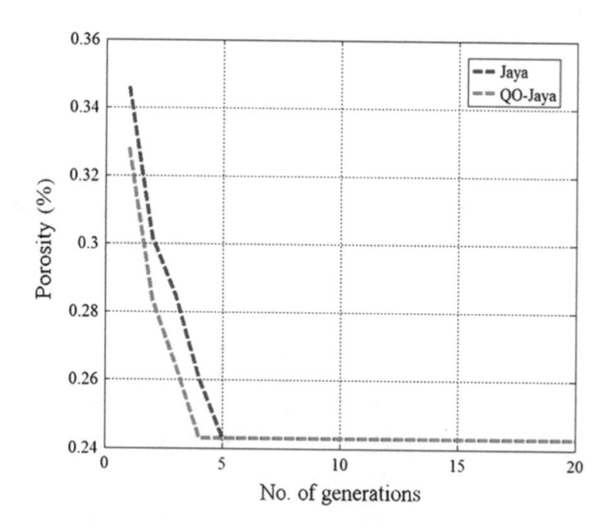

Fig. 9.4 Convergence graphs for Jaya and QO-Jaya algorithms for porosity

Table 9.8 The best, mean and standard deviation of porosity values obtained by TLBO, Jaya and QO-Jaya algorithms

	TLBO (Rao et al. 2014)	Jaya	QO-Jaya
Best	0.2430	**0.2430**	**0.2430**
Mean	0.2496	**0.2430**	**0.2430**
SD	0.02	**0**	**0**

Values in bold indicate better performance of the algorithm

9.2.4 Process Parameters Optimization Model for Green Sand Casting Process

The optimization problem formulated in this work is based on the empirical models developed by Surekha et al. (2012) for green sand casting process. The objectives considered are: green compression strength '*GCS*', permeability '*P*', hardness '*H*' and bulk density '*BD*' of green sand mould. The process input parameters considered are: grain fitness number '*G*', clay content '*C*' (%), water content '*W*' (%) and number of strokes '*S*'.

Objective functions

The objective functions are expressed by Eqs. (9.45)–(9.49)

$$\begin{aligned} \textit{Maximize } GCS = {} & 17.2527 - 1.7384 \times G - 2.7463 \times C + 32.3203 \times W \\ & + 6.575 \times S + 0.014 \times G^2 + 0.0945 \times C^2 - 7.7857 \times W^2 \\ & - 1.2079 \times S^2 + 0.0468 \times G \times C - 0.1215 \times G \times W \\ & - 0.0451 \times G \times S + 0.5516 \times G \times W + 0.6378 \times C \times S \\ & + 2.689 \times W \times S \end{aligned} \tag{9.45}$$

$$\begin{aligned} \textit{Maximize } P = {} & 1192.51 - 15.98 \times G - 35.66 \times C + 9.51 \times W - 105.66 \times S \\ & + 0.07 \times G^2 + 0.45 \times C^2 - 4.13 \times W^2 + 4.22 \times S^2 + 0.11 \times G \times C \\ & + 0.2 \times G \times W + 0.52 \times G \times S + 1.19 \times C \times W + 1.99 \times C \times S \end{aligned} \tag{9.46}$$

$$\begin{aligned} \textit{Maximize } H = {} & 38.2843 - 0.0494 \times G + 2.4746 \times C + 7.8434 \times W + 7.77 \times G \times S \\ & + 0.001 \times G^2 - 0.00389 \times C^2 - 1.6988 \times W^2 - 0.6556 \times S^2 \\ & - 0.0015 \times G \times C - 0.0151 \times G \times W - 0.0006 \times G \times S \\ & - 0.075 \times C \times W - 0.1938 \times C \times S + 0.65 \times W \times S \end{aligned} \tag{9.47}$$

$$\begin{aligned} \textit{Maximize } BD = {} & 1.02616 + 0.01316 \times G - 0.00052 \times C - 0.06845 \times W \\ & + 0.0083 \times S - 0.00008 \times G^2 + 0.0009 \times C^2 + 0.0239 \times W^2 \\ & - 0.00107 \times S^2 - 0.00004 \times G \times C - 0.00018 \times G \times W \\ & + 0.00029 \times G \times S - 0.00302 \times C \times W - 0.00019 \times C \times S \\ & - 0.00186 \times W \times S \end{aligned} \tag{9.48}$$

The combined objective function used for simultaneous optimization of the above mentioned four objectives is expressed as follows.

$$Maximize\ Z = w_1\left(\frac{GCS}{GCS^*}\right) + w_2\left(\frac{P}{P^*}\right) + w_3\left(\frac{H}{H^*}\right) + w_4\left(\frac{BD}{BD^*}\right) \quad (9.49)$$

where, w_1, w_2, w_3 and w_4 are weight assigned to the objectives between [0, 1], while, their summation should be equal to one. GCS^*, P^*, H^* and BD^* are respective maximum values of *GCS*, *P*, *H* and *BD* obtained by solving each objective separately.

Process input parameter bounds

The bounds on the process parameters are expressed as follows

$$52 \leq G \leq 94 \quad (9.50)$$

$$8 \leq C \leq 12 \quad (9.51)$$

$$1.5 \leq W \leq 3 \quad (9.52)$$

$$3 \leq S \leq 5 \quad (9.53)$$

This problem was solved by Surekha et al. (2012) using GA and PSO algorithms. Surekha et al. (2012) used a priori approach to optimize *GCS*, *P*, *H* and *BD*, simultaneously, by formulating a combined objective function using weighted sum approach. Surekha et al. (2012) considering five cases each with a distinct set of weights and combined objective function was optimized using GA and PSO algorithms. The maximum function evaluations equal to 13,000 and 1500 was considered for GA and for PSO algorithm, respectively, by Surekha et al. (2012).

Now, Jaya algorithm and QO-Jaya algorithm are applied to solve the same problem. For fair comparison of results, in the present work a priori approach is used to optimize *GCS*, *P*, *H* and *BD*, simultaneously. For this purpose the weighted sum approach is used to formulate the combined objective function and the same is optimized using Jaya and QO-Jaya algorithms. The maximum number of function evaluations for Jaya and QO-Jaya algorithms is maintained as 1500, which is same as the maximum function evaluations used by PSO algorithm and is very less than the maximum function evaluations used by GA. The results of GA, PSO, Jaya and QO-Jaya algorithms for all the cases and are reported in Table 9.9.

It is observed that in all five cases the Jaya algorithm and QO-Jaya algorithm could achieve a higher value of combined objective function (*Z*) as compared to GA and PSO algorithm. Table 9.10 shows the values of process input parameters, function evaluations required to achieve convergence and computational required by Jaya algorithm and QO-Jaya algorithm.

It is observed that GA required 11,050 function evaluations to achieve convergence (for case 1); PSO algorithm required 3000 function evaluations to achieve

Table 9.9 The optimal values of combined objective function obtained by GA, PSO, Jaya and QO-Jaya algorithms for green sand casting process

Case	Weights (w_1; w_2; w_3; w_4)	Combined objective function (Z)			
		GA (Surekha et al. 2012)	PSO (Surekha et al. 2012)	Jaya	QO-Jaya
1	0.25; 0.25; 0.25; 0.25	0.6654	0.6647	**0.8152**	**0.8152**
2	0.7; 0.1; 0.1; 0.1	0.7620	0.7609	**0.9224**	**0.9224**
3	0.1; 0.7; 0.1; 0.1	0.8522	0.8719	**0.9162**	**0.9162**
4	0.1; 0.1; 0.7; 0.1	0.7529	0.7630	**0.9224**	**0.9224**
5	0.1; 0.1; 0.1; 0.7	0.6072	0.6066	**0.9075**	**0.9075**

Values in bold indicate better performance of the algorithm

Table 9.10 The optimal values of process input parameters obtained by Jaya and QO-Jaya algorithms for green sand casting process

Case	Algorithm	Process parameters				Z	FEs	CT (s)
		G	C	W	S			
1	Jaya	52	12	2.8483	4.8879	0.8152	625	**0.231**
	QO-Jaya	52	12	2.8483	4.8879	0.8152	**400**	0.318
2	Jaya	94	12	2.6457	5	0.9224	750	**0.234**
	QO-Jaya	94	12	2.6457	5	0.9224	**525**	0.354
3	Jaya	52	8	2.4156	3	0.9162	600	**0.230**
	QO-Jaya	52	8	2.4156	3	0.9162	**450**	0.344
4	Jaya	94	12	2.6513	5	0.9224	725	**0.209**
	QO-Jaya	94	12	2.6513	5	0.9224	**500**	0.351
5	Jaya	94	12	2.6509	5	0.9075	625	**0.206**
	QO-Jaya	94	12	2.6509	5	0.9075	**550**	0.343

Values in bold indicate better performance; FEs: function evaluations; CT: computational time

convergence (for case 1), which in fact exceeds the maximum generation limit for PSO algorithm set of Surekha et al. (2012). On the other hand, Jaya algorithm and QO-Jaya algorithm required only 625 function evaluations and 400 function evaluations, respectively, to achieve convergence (for case 1). The number of generations required by GA and PSO algorithm for case 2, case 3, case 4 and case 5 were not reported by Surekha et al. (2012). In order to demonstrate the progress of Jaya algorithm and QO-Jaya algorithm the convergence for case 1 is shown in Fig. 9.5.

It is observed that in the case of squeeze casting process, the objective is to maximize the average hardness and tensile strength of the parts. The results of Jaya and QO-Jaya algorithms are compared with the results of desirability function approach and TLBO algorithm. The Jaya and QO-Jaya algorithm achieved a higher value of hardness and tensile strength as compared to desirability function approach

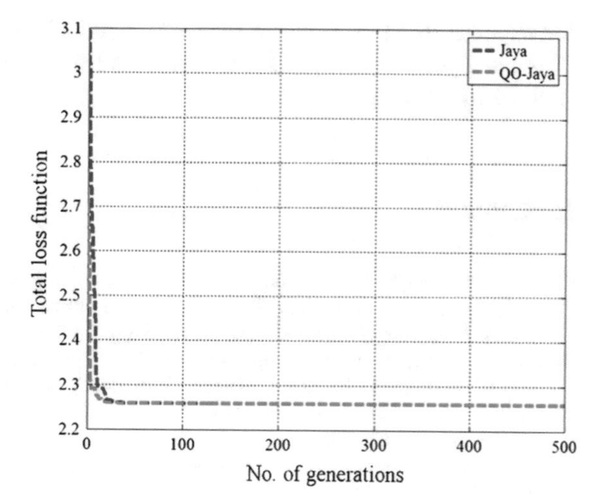

Fig. 9.5 Convergence graphs for Jaya and QO-Jaya algorithms for combined objective function for case 1 of green sand casting process

and TLBO algorithm within less number of function evaluations and less computational time. The Jaya and QO-Jaya algorithm have achieved a 28.5% improvement in average hardness and 4.22% improvement in tensile strength as compared to desirability function approach.

In the case of continuous casting process, the objective is to minimize the loss function. The results of Jaya and QO-Jaya algorithms are compared with the results of SA and TLBO algorithm. The Jaya and QO-Jaya algorithms obtained the 46.34 and 10.63% reduction in loss function as compared to SA and TLBO algorithms, respectively, with less number of function evaluations and computational time. The QO-Jaya algorithm showed a higher convergence speed as compared to the Jaya algorithm. However, the Jaya algorithm required less computational time to complete one run of simulation as compared to QO-Jaya algorithm.

In the case of pressure die casting process, the objective is to minimize the porosity. The results of Jaya and QO-Jaya algorithms are compared with the results of GA and TLBO algorithm which were used by the previous researchers. The Jaya and QO-Jaya algorithms achieved a 3.19% reduction in porosity as compared to GA. The number of function evaluations required by Jaya and QO-Jaya algorithms is remarkably less as compared to TLBO algorithm. The QO-Jaya algorithm showed a higher convergence speed as compared to Jaya algorithm.

In the case of green sand casting process, the objectives are to maximize the green compressive strength, permeability, hardness and bulk density. A combined objective function is formed by using the weighted sum approach. Five cases are formulated by considering five different sets of weights assigned to the objectives. The results of Jaya and QO-Jaya algorithms are compared with the results of GA and PSO algorithm. The Jaya and QO-Jaya algorithms achieved a 22.51, 21.05, 7.51, 22.51 and 49.46% improvement in combined objective function as compared to GA and a 22.64, 21.22, 5.08, 20.89 and 49.60% improvement in combined objective function as compared to PSO algorithm in all five cases with high convergence speed. In all five cases the number of function evaluations required by QO-Jaya algorithm is significantly lower than the number of function evaluations required by Jaya algorithm.

References

Chiang, K. T., Liu, N. M., & Chou, C. C. (2008). Machining parameters optimization on the die casting process of magnesium alloy using the grey-based fuzzy algorithm. *The International Journal of Advanced Manufacturing Technology, 38,* 229–237.

Guharaja, S., Haq, A. N., & Karuppannan, K. M. (2006). Optimization of green sand casting process parameters by using Taguchi's method. *International Journal of Advanced Manufacturing Technology, 30,* 1040–1048.

Kittur, Jayant K., Choudhari, M. N., & Parappagoudar, M. B. (2014). Modeling and multi-response optimization of pressure die casting process using response surface methodology. *International Journal of Advanced Manufacturing Technology, 77,* 211–224.

Kulkarni, M. S., & Babu, A. S. (2005). Managing quality in continuous casting process using product quality model and simulated annealing. *Journal of Materials Processing Technology, 166,* 294–306.

Kumar, S., Satsangi, P. S., & Prajapati, D. R. (2011). Optimization of green sand casting process parameters of a foundry by using Taguchi's method. *International Journal of Advanced Manufacturing Technology, 55,* 23–34.

Kumaravadivel, A., & Natarajan, U. (2013). Optimization of sand-casting process variables—A process window approach. *International Journal of Advanced Manufacturing Technology, 66,* 695–709.

Li, Y., Yang, H., & Xing, Z. (2017). Numerical simulation and process optimization of squeeze casting process of an automobile control arm. *International Journal of Advanced Manufacturing Technology, 88,* 941–947.

Park, Y. K., & Yang, J. M. (2011). Maximizing average efficiency of process time for pressure die casting in real foundries. *International Journal of Advanced Manufacturing Technology, 53,* 889–897.

Patel, M. G. C., Prasad, K., & Parappagoudar, M. B. (2016). An intelligent system for squeeze casting process—Soft computing based approach. *International Journal of Advanced Manufacturing Technology, 86,* 3051–3065.

Rao, R. V., Kalyankar, V. D., & Waghmare, G. G. (2014). Parameters optimization of selected casting processes using teaching-learning-based optimization algorithm. *Applied Mathematical Modelling, 38,* 5592–5608.

Rao, R. V., & Rai, D. P. (2017). Optimization of selected casting processes using Jaya algorithm. *Materials Today: Proceedings, 4,* 11056–11067.

Senthil, P., & Amirthagadeswaran, K. S. (2012). Optimization of squeeze casting parameters for non symmetrical AC2A aluminium alloy castings through Taguchi method. *Journal of Mechanical Science and Technology, 26,* 1141–1147.

Senthil, P., & Amirthagadeswaran, K. S. (2014). Experimental study and squeeze casting process optimization for high quality AC2A aluminium alloy castings. *Arabian Journal for science and Engineering, 39*(3), 2215–2225.

Surekha, B., Kaushik, L. K., Panduy, A. K., Vundavilli, P. R., & Parappagoudar, M. B. (2012). Multi-objective optimization of green sand mould system using evolutionary algorithms. *International Journal of Advanced Manufacturing Technology, 58,* 9–17.

Tsoukalas, V. D. (2008). Optimization of porosity formation in AlSi9Cu3 pressure die castings using genetic algorithm analysis. *Materials and Design, 29*(10), 2027–2033.

Vijian, P., & Arunachalam, V. P. (2006). Optimization of squeeze casting process parameters using Taguchi analysis. *International Journal of Advanced Manufacturing Technology, 33,* 1122–1127.

Wang, X., Wang, Z., Liu, Y., Du, F., Yao, M., & Zhang, X. (2016). A particle swarm approach for optimization of secondary cooling process in slab continuous casting. *International Journal of Heat and Mass Transfer, 93,* 250–256.

Zhang, H. G., & Hu, Q. X. (2016). Study of the filling mechanism and parameter optimization method for vacuum casting. *International Journal of Advanced Manufacturing Technology, 83,* 711–720.

Zhang, L., & Wang, R. (2013). An intelligent system for low-pressure die-cast process parameters optimization. *International Journal of Advanced Manufacturing Technology, 65,* 517–524.

Zheng, J., Wang, Q., Zhao, P., & Wu, C. (2009). Optimization of high-pressure die-casting process parameters using artificial neural network. *International Journal of Advanced Manufacturing Technology, 44,* 667–674.

Chapter 10
Applications of Jaya Algorithm and Its Modified Versions to Different Disciplines of Engineering and Sciences

Abstract After its introduction in 2016 by Rao (Int J Ind Eng Comput 7:19–34, 2016), the Jaya algorithm is quickly finding a large number of applications in different fields of engineering and science. This chapter presents an overview of the *applications* of the Jaya algorithm and its modifications published in reputed international journals since 2016 till March 2018.

10.1 Applications of Jaya Algorithm and Its Modified Versions

Rao (2016) proposed Jaya algorithm for solving the constrained and unconstrained optimization problems. The algorithm was based on the concept that the solution obtained for a given problem should move towards the best solution and should avoid the worst solution. This algorithm requires only the common control parameters and does not require any algorithm-specific control parameters. The performance of the proposed algorithm was investigated by implementing it on 24 constrained benchmark functions having different characteristics given in Congress on Evolutionary Computation (CEC 2006) and the performance was compared with that of other well-known optimization algorithms. The results had proved the better effectiveness of the proposed algorithm.

Zhang et al. (2016) proposed a tea-category identification (TCI) system, which can automatically determine tea category from images captured by a 3 charge-coupled device (CCD) digital camera. Three-hundred tea images were acquired as the dataset. Apart from the 64 traditional color histogram features that were extracted, we also introduced a relatively new feature as fractional Fourier entropy (FRFE) and extracted 25 FRFE features from each tea image. Furthermore, the kernel principal component analysis (KPCA) was harnessed to reduce $64 + 25 = 89$ features. The four reduced features were fed into a feed forward neural network (FNN). Its optimal weights were obtained by Jaya algorithm. The 10×10-fold stratified cross-validation (SCV) showed that the TCI system obtains an overall average sensitivity rate of 97.9%, which was higher than seven existing

R. Venkata Rao, *Jaya: An Advanced Optimization Algorithm and its Engineering Applications*, https://doi.org/10.1007/978-3-319-78922-4_10

approaches. The contributions covered the following points: (i) a novel feature—fractional Fourier entropy (FRFE)—was introduced and its effectiveness in extracting features for tea images was proved; (ii) a novel classifier—Jaya-FNN—was proposed by combining feed forward neural network with Jaya algorithm; (iii) The TCI system was proved better than seven state-of-the-art approaches.

Rao and Waghmare (2017) presented the performance of Jaya algorithm on a class of constrained design optimization problems. The performance of the Jaya algorithm was tested on 21 benchmark problems related to constrained design optimization. In addition to the 21 benchmark problems, the performance of the algorithm was investigated on four constrained mechanical design problems, i.e. robot gripper, multiple disc clutch brake, hydrostatic thrust bearing and rolling element bearing. The computational results revealed that the Jaya algorithm was superior to or competitive with other optimization algorithms for the problems considered.

Rao and Saroj (2016) proposed an elitist-Jaya algorithm for multi-objective design optimization of shell-and-tube and plate-fin heat exchangers. Elitist version of the Jaya algorithm was proposed to simultaneously optimize the total annual cost and effectiveness of the heat exchangers. Two heat exchanger design problems were considered for investigating the performance of Jaya algorithm. The same problems were earlier optimized by other researchers using genetic algorithm and modified teaching-learning-based optimization algorithm. The consequences of common controlling parameters e.g. number of iterations, population size and elite size, on the proposed algorithm's performance was tested with various combinations of the same. The results of computational experiments proved the superiority of the proposed algorithm over the reported approaches used for the design optimization problems of heat exchangers.

Rao et al. (2016a) used Jaya algorithm for the dimensional optimization of a micro-channel heat sink. Two case studies were considered and in both the case studies two objective functions related to the heat transfer and pressure drop, i.e. thermal resistance and pumping power, were formulated to examine the performance of the micro-channel heat sink. In the first case study, two non-dimensional design variables related to the micro-channel depth, width and fin width were chosen and their ranges were decided through preliminary calculations of three-dimensional Navier–Stokes and energy equations. In the second case study, three design variables related to the width of the micro-channel at the top and bottom, depth of the micro-channel and width of fin were considered. The objective functions in both the case studies were expressed in terms of the respective design variables and the two objectives were optimized simultaneously using Jaya algorithm. The results were compared with the results obtained by the TLBO algorithm and a hybrid multi-objective evolutionary algorithm (MOEA) and numerical analysis. The results obtained by the application of Jaya algorithm were found much better than those reported by the other approaches.

Rao et al. (2016b) considered the multi-objective optimization aspects of plasma arc machining (PAM), electro-discharge machining (EDM), and micro electro-discharge machining (μ-EDM) processes. Experiments were performed and actual experimental data was used to develop regression models for the considered machining processes. A posteriori version of Jaya algorithm (MO-Jaya algorithm) was proposed to solve the multi-objective optimization models in a single simulation run. The PAM, EDM and μ-EDM processes were optimized using MO-Jaya algorithm and a set of Pwereto-efficient solutions was obtained for each of the considered machining processes. This Pwereto optimal set of solutions provided flexibility to the process planner to choose the best setting of parameters depending on the application.

Rao et al. (2016c) studied the optimization problem of an important traditional machining process namely surface grinding. The comparison of results of optimization showed that the results of Jaya algorithm were better than the results reported by previous researchers using genetic algorithm, simulated annealing, artificial bee colony, harmony search, particle swarm optimization, ant colony optimization and TLBO algorithms.

Sinha and Ghosh (2016) considered identification and monitoring of electroencephalogram-based brain-computer interface (BCI) for motor imagery (MI) task and proposed an efficient adaptive neuro-fuzzy classifier (NFC). The brain–computer interface (BCI) identifies brain patterns to translate thoughts into action. The identification relies on the performance of the classifier. The Jaya optimization algorithm was integrated with adaptive neuro-fuzzy inference systems to enhance classification accuracy. The linguistic hedge (LH) was used for proper elicitation and pruning of the fuzzy rules and network was trained using scaled conjugate gradient (SCG) and speeding up SCG (SSCG) techniques. Jaya-based k-means was applied to divide the feature set into two mutually exclusive clusters and fire the fuzzy rule. The performance of the proposed classifier, Jaya-based NFC using SSCG as training algorithm and was powered by LH (JayaNFCSSCGLH), was compwered with four different NFCs for classifying two class MI-based tasks. It was observed a shortening of computation time per iteration by 57.78% in the case of SSCG as compwered with the SCG technique of training. LH-based feature selecting capability of the proposed classifier not only reduced computation time but also improved the accuracy by discarding irrelevant features. Lesser computation time with fast convergence and high accuracy among considered NFCs made it a suitable choice for the real-time application. Supremacy of Jaya-NFCSSCGLH among the considered classifiers was validated through Friedman test.

Azizipanah-Abarghooee et al. (2016a) proposed a new optimization toolset for the bi-objective multi-werea economic dispatch problem to determine the transmission power flow and power output of units while satisfying system demand and security constraints at each werea. The proposed architecture was built on an improved gradient-based Jaya algorithm to generate a feasible set of Pwereto-optimal solutions corresponding to the operation cost and emission calculated through a new robust bi-objective gradient-based method. The projected algorithm was proved to be capable of finding the robust global or near-global Pwereto solutions fast and accurate.

Azizipanah-Abarghooee et al. (2016b) proposed a probabilistic unit commitment problem with incentive-based demand response and high level of wind power.

The novel formulation provided an optimal allocation of up/down spinning reserve. A more efficient unit commitment algorithm based on operational cycles was developed. A multi-period elastic residual demand economic model based on the self- and cross-price elasticities and customers' benefit function was used. In the proposed scheme, the probability of residual demand falling within the up/down spinning reserve imposed by $n - 1$ security criterion was considered as a stochastic constraint. A chance-constrained method, with a new iterative economic dispatch correction, wind power curtailment, and commitment of cheaper units, was applied to guarantee that the probability of loss of load was lower than a pre-defined risk level. The developed architecture was built upon an improved Jaya algorithm to generate feasible, robust and optimal solutions corresponding to the operational cost. The proposed framework was applied to a small test system with 10 units and also to the IEEE 118-bus system to illustrate its advantages in efficient scheduling of generation in the power systems.

Mishra and Ray (2016) proposed Jaya algorithm to optimize the coefficients of proportional plus integral controller and filter parameters of photovoltaic fed distributed static compensator (PV-DSTATCOM). From the obtained simulation and experimental results, the performance of JAYA optimized PV-DSTATCOM was found quite satisfactory and stimulating as compared to grenade explosion method and teaching-learning based optimized PV-DSTATCOM.

Warid et al. (2016) presented the application of Jaya algorithm to deal with different optimum power flow (OPF) problems. In this work, three goal functions were considered for the OPF solution: generation cost minimization, real power loss reduction, and voltage stability improvement. In addition, the effect of distributed generation (DG) was incorporated into the OPF problem using a modified formulation. For best allocation of DG unit(s), a sensitivity-based procedure was introduced. Simulations were carried out on the modified IEEE 30-bus and IEEE 118-bus networks to determine the effectiveness of the Jaya algorithm. The single objective optimization cases were performed both with and without DG. For all considered cases, results had shown that the Jaya algorithm could produce an optimum solution with rapid convergence. The optimal solution obtained by the Jaya algorithm was compwered with different stochastic algorithms, and demonstrably outperformed them in terms of solution optimality and solution feasibility, proving its effectiveness and potential.

Abhishek et al. (2017) highlighted the application potential of a multi-response optimization route by integrating nonlinear regression modelling, fuzzy inference system (FIS) in combination with the Jaya optimization algorithm, for the selection of optimal process parameter setting during the machining (turning) of carbon fibre-reinforced (epoxy) composites. Experiments were carried out in consideration with spindle speed, feed rate, and depth of cut as process control parameters, whereas material removal rate (MRR), roughness average (R_a), and net cutting force were treated as machining performance characteristics. Attempt was made to identify the best setting of process parameters for optimizing the output responses, simultaneously. The result of the Jaya algorithm was compared to that of TLBO (teaching–learning-based optimization) algorithm. In addition to this, the results obtained were also compared to that of two evolutionary optimization algorithms viz., GA (genetic algorithm) and ICA (imperialist competitive algorithm). Good agreement was observed amongst the obtained results.

Rao and Saroj (2017a) proposed a self-adaptive multi-population based Jaya (SAMP-Jaya) algorithm for solving the constrained and unconstrained numerical and engineering optimization problems. The search mechanism of the Jaya algorithm was upgraded by using the multi-population search scheme. The scheme used an adaptive scheme for dividing the population into sub-populations which controlled the exploration and exploitation rates of the search process based on the problem landscape. The robustness of the proposed SAMP-Jaya algorithm was tested on 15 CEC 2015 unconstrained benchmark problems in addition to 15 unconstrained and 10 constrained standard benchmark problems taken from the literature. The Friedman rank test was conducted in order to compare the performance of the algorithms. SAMP-Jaya had obtained first rank among six algorithms for 15 CEC 2015 unconstrained problems with the average scores of 1.4 and 1.9 for 10-dimension and 30-dimension problems respectively. Also, the proposed algorithm had obtained first rank for 15 unimodal and multimodal unconstrained benchmark problems with the average scores of 1.7667 and 2.2667 with 50,000 and 200,000 function evaluations respectively. The performance of the proposed algorithm was further compared with the other latest algorithms such as across neighborhood search (ANS) optimization algorithm, multi-population ensemble of mutation differential evolution (MEMDE), social learning particle swarm optimization algorithm (SL-PSO), competitive swarm optimizer (CSO) and it was found that the performance of the proposed algorithm was better in more than 65% cases. Furthermore, the proposed algorithm was used for solving a case study of the entropy generation minimization of a plate-fin heat exchanger (PFHE). It was found that the number of entropy generation units was reduced by 12.73, 3.5 and 9.6% using the proposed algorithm as compared to the designs given by genetic algorithm (GA), particle swarm optimization (PSO) and cuckoo search algorithm (CSA) respectively. Thus the computational experiments had proved the effectiveness of the proposed algorithm for solving the engineering optimization problems.

Rao and Saroj (2017b) explored the use of an elitist-Jaya algorithm for economic optimization of shell-and-tube heat exchanger (STHE) design. Three different optimization problems of STHE were considered in this work. The same problems were earlier attempted by other researchers using genetic algorithm (GA), simulated annealing (SA), non-dominated sorting algorithm (NSGA-II) and design approaches motivated by constructional theory. Elitist version of the algorithm was proposed to optimize the setup cost and operational cost of STHE simultaneously. Discrete variable optimization was also carried out in order to take the advantage of the standard design available for heat exchanger parts. The effect of common controlling parameters e.g. population size, number of iterations and elite size, was tested by considering different combinations of the same. Furthermore, two other case studies of STHE design were considered from the literature and the performance of the elitist-Jaya algorithm is compared with the recently published results.

The results of computational experiments proved the superiority of the proposed algorithm over the latest reported methods used for the optimization of the same problems.

Rao and Saroj (2017c) explored the use of the Jaya algorithm for economic optimization of shell-and-tube heat exchanger (STHE) design. Consistency and maintenance because of fouling were also considered. Two different optimization problems of STHE were considered in this work. The same problems were earlier attempted by other researchers using genetic algorithm (GA), particle swarm optimization (PSO) and civilized swarm optimization (CSO). Eleven design parameters related with STHE configuration were used as decision variables. The results obtained by the computational studies were better and the superiority of the Jaya algorithm was proved over the latest reported methods used for the optimization of the same problems.

Rao et al. (2017a) used Jaya algorithm for parameter optimization of nano-finishing processes. The results showed the better performance of the Jaya algorithm over the other approaches attempted by the previous researchers such as genetic algorithm and desirability function approach for the same nano-finishing processes.

Rao et al. (2017b) explored the use of Jaya algorithm for the single- and multi-objective design optimization of plate-fin heat exchangers (PFHEs). Design of PFHEs involves a number of geometric and physical parameters with high complexity. The general design approaches were based on trial and error and become tedious and time consuming and do not guarantee the achievement of an optimal design. The Jaya algorithm was proposed for the design optimization of PFHEs by minimizing the total surface area of heat transfer, total annual cost, and total pressure drop of the system and maximizing the effectiveness. Seven design parameters were considered which were imposed by constraints on the design. Single- as well as multi-objective design optimization was carried out using the proposed algorithm. The results obtained by Jaya algorithm were compared with the results of latest reported algorithms. These comparisons revealed that the Jaya algorithm can be successfully applied for the design optimization of PFHEs.

Rao et al. (2016a) considered multi-objective optimization aspects of four modern machining processes namely, wire-electro discharge machining (WEDM) process, laser cutting process, electrochemical machining (ECM) process and focused ion beam (FIB) micro-milling process. In order to handle multiple objectives simultaneously a new posteriori multi-objective optimization algorithm named as multi-objective Jaya (MO-Jaya) algorithm was proposed which could provide multiple optimal solutions in a single simulation run. The regression models for the machining processes which were developed by previous researchers were used as fitness functions for MO-Jaya algorithm.

Rao et al. (2017d) considered a solar-powered Stirling engine for optimization with multiple criteria. The authors had explored the use of self-adaptive Jaya algorithm and the basic Jaya algorithm for maximizing output power and thermal efficiency and minimizing pressure losses of the entire Stirling system. The comparison of the proposed algorithm was made with those obtained by using the non-dominated sorting genetic-algorithm-II (NSGA-II), an optimizer inbuilt in MATLAB (function gamultiobj), Front-based Yin-Yang-Pair Optimization (FYYPO), Multi-Objective Grey Wolf Optimizers (MOGWOs), Teaching–Learning-Based Optimization (TLBO), tutorial training and self-learning inspired teaching-learning based optimization (TS-TLBO), the decision-making methods like linear programming technique for multi-dimensional analysis of preference (LINMAP), technique for order of preference by similarity to ideal solution (TOPSIS), and Bellman-Zadeh, and experimental results. The results achieved by using Jaya and self-adaptive Jaya algorithms were compared with the results achieved by using the NSGA-II, gamultiobj, FYYPO, MOGWO, TLBO, TS-TLBO, LINMAP, TOPSIS, Bellman-Zadeh, and experimental results. The Jaya algorithm and its self-adaptive population version were proved to be better compared to other optimization algorithms with respect to the computational effort and function evaluations.

In the case of WEDM process the optimization problem was an unconstrained, linear and parameter bounded. In the case of laser cutting process the optimization problem was a non-linear, unconstrained, quadratic and parameter bounded. In the ECM process the optimization problem was a non-linear, unconstrained, quadratic and parameter bounded. The second case study of ECM process the optimization problem was a non-linear, constrained, non-quadratic and parameter bounded. In the case of FIB micro-milling process, the optimization problem was a non-linear, unconstrained, quadratic and parameter bounded. In addition, the performance of MO-Jaya algorithm was also tested on a non-linear, non-quadratic unconstrained multi-objective benchmark function of CEC2009. In order to handle the constraints effectively a heuristic approach for handling constraints known as the constrained-dominance concept was used in MO-Jaya algorithm. In order to ensure that the newly generated solutions were within the parameter bounds a parameter-bounding strategy was used in MO-Jaya algorithm. The results of MO-Jaya algorithm were compwered with the results of GA, NSGA, NSGA-II, BBO, NSTLBO, PSO, sequential quadratic programming (SQP) and Monte Carlo simulations. The results had shown the better performance of the MO-Jaya algorithm.

Rao et al. (2017e) solved the process parameters optimization problems of abrasive waterjet machining process using Jaya algorithm and its posteriori version named as multi-objective Jaya (MO-Jaya) algorithm. The results of Jaya and MO-Jaya algorithms were compared with the results obtained by other well-known optimization algorithms such as simulated annealing, particle swam optimization, firefly algorithm, cuckoo search algorithm, blackhole algorithm and bio-geography based optimization. A hypervolume performance metric was used to compare the results of MO-Jaya algorithm with the results of non-dominated sorting genetic algorithm and non-dominated sorting teaching–learning-based optimization algorithm. The results of Jaya and MO-Jaya algorithms were found to be better as compared to the other optimization algorithms. In addition, a multi-objective decision making method named PROMETHEE method was applied in order to select a particular solution out-of the multiple Pareto-optimal solutions provided by MO-Jaya algorithm which best suits the requirements of the process planer.

Rao and More (2017a) explored the use of self-adaptive Jaya algorithm for optimal design of selected thermal devices viz; heat pipe, cooling tower, honeycomb heat sink and thermo-acoustic prime mover. Four different optimization case studies of the selected thermal devices were presented. The researchers had attempted the same design problems in the past using niched pareto genetic algorithm (NPGA), response surface method (RSM), leap-frog optimization program with constraints (LFOPC) algorithm, teaching-learning based optimization (TLBO) algorithm, grenade explosion method (GEM) and multi-objective genetic algorithm (MOGA). The results achieved by using self-adaptive Jaya algorithm were compared with those achieved by using the NPGA, RSM, LFOPC, TLBO, GEM and MOGA algorithms. The self-adaptive Jaya algorithm was proved superior as compared to the other optimization methods in terms of the results, computational effort and function evaluations.

Rao and More (2017b) explored the use of self-adaptive Jaya algorithm for optimal design of cooling tower from economic facets. In this work, six different examples were considered in the design optimization of mechanical draft cooling tower. Various researchers had attempted the same mathematical models by using different methods like Merkel method, Poppe method and artificial bee colony (ABC) algorithm. The results achieved by using the proposed self-adaptive Jaya algorithm were compared with the results achieved by using the Merkel method, Poppe method, ABC algorithm and basic Jaya algorithm. The proposed self-adaptive Jaya algorithm was proved better as compared to the other optimization methods with respect to achieving the optimal value of the objective function at less computational effort.

Rao and Rai (2017a) optimized the performance of submerged arc welding (SAW) process, electron beam welding (EBW) process, friction stir welding (FSW) process and gas tungsten arc welding (GTAW) process through parameter optimization using 'Quasi-oppositional-based Jaya algorithm' (QO-Jaya). The optimization case studies for each of the above-mentioned welding processes were considered and the results obtained by the QO-Jaya algorithm were compared with the results obtained by well-known optimization algorithms such as genetic algorithm (GA), simulated annealing (SA), teaching-learning-based optimization (TLBO) and basic Jaya algorithm.

Rao and Rai (2017b) optimized the submerged arc welding (SAW) process parameters using "Quasi-oppositional based Jaya algorithm" (QO-Jaya). Three optimization case studies were considered and the results obtained by Jaya algorithm and QO-Jaya algorithm were compared with the results obtained by well-known optimization algorithms such as Genetic algorithm (GA), Particle swarm optimization (PSO), Imperialist competitive algorithm (ICA) and TLBO algorithm.

Singh et al. (2017) explored the design and performance analysis of a proportional-integral-derivative (PID) controller employing Jaya algorithm for automatic generation control (AGC) of an interconnected power system. A filter with derivative term is used with PID controller to minimize the effect of noise in the input signal. The different performance measures considered were integral-time-multiplied-absolute-error (ITAE), the minimization of peak overshoots and the settling times of the deviations in the frequencies and tie-line power. Furthermore, these performance measures i.e. objectives were combined to form a single objective problem using analytic hierarchy process (AHP). Jaya algorithm was used for minimizing this single objective. The two-area interconnected linear power system model was considered for the simulation process. The simulation studies were carried out under five different cases with diverse sets of disturbances. The efficacy and superiority of the proposed Jaya algorithm based PID controller was shown by comparing simulation results with other algorithms like particle swarm optimization (PSO), differential evolution (DE), Nelder-Mead simplex (NMS), elephant herding optimization (EHO) and TLBO algorithm. Time-domain simulations and statistical analysis were presented to validate the effectiveness of the proposed controller.

Wang et al. (2017) proposed a novel computer-aided diagnostic system for detecting abnormal breasts in mammogram images. The region of interest was segmented first. Then the weighted-type fractional Fourier transform (WFRFT) was employed to obtain the unified time-frequency spectrum. The principal component analysis (PCA) was used to reduce the spectrum to only 18 principal components. A feed-forward neural network (FNN) was then utilized to generate the classifier. Finally, Jaya algorithm was employed to train the classifier. The proposed WFRFT + PCA + Jaya-FNN achieved better sensitivity, specificity and accuracy.

Kumar et al. (2017a, b) introduced a hybrid of 'Jaya' and 'differential evolution (DE)' (JayaDE) technique for maximum power point tracking (MPPT) in the highly fluctuating atmospheric conditions. This JayaDE algorithm was tested on MATLAB simulator and was verified on a developed hardware of the solar photovoltaic (PV) system, which consisted of a single peak and many multiple peaks in the voltage-power curve. Moreover, the tracking ability was compared with the recent state of the art methods. The satisfactory steady-state and dynamic performances of this new hybrid technique under variable irradiance and temperature levels showed the superiority over the state of the art control methods.

Mohebian et al. (2017) carried out simultaneous optimization of the size of components and the arrangement of connections for performance-based seismic design of low-rise steel plate shear walls (SPSWs). Design variables included the size of beams and columns, the thickness of the infill panels, the type of each beam-to-column connection and the type of each infill-to-boundary frame connection. The objective function was considered to be the sum of the material cost and rigid connection fabrication cost. For comparison purposes, the SPSW model was also optimized with regard to two fixed connection arrangements. To fulfill the optimization task a new hybrid optimization algorithm called colliding bodies optimization—Jaya was proposed. The performance of the hybrid optimization algorithm was assessed by two benchmark optimization problems. The results of the application of the hybrid algorithm to the benchmark problems indicate the efficiency, robustness, and the fast convergence of the proposed algorithm compared with other meta-heuristic algorithms. The achieved results for the SPSWs demonstrated that incorporating the optimal arrangement of beam-to-column and infill-to-boundary frame connections into the optimization procedure resulted in considerable reduction of the overall cost.

Kumar and Mishra (2017) presented a novel adaptive Jaya based functional link artificial neural network (Jaya-FLANN) filter for suppressing different noise present in ultrasound (US) images. Jaya was the optimization algorithm employed to assist in updating weights of FLANN. The target function for Jaya was the minimum error between noisy and contextual pixels of reference images. Compared to Wiener, Multi-Layer Perceptron (MLP), Cat Swarm Optimization based FLANN (CSOFLANN) and Particle Swarm Optimization based FLANN (PSO-FLANN), Jaya-FLANN filter was observed to be superior in terms of Peak Signal to Noise Ratio (PSNR), computational time.

Malik and Kumar (2017) designed a technique to optimize the phase noise compensation in the multiple inputs multiple output orthogonal frequency division multiplexing by using the neural network. The authors had presented a neural network and Jaya algorithm based technique for the estimation of phase noise. The neural network was used to remove the common phase error while inter-carrier interference was removed by the Jaya algorithm. The proposed technique was compared with three states of art techniques on different block size by using bit error rate.

Gao et al. (2017) studied a large-scale urban traffic light scheduling problem (LUTLSP). A centralized model was developed to describe the LUTLSP, where each outgoing flow rate was described as a nonlinear mixed logical switching function over the source link's density, the destination link's density and capacity, and the driver's potential psychological response to the past traffic light signals. The objective was to minimize the total network-wise delay time of all vehicles in a time window. Three metaheuristic optimization algorithms, named as Jaya algorithm, harmony search (HS) and water cycle algorithm (WCA) were implemented to solve the LUTLSP. Since the authors had adopted a discrete-time formulation of LUTLSP, discrete versions of Jaya and WCA were developed first. Secondly, some improvement strategies were proposed to speed up the convergence of applied optimizers. Thirdly, a feature based search operator was utilized to improve the search performance of reported optimization methods. Finally, experiments were carried out based on the real traffic data in Singapore. The HS, WCA, Jaya, and their variants were evaluated by solving 11 cases of traffic networks. The comparisons and discussions verified that the considered metaheuristic optimization methods could effectively solve the LUTLSP considerably surpassing the existing traffic light control strategy.

Prakash et al. (2017) proposed a binary Jaya algorithm for optimal placement of phasor measurement units(PMUs) for power system observability. The solution of optimal PMUs placement problem (OPPP) was addressed using a binary Jaya algorithm. The complete power system observability with maximum measurement redundancy was considered as the performance index in OPPP. To test the supremacy and accuracy of the algorithm, different standard test systems were examined for solving OPPP and the obtained results were compared with other state-of-art algorithms reported in the literature. The analysis of the results showed that the algorithm was equally good or better in solving the problem when compared to the other reported algorithms.

Zamli et al. (2017) described an experimental study of hyper-heuristic selection and acceptance mechanism for combinatorial t-way test suite generation. Additionally, a new hyper-heuristic selection acceptance mechanism, called FIS, was proposed based on the fuzzy inference system and compared with other hyper-heuristic and meta-heuristic approaches. The authors had reported the simplicity of the Jaya search operator as compared to complex crossover, peer learning, and global pollination operators. The Jaya code was reported to run faster.

Choudhary et al. (2018) investigated the effects of advanced submerged arc welding process parameters on weld bead geometry. Single and multiobjective optimization of process parameters were performed using desirability approach and Jaya algorithm. The results revealed that a smaller bead width and lower dilution could be achieved with the developed torch by allowing the use of higher preheat current values.

Michailidis (2017) proposed a hybrid parallel Jaya optimization algorithm for a multi-core environment with the aim of solving large-scale global optimization problems. The proposed algorithm was called HHCPJaya, and combined the hyper-population approach with the hierarchical cooperation search mechanism. The HHCPJaya algorithm divided the population into many small subpopulations, each of which focused on a distinct block of the original population dimensions. In the hyper-population approach, the small subpopulations were increasing by assigning more than one subpopulation to each core, and each subpopulation evolved independently to enhance the explorative and exploitative nature of the population. This hyper-population approach was combined with the two-level hierarchical cooperative search scheme to find global solutions from all subpopulations. Furthermore, an additional updating phase on the respective subpopulations based on global solutions was incorporated, with the aim of further improving the convergence rate and the quality of solutions. Several experiments applying the proposed parallel algorithm in different settings proved that it demonstrated sufficient promise in terms of the quality of solutions and the convergence rate. Furthermore, a relatively small computational effort was required to solve complex and large-scale optimization problems.

Das et al. (2017) established a hybridized intelligent machine learning based currency exchange forecasting model using Extreme Learning Machines (ELMs) and the Jaya optimization technique. This model can very well forecast the exchange price of USD (US Dollar) to INR (Indian Rupee) and USD to EURO based on statistical measures, technical indicators and combination of both measures over a time frame varying from 1 day to 1 month ahead. The proposed ELM-Jaya model was compared with existing optimized Neural Network and Functional Link Artificial Neural Network based predictive models. Finally, the model was validated using various performance measures such as; MAPE, Theil's U, ARV and MAE. The comparison of different features demonstrated that the technical indicators outperformed both the statistical measures and a combination of statistical measures and technical indicators in ELM-Jaya forecasting model.

Li and Yang (2017) proposed a novel pattern-recognition-based transient stability assessment (PRTSA) approach based on an ensemble of OS-extreme learning machine (EOSELM) with binary Jaya (BinJaya)-based feature selection with the use of phasor measurement units (PMUs) data. Furthermore, a BinJaya-based feature selection approach was put forward for selecting an optimal feature subset from the entire feature space constituted by a group of system-level classification features extracted from PMU data. The application results on the IEEE 39-bus system and a real provincial system showed that the proposed approach had superior computation speed and prediction accuracy than other state-of-the-art sequential learning algorithms. In addition, without sacrificing the classification performance, the dimension of the input space had been reduced to about one-third of its initial value.

Nayak et al. (2017) introduced an efficient Pathological brain detection systems (PBDSs) based on MR images that markedly improved the recent results. The proposed system made use of contrast limited adaptive histogram equalization (CLAHE) and orthogonal discrete ripplet-II transform (O-DR2T) with degree 2 to enhance the quality of the input MR images and extract the features respectively. Subsequently, relevant features were obtained using PCA + LDA approach. Finally, a novel learning algorithm called IJaya-ELM was proposed that combined improved Jaya algorithm (IJaya) and extreme learning machine (ELM) for segregation of MR images as pathological or healthy. The improved Jaya algorithm was utilized to optimize the input weights and hidden biases of single-hidden-layer feedforward neural networks (SLFN), whereas one analytical method was used for determining the output weights. The proposed algorithm performed optimization according to both the root mean squared error (RMSE) and the norm of the output weights of SLFNs. Extensive experiments were carried out using three benchmark datasets and the results were compared against other competent schemes. The experimental results demonstrated that the proposed scheme brings potential improvements in terms of classification accuracy and number of features. Moreover, the proposed IJaya-ELM classifier achieved higher accuracy and obtained compact network architecture compared to conventional ELM and BPNN classifier.

Du et al. (2017) used Jaya algorithm for solving the optimization-based damage identification problem. The vector of design variables represented the damage extent of elements discretized by the finite element model, and a hybrid objective function was proposed by combining two different objective functions to determine the sites and extent of damage. The first one was based on the multiple damage location assurance criterion and the second one was based on modal flexibility change. The robustness and efficiency of the proposed damage detection method were verified through three specific structures. The obtained results indicated that even under relatively high noise level, the proposed method not only successfully detected and quantified damage in engineering structures, but also showed better efficiency in terms of computational cost.

Huang et al. (2018) proposed a novel model-free solution algorithm, the natural cubic-spline-guided Jaya algorithm (S-Jaya), for efficiently solving the maximum power point tracking (MPPT) problem of PV systems under partial shading conditions. A photovoltaic (PV) system which controls the power generation with its operating voltage was considered. A natural cubic-spline-based prediction model was incorporated into the iterative search process to guide the update of candidate solutions (operating voltage settings) in the S-Jaya and such extension was capable of improving the tracking performance. Simulation studies and experiments were conducted to validate the effectiveness of the proposed S-Jaya algorithm for better addressing PV MPPT problems considering a variety of partial-shading conditions. Results of simulation studies and experiments had demonstrated that the S-Jaya algorithm converged faster and provided a higher overall tracking efficiency.

Ocłoń et al. (2018) presented a modified Jaya algorithm for optimizing the material costs and electric-thermal performance of an Underground Power Cable System (UPCS). A High Voltage (HV) underground cable line with three 400 kV AC cables arranged in flat formation in an exemplary case study was considered. When buried underground, three XLPE high voltage cables were situated in thermal backfill layer for ensuring the optimal thermal performance of the cable system. The study discussed the effect of thermal conductivities of soil and cable backfill material on the UPCS total investment costs. The soil thermal conductivity was assumed constant and equal to 0.8 W/(m K). The cable backfills considered in the study were as follows: sand and cement mix, Fluidized Thermal Backfill™ (FTB) and Powercrete™ a product of Heidelberg Cement Group. Constant thermal conductivities of the backfills in the dry state are assumed, respectively, 1.0, 1.54 and 3.0 W/(m K). The cable backfill dimensions and cable conductor area were selected as design variables in the optimization problem. The modified JAYA algorithm was applied to minimize material costs of UPCS under the constraint that the cable conductor temperature should not exceed its optimum value of 65 °C. The cable temperature was determined from the two-dimensional steady state heat conduction equation discretized using the Finite Element Method (FEM). The performance of the modified Jaya algorithm was compared with classical Jaya and PSO algorithms. The modified Jaya algorithm, for the presented case study, obtained lower values of the cost function.

Ghadivel et al. (2018) proposed an efficient improved hybrid Jaya algorithm based on time-varying acceleration coefficients (TVACs) and the learning phase introduced in teaching–learning-based optimization (TLBO), named the LJaya-TVAC algorithm, for solving various types of nonlinear mixed-integer reliability–redundancy allocation problems (RRAPs) and standard real-parameter test functions. RRAPs included series, series–parallel, complex (bridge) and overspeed protection systems. The search power of the proposed LJaya-TVAC algorithm for finding the optimal solutions was first tested on the standard real-parameter unimodal and multi-modal functions with dimensions of 30–100, and then tested on various types of nonlinear mixed-integer RRAPs. The results were compared with the original Jaya algorithm and the best results reported in the literature. The optimal results obtained with the proposed LJaya-TVAC algorithm provided evidence for its better and acceptable performance.

Wang et al. (2018a, b) had developed an improved emotion recognition system over facial expression images. The algorithm chosen was a combination of stationary wavelet transform, single hidden layer feed forward neural network, and Jaya algorithm. The proposed approach performed better than five state-of-the-art approaches in terms of overall accuracy.

Degertekin et al. (2017) used Jaya algorithm for sizing and layout optimization of truss structures. The original Jaya algorithm formulation was modified in order to improve convergence speed and reduce the number of structural analyses required in the optimization process. The suitability of Jaya algorithm for truss optimization was investigated by solving six classical weight minimization problems of truss structures including sizing, layout and large-scale optimization problems with up to 204 design variables. Discrete sizing/layout variables and simplified topology optimization also were considered. The test problems solved in this study were very common benchmarks in structural optimization and practically described all scenarios that might be faced by designers. The results demonstrated that the Jaya algorithm could obtain better designs than those of the other state-of-the-art metaheuristics and gradient-based optimization methods in terms of optimized weight, standard deviation and number of structural analyses.

Gambhir and Gupta (2018) applied meta-heuristic algorithms of cuckoo search, TLBO, PSO and Jaya algorithms to the analytical model of long period fibre grating (LPFG) sensors. Peak transmission loss of −76 dB was obtained by optimizing the grating parameters using Jaya and cuckoo search algorithms. Jothi et al. (2018) described the application of graphical user interface technique for the differentiation of acute lymphoblastic leukemia nucleus from healthy lymphocytes in a medical image is described. This method employed a backtrack search optimization algorithm for clustering. Five different categories of features were extracted from the segmented nucleus images, i.e., morphological, wavelet, color, texture and statistical features. Three different kinds of hybrid supervised feature selection algorithms such as tolerance rough set particle swarm optimization-based quick reduct, tolerance rough set particle swarm optimization-based relative reduct and tolerance rough set firefly-based quick reduct were applied for selecting prominent features. The redundant features were eliminated to generate the reduced set. Jaya algorithm was applied for optimizing the rules generated from the classification algorithms. Classification algorithms such as Naïve Bayes, linear discriminant analysis, *K*-nearest neighbor, support vector machine, decision tree and ensemble random undersampling boost were applied on leukemia dataset. Experimental results depicted that the classification algorithms after optimizing with Jaya algorithm had improved the classification accuracy compared to the results obtained before optimizing with Jaya algorithm.

Warid et al. (2018) introduced quasi-oppositional modified Jaya (QOMJaya) to solve different multi-objective optimal power flow (MOOPF) problems. Significant modifications to the basic Jaya algorithm were done to create a modified Jaya (MJaya) algorithm that could handle the MOOPF problem. A fuzzy decision-making strategy was proposed and incorporated into the Jaya algorithm as selection criteria for best and worst solutions. A new criterion for comparing updated and current candidate solutions was proposed. The concept of Pareto optimality was used to extract a set of non-dominated solutions. A crowding distance measure approach was utilized to maintain the diversity of Pareto optimality. In addition, a novel external elitist repository was developed to conserve discovered non-dominated solutions and to produce true and well-distributed Pareto optimal fronts. The proposed algorithm was scrutinized and validated using the modified IEEE 30-bus test system. Simulation results revealed the proposed algorithm's ability to produce real and well-distributed Pareto optimum fronts for all considered multi-objective optimization cases.

Wang et al. (2018a, b) proposed a parallel Jaya algorithm implemented on the graphics processing unit (GPU-Jaya) to estimate parameters of the Li-ion battery model. Similar to the generic Jaya algorithm, the GPU-Jaya was free of tuning algorithm-specific parameters. Compared with the generic Jaya algorithm, three main procedures of the GPU-Jaya, the solution update, fitness value computation, and the best/worst solution selection were all computed in parallel on GPU via a compute unified device architecture (CUDA). Two types of memories of CUDA, the global memory and the shared memory were utilized in the execution. The effectiveness of the proposed GPU-Jaya algorithm in estimating model parameters of two Li-ion batteries was validated via real experiments while its high efficiency was demonstrated by comparing with the generic Jaya and other considered benchmarking algorithms. The experimental results reflected that the GPU-Jaya algorithm could accurately estimate battery model parameters while tremendously reducing the execution time using both entry-level and professional GPUs.

Rao and Saroj (2018a) proposed an elitist-based self-adaptive multi-population (SAMPE) Jaya algorithm to solve the constrained and unconstrained problems related to numerical and engineering optimization. The search mechanism of the Jaya algorithm was improved by using the subpopulation search scheme with elitism. The algorithm had used an adaptive scheme for dividing the population into subpopulations. The effectiveness of the proposed SAMPE-Jaya algorithm was verified on CEC 2015 benchmark problems in addition to fifteen unconstrained, six constrained standard benchmark problems and four constrained mechanical design optimization problems considered from the literature. The Friedman rank test was also done for comparing the performance of the SAMPE-Jaya algorithm with other algorithms. It was also tested on three large-scale problems with the dimensions of 100, 500 and 1000. Furthermore, the proposed SAMPE-Jaya algorithm was used for solving a case study of design optimization of a micro-channel heat sink. The computational experiments had proved the effectiveness of the proposed SAMPE-Jaya algorithm.

Rao and Saroj (2018b) explored the use of a self-adaptive multi-population elitist (SAMPE) Jaya algorithm for the economic optimization of shell-and-tube heat exchanger (STHE) design. Three different optimization problems of STHE were considered in this work. The same problems were earlier attempted by other researchers using genetic algorithm (GA), particle swarm optimization (PSO) algorithm, biogeography-based optimization (BBO), imperialist competitive algorithm (ICA), artificial bee colony (ABC), cuckoo-search algorithm (CSA), intelligence-tuned harmony search (ITHS), and cohort intelligence (CI) algorithm. The SAMPE-Jaya algorithm was proposed to optimize the setup cost and operational cost of STHEs simultaneously. The performance of the SAPME-Jaya algorithm was tested on four well-known constrained, ten unconstrained standard benchmark problems, and three STHE design optimization problems. The results of computational experiments proved the superiority of the method over the latest reported methods used for the optimization of the same problems.

Rao et al. (2018) Used SAMPE-Jaya algorithm for the design optimization of heat pipes. Minimization of the total mass of heat pipe was considered as a single objective problem with an array of working fluids namely methanol, ethanol and ammonia. Furthermore, maximization of heat transfer rate and minimization of thermal resistance of a heat pipe were carried out simultaneously by using a priori approach. The results obtained by the SAMPE-Jaya algorithm were found to be superior as compared to those given by Teaching-Learning-Based-Optimization (TLBO) algorithm, Global Best Algorithm (GBA), Jaya and SAMP-Jaya algorithms in the case of single objective optimization. In the case of multi-objective optimization, the results of SAMPE-Jaya algorithm were found superior as compared to those given by Niched Pareto Genetic Algorithm (NPGA), Grenade Explosion Method (GEM), TLBO, Jaya, self-adaptive Jaya and SAMP-Jaya algorithms.

In addition to the above research works published in reputed international journals, a number of other research works have been published in various reputed international conference proceedings.

10.2 Concluding Remarks

The Jaya algorithm has carved a niche for itself in the field of advanced optimization and many researchers have understood the potential of the algorithm. Like TLBO algorithm, the Jaya algorithm does not require tuning of algorithmic-specific parameters. This concept of the algorithm is one of the attracting features in addition to its *simplicity*, *robustness* and the *ability to provide the global or near global optimum solutions* in *comparatively less number of function evaluations*. Self-adaptive Jaya algorithm has eliminated tuning of population size which is one of the common control parameters. Few researchers have made modifications to the basic Jaya algorithm and proved the effectiveness of the modified versions. However, to make the implementation of Jaya algorithm simpler, it may be desirable to maintain the algorithm-specific parameter-less concept in the modified versions also. Otherwise, the user will be faced with the burden of tuning the algorithm-specific parameters in addition to the common control parameters.

Like TLBO algorithm, the logic of the Jaya algorithm is such that the algorithm can be used with equal ease for the maximization as well as minimization problems. Single objective as well as multi-objective optimization problems can be easily attempted. Researchers have proved that this algorithm is successful in solving small as well as large scale optimization problems. One may argue that this algorithm has origin bias while solving the standard benchmark functions but it has been already proved that the algorithm can work well on shifted, shifted rotated expanded and composite benchmark functions. Various types of real-world optimization problems have also been successfully solved and reported by the researchers.

The working of the Jaya algorithm and its variants demonstrated step-by-step by means of examples in Chap. 2 is expected to be useful to the beginners to understand the basics of the algorithm. One can understand the *ease* of the algorithm after reading through the steps. The potential and effectiveness of the Jaya algorithm and its variants are demonstrated in this book by reporting the results of application on the complex and composite benchmark functions, constrained and unconstrained single objective as well as multi-objective practical design problems. The performance of the algorithm is compared with the other well known evolutionary algorithms, swarm intelligence based algorithms, and other algorithms. In majority of the cases the results obtained by the Jaya algorithm and its variants are found superior or competitive to the other optimization algorithms. Statistical tests have also been conducted to validate the Jaya algorithm and its variants.

It cannot be concluded that the Jaya algorithm is the 'best' algorithm among all the optimization algorithms available in the literature. In fact, there may not be any such 'best' algorithm existing! In recent years, the field of combinatorial optimization has witnessed a true tsunami of "novel" metaheuristic methods, most of them based on a metaphor. The behavior of virtually any species of insects, animals, birds, natural phenomena, etc. is serving as inspiration to launch yet another metaheuristic. Few researchers are focusing all their efforts on producing algorithms that appear to achieve good performance on the standard set of benchmark instances, even though these are often not representative of realistic instances of the optimization problem at hand. Those researchers manage to tune their algorithm-specific parameters to just the right values for it to perform well on the standard benchmark sets to get published (Sorensen 2015). However, Jaya algorithm is different and is not having any such algorithm-specific parameters to tune and the algorithm is not based on any metaphor.

The Jaya algorithm is relatively a new algorithm and has strong potential to solve the optimization problems. If the algorithm is found having certain limitations then the researchers are requested to find the ways to overcome the limitations and to further strengthen the algorithm. The efforts should not be in the form of destructive criticism. What can be said with more confidence at present about the Jaya algorithm is that it is simple to apply, it has no algorithm-specific parameters and it provides the optimum results in comparatively less number of function evaluations. There is no need to make any undue criticism on these aspects, as the algorithm has established itself and has set itself distinct. Researchers are encouraged to make improvements to the Jaya algorithm so that the algorithm will become much more powerful with much improved performance. It is hoped that the researchers belonging to different fields will find the Jaya algorithm as a powerful tool to optimize the systems and processes.

References

Abhishek, K., Kumar, V. R., Datta, S., & Mahapatra, S. S. (2017). Application of JAYA algorithm for the optimization of machining performance characteristics during the turning of CFRP (epoxy) composites: comparison with TLBO, GA, and ICA. *Engineering with Computers, 33* (3), 457–475

Azizipanah-Abarghooee, R., Dehghanian, P., & Terzija, V. (2016a). Practical multi-area bi-objective environmental economic dispatch equipped with a hybrid gradient search method and improved Jaya algorithm. *IET Generation, Transmission and Distribution, 10*(14), 3580–3596.

Azizipanah-Abarghooee, R., Golestaneh, F., Gooi, H. B., Lin, J., Bavafa, F., & Terzija, V. (2016b). Corrective economic dispatch and operational cycles for probabilistic unit commitment with demand response and high wind power. *Applied Energy, 182,* 634–651.

Choudhary, A., Kumar, M., & Unune, D. R. (2018). Investigating effects of resistance wire heating on AISI 1023 weldment characteristics during ASAW. *Materials and Manufacturing Processes, 33*(7), 759–769.

Das, S. R., Mishra, D., & Rout, M. (2017). A hybridized ELM-Jaya forecasting model for currency exchange prediction, *Journal of King Saud University-Computer and Information Sciences*. https://doi.org/10.1016/j.jksuci.2017.09.006.

Degertekin, S. O., Lamberti, L., & Ugur, I. B. (2017). Sizing, layout and topology design optimization of truss structures using the Jaya algorithm. *Applied Soft Computing*. https://doi.org/10.1016/j.asoc.2017.10.001.

Du, D. C., Vinh, H. H., Trung, V. D., Hong Quyen, N. T., & Trung, N. T. (2017). Efficiency of Jaya algorithm for solving the optimization-based structural damage identification problem based on a hybrid objective function. *Engineering Optimization*. https://doi.org/10.1080/0305215X.2017.1367392.

Gambhir, M., & Gupta, S. (2018). Advanced optimization algorithms for grating based sensors: A comparative analysis. *Optik*. https://doi.org/10.1016/j.ijleo.2018.03.062.

Gao, K., Zhang, Y., Sadollah, A., Lentzakis, A., & Su, R. (2017). Jaya, harmony search and water cycle algorithms for solving large-scale real-life urban traffic light scheduling problem. *Swarm and Evolutionary Computation, 37,* 58–72.

Ghadivel, S., Azizivahed, A., & Li, L. (2018). A hybrid Jaya algorithm for reliability–redundancy allocation problems. *Engineering Optimization, 50*(4), 698–715.

Huang, C., Wang, L., Yeung, R. S. C., Zhang, Z., Chung, H. S. H., & Bensoussan, A. (2018). A prediction model-guided Jaya algorithm for the PV system maximum power point tracking. *IEEE Transactions on Sustainable Energy, 9*(1), 45–55.

Jothi, G., Inbarani, H. H., Azar, A. T., & Devi, K. R. (2018). Rough set theory with Jaya optimization for acute lymphoblastic leukemia classification. *Neural Computing and Applications*. https://doi.org/10.1007/s00521-018-3359-7.

Kumar, A., Datta, S., & Mahapatra, S. S. (2017a). Application of JAYA algorithm for the optimization of machining performance characteristics during the turning of CFRP (epoxy) composites: Comparison with TLBO, GA, and ICA. *Engineering with Computers, 33*(3), 457–475.

Kumar, N., Hussain, I., Singh, B., & Panigrahi, B. K. (2017b). Rapid MPPT for uniformly and partial shaded PV system by using JayaDE algorithm in highly fluctuating atmospheric conditions. *IEEE Transactions on Industrial Informatics, 13*(5), 2406–2416.

Kumar, M., & Mishra, S. K. (2017). Jaya-FLANN based adaptive filter for mixed noise suppression from ultrasound images. *Biomedical Research, 29*(5) (in press).

Li, Y., & Yang, Z. (2017). Application of EOS-ELM with binary Jaya-based feature selection to real-time transient stability assessment using PMU data. *IEEE Access, 5,* 23092–23101.

Malik, S., & Kumar, S. (2017). Optimized phase noise compensation technique using neural network. *Indian Journal of Science and Technology, 10*(5). https://doi.org/10.17485/ijst/2017/v10i5/104348.

Michailidis, P. D. (2017). An efficient multi-core implementation of the Jaya optimisation algorithm. *International Journal of Parallel, Emergent and Distributed Systems,* https://doi.org/10.1080/17445760.2017.141638.

Mishra, S., & Ray, P. K. (2016). Power quality improvement using photovoltaic fed DSTATCOM based on JAYA optimization. *IEEE Transactions on Sustainable Energy, 7*(4), 1672–1680.

Mohebian, P., Mousavi, M., & Rahami, H. (2017). Simultanous optimization of size and connection arrangement for low-rise steel plate shear wall systems. *Iran University of Science & Technology, 7*(2), 269–290.

Nayak, D. R., Dash, R., & Majhi, B. (2017). Development of pathological brain detection system using Jaya optimized improved extreme learning machine and orthogonal ripplet-II transform. *Multimedia Tools and Applications*. https://doi.org/10.1007/s11042-017-5281-x.

Ocłoń, P., Cisek, P., Rerak, M., Taler, D., Rao, R. V., Vallati, A., et al. (2018). Thermal performance optimization of the underground power cable system by using a modified Jaya algorithm. *International Journal of Thermal Sciences, 123,* 162–180.

Prakash, T., Singh, V. P., Singh, S. P., & Mohanty, S. R. (2017). Binary Jaya algorithm based optimal placement of phasor measurement units for power system observability. *Energy Conversion and Management, 140,* 34–35.

Rao, R. V. (2016). Jaya: A simple and new optimization algorithm for solving constrained and unconstrained optimization problems. *International Journal of Industrial Engineering Computations, 7,* 19–34.

Rao, R. V., & More, K. C. (2017a). Design optimization and analysis of selected thermal devices using self-adaptive Jaya algorithm. *Energy Conversion and Management, 140,* 24–35.

Rao, R. V., & More, K. C. (2017b). Optimal design and analysis of mechanical draft cooling tower using improved Jaya algorithm. *International Journal of Refrigeration, 82,* 312–324.

Rao, R. V., More, K. C., Coelho, L. S., & Mariani, V. C. (2017a). Multi-objective optimization of the Stirling heat engine through self-adaptive Jaya algorithm. *Journal of Renewable and Sustainable Energy, 9*(3), 033703. https://doi.org/10.1063/1.4987149.

Rao, R. V., More, K. C., Taler, J., & Ocłoń, P. (2016a). Dimensional optimization of a micro-channel heat sink using Jaya algorithm. *Applied Thermal Engineering, 103,* 572–582.

Rao, R. V., & Rai, D. P. (2017a). Optimisation of welding processes using quasi-oppositional-based Jaya algorithm. *Journal of Experimental & Theoretical Artificial Intelligence, 29*(5), 1099–1117.

Rao, R. V., & Rai, D. P. (2017b). Optimization of submerged arc welding process parameters using quasi-oppositional based Jaya algorithm. *Journal of Mechanical Science and Technology, 31*(5), 2513–2522.

Rao, R. V., Rai, D. P., & Balic, J. (2016b). A new optimization algorithm for parameter optimization of nano-finishing processes. *Scientia Iranica. Transaction E, Industrial Engineering, 24*(2), 868–875.

Rao, R. V., Rai, D. P., & Balic, J. (2016c). Surface grinding process optimization using Jaya algorithm. In *Computational intelligence in data mining* (Vol. 2, pp. 487–495). New Delhi: Springer.

Rao, R. V., Rai, D. P. & Balic, J. (2017a). A new algorithm for parameter optimization of nano-finishing processes. *Scientia Iranica: Transactions on Industrial Engineering, 24*(2), 868–875.

Rao, R. V., Rai, D. P., & Balic, J. (2017b). A multi-objective algorithm for optimization of modern machining processes. *Engineering Applications of Artificial Intelligence, 61,* 103–125.

Rao, R. V., Rai, D. P., Ramkumar, J., & Balic, J. (2016). A new multi-objective Jaya algorithm for optimization of modern machining processes. *Advances in Production Engineering & Management, 11*(4), 271.

Rao, R. V., & Saroj, A. (2016). Multi-objective design optimization of heat exchangers using elitist-Jaya algorithm. *Energy Systems*. https://doi.org/10.1007/s12667-016-0221-9.

Rao, R. V., & Saroj, A. (2017a). A self-adaptive multi-population based Jaya algorithm for engineering optimization. *Swarm and Evolutionary Computation, 37,* 1–26.

Rao, R. V., & Saroj, A. (2017b). Constrained economic optimization of shell-and-tube heat exchangers using elitist-Jaya algorithm. *Energy, 128,* 785–800.

Rao, R. V., & Saroj, A. (2017c). Economic optimization of shell-and-tube heat exchanger using Jaya algorithm with maintenance consideration. *Applied Thermal Engineering, 116,* 473–487.

Rao, R. V., & Saroj, A. (2018a). Economic optimization of shell-and-tube heat exchanger using Jaya algorithm with maintenance consideration. *ASME Transactions: Journal of Thermal Science and Engineering Applications, 10,* 041001-1–041001-12.

Rao, R. V., & Saroj, A. (2018b). An elitism-based self-adaptive multi-population Jaya algorithm and its applications. *Soft Computing*. https://doi.org/10.1007/s00500-018-3095-z.

Rao, R. V., Saroj, A., & Bhattacharyya, S. (2018). Design optimization of heat pipes using elitism-based self-adaptive multipopulation Jaya algorithm. *Journal of Thermophysics and Heat Transfer*. https://doi.org/10.2514/1.T5348.

Rao, R. V., Saroj, A., Ocłoń, P., Taler, J., & Taler, D. (2017b). Single-and multi-objective design optimization of plate-fin heat exchangers using Jaya algorithm. *Heat Transfer Engineering*. https://doi.org/10.1080/01457632.2017.1363629.

Rao, R. V., & Waghmare, G. G. (2017). A new optimization algorithm for solving complex constrained design optimization problems. *Engineering Optimization, 49*(1), 60–83.

Singh, S. P., Prakash, T., Singh, V. P., & Babu, M. G. (2017). Analytic hierarchy process based automatic generation control of multi-area interconnected power system using Jaya algorithm. *Engineering Applications of Artificial Intelligence, 60,* 35–44.

Sinha, R. K., & Ghosh, S. (2016). Jaya based ANFIS for monitoring of two class motor imagery task. *IEEE Access, 4,* 9273–9282.

Sorensen, K. (2015). Metaheuristics—The metaphor exposed. *International Transactions in Operational Research, 22,* 3–18.

Wang, S. H., Phillips, P., Dong, Z. C., & Zhang, Y. D. (2018a). Intelligent facial emotion recognition based on stationary wavelet entropy and Jaya algorithm. *Neurocomputing, 272,* 668–676.

Wang, S., Rao, R. V., Chen, P., Zhang, Y., Liu, A., & Wei, L. (2017). Abnormal breast detection in mammogram images by feed-forward neural network trained by Jaya algorithm. *Fundamenta Informaticae, 151*(1–4), 191–211.

Wang, L., Zhang, Z., Huang, C., & Tsui, K. L. (2018b). A GPU-accelerated parallel Jaya algorithm for efficiently estimating Li-ion battery model parameters. *Applied Soft Computing, 65,* 12–20.

Warid, W., Hizam, H., Mariun, N., & Abdul-Wahab, N. I. (2016). Optimal power flow using the Jaya algorithm. *Energies, 9*(9), 678.

Warid, W., Hizam, H., Mariun, N., & Wahab, N. I. A. (2018). A novel quasi-oppositional modified Jaya algorithm for multi-objective optimal power flow solution. *Applied Soft Computing, 65,* 360–373.

Zamli, K. Z., Din, F., Kendall, G., & Ahmed, B. S. (2017). An experimental study of hyper-heuristic selection and acceptance mechanism for combinatorial t-way test suite generation. *Information Sciences, 399,* 121–153.

Zhang, Y., Yang, X., Cattani, C., Rao, R. V., Wang, S., & Phillips, P. (2016). Tea category identification using a novel fractional Fourier entropy and Jaya algorithm. *Entropy, 18*(3), 77.

Appendix
Codes of Jaya Algorithm and Its Variants

The codes for unconstrained and constrained optimization problems for sample single- and multi-objective optimization functions are given below. The user has to create separate MATLAB files but the files are to be saved in a single folder. These codes may be used for reference and the user may define the objective function(s), design variables and their ranges as per his or her own requirements.

A.1 Jaya Code for Unconstrained Rosenbrock Function

This is a complete program for solving Rosenbrock function using Jaya algorithm. In order to run the program, the following code may be copied and pasted, as it is, into the MATLAB editor file and the same may be executed. This program is only for demonstration purpose. The numbers assigned to the population size, generations, design variables and maximum function evaluations in this program need not be taken as default values.

```
%% Jaya algorithm
%% Rosenbrock function
function Jaya()
clc;
clear all;
RUNS=30;
runs=0;
while(runs<RUNS)
pop=25; % population size
var=30; % no. of design variables
maxFes=500000;
maxGen=floor(maxFes/pop);
mini=-30*ones(1,var);
maxi=30*ones(1,var);
```


R. Venkata Rao, *Jaya: An Advanced Optimization Algorithm and its Engineering Applications*, https://doi.org/10.1007/978-3-319-78922-4

```
[row,var]=size(mini);
x=zeros(pop,var);
for i=1:var
x(:,i)=mini(i)+(maxi(i)-mini(i))*rand(pop,1);
end
ch=1;
gen=0;
f=myobj(x);
while(gen<maxGen)
xnew=updatepopulation(x,f);
xnew=trimr(mini,maxi,xnew);
fnew=myobj(xnew);
for i=1:pop
if(fnew(i)<f(i))
x(i,:)=xnew(i,:);
f(i)=fnew(i);
end
end
disp('%%%%%%%% Final population %%%%%%%%');
disp([x,f]);
fnew=[];xnew=[];
gen=gen+1;
fopt(gen)=min(f);
end
runs=runs+1;
[val,ind]=min(fopt);
Fes(runs)=pop*ind;
best(runs)=val;
end
bbest=min(best);
mbest=mean(best);
wbest=max(best);
stdbest=std(best);
mFes=mean(Fes);
stdFes=std(Fes);
fprintf('\n best=%f',bbest);
fprintf('\n mean=%f',mbest);
fprintf('\n worst=%f',wbest);
fprintf('\n std. dev.=%f',stdbest);
fprintf('\n mean function evaluations=%f',mFes);
end
function[z]=trimr(mini,maxi,x)
[row,col]=size(x);
for i=1:col
x(x(:,i)<mini(i),i)=mini(i);
```

```
x(x(:,i)>maxi(i),i)=maxi(i);
end
z=x;
end
function [xnew]=updatepopulation(x,f)
[row,col]=size(x);
[t,tindex]=min(f);
Best=x(tindex,:);
[w,windex]=max(f);
worst=x(windex,:);
xnew=zeros(row,col);
for i=1:row
for j=1:col
r=rand(1,2);
xnew(i,j)=x(i,j)+r(1)*(Best(j)-abs(x(i,j)))-r(2)*(worst(j)-abs(x(i,j)));
end
end
end
function [f]=myobj(x)
[r,c]=size(x);
for i=1:r
y=0;
for j=1:c-1
y=y+(100*(x(i,j)^2-x(i,j+1))^2+(1-x(i,j))^2);
end
z(i)=y;
end
f=z';
end
```

A.2 Jaya Code for Constrained Himmelblau Function

This is a complete program for solving constrained Himmelblau function using Jaya algorithm. In order to run the program, the following code may be copied and pasted, as it is, into the Matlab editor file and the same may be executed. This program is only for demonstration purpose. The numbers assigned to the population size, generations, design variables and maximum function evaluations in this program need not be taken as default values.

```
%% Jaya algorithm
%% Constrained optimization
%% Himmelblau function
function Jaya()
```

```
clc;
clear all;
RUNS=10;
runs=0;
while(runs<RUNS)
pop=5; % population size
var=2; % no. of design variables
maxFes=150000;
maxGen=floor(maxFes/pop);
mini=[-5 -5];
maxi=[5 5];
[row,var]=size(mini);
x=zeros(pop,var);
for i=1:var
x(:,i)=mini(i)+(maxi(i)-mini(i))*rand(pop,1);
end
ch=1;
gen=0;
f=myobj(x);
while(gen<maxGen)
xnew=updatepopulation(x,f);
xnew=trimr(mini,maxi,xnew);
fnew=myobj(xnew);
for i=1:pop
if(fnew(i)<f(i))
x(i,:)=xnew(i,:);
f(i)=fnew(i);
end
end
disp('%%%%%% Final population%%%%%%%');
disp([x,f]);
fnew=[];xnew=[];
gen=gen+1;
fopt(gen)=min(f);
end
runs=runs+1;
[val,ind]=min(fopt);
Fes(runs)=pop*ind;
best(runs)=val;
end
bbest=min(best);
mbest=mean(best);
wbest=max(best);
stdbest=std(best);
mFes=mean(Fes);
```

```
fprintf('\n best=%f',bbest);
fprintf('\n mean=%f',mbest);
fprintf('\n worst=%f',wbest);
fprintf('\n std. dev.=%f',stdbest);
fprintf('\n mean Fes=%f',mFes);
end
function[z]=trimr(mini,maxi,x)
[row,col]=size(x);
for i=1:col
x(x(:,i)<mini(i),i)=mini(i);
x(x(:,i)>maxi(i),i)=maxi(i);
end
z=x;
end
function [xnew]=updatepopulation(x,f)
[row,col]=size(x);
[t,tindex]=min(f);
Best=x(tindex,:);
[w,windex]=max(f);
worst=x(windex,:);
xnew=zeros(row,col);
for i=1:row
for j=1:col
r=rand(1,2);
xnew(i,j)=x(i,j)+r(1)*(Best(j)-abs(x(i,j)))-r(2)*(worst(j)-abs(x(i,j)));
end
end
end
function [f]=myobj(x)
[r,c]=size(x);
Z=zeros(r,1);
for i=1:r
x1=x(i,1);
x2=x(i,2);
z=(((x1^2)+x2-11)^2)+((x1+x2^2-7)^2);
g1=26-((x1-5)^2)-((x2)^2);
g2=20-(4*x1)-x2;
p1=10*((min(0,g1))^2); % penalty if constraint 1 is violated
p2=10*((min(0,g2))^2); % penalty if constraint 2 is violated
Z(i)=z+p1+p2; % penalized objective function value
end
f=Z;
end
```

A.3 SAMPE Jaya Code for Unconstrained Himmelblau Function

This is a complete program for solving unconstrained Himmelblau functionusing self-adaptive multi-population elitist-Jaya algorithm. The user has to create separate MATLAB files (all saved in a single folder). He or she can define the design variables, ranges and the objective function(s) as per his or her requirements.

```
%%%%%%%%%%%%%%%% objective%%%%%%%%%%%%%%%%%
function [ Z ] = objective( x )
% Unconstrained optimization
% Himmelblau function
% Detailed explanation goes here
x1=x(1);
x2=x(2);
Z=(((x1^2)+x2-11)^2)+((x1+x2^2-7)^2);
end
%%%%%%%%Algorithm_MP_ES%%%%%%%%%%%%%
clc
clearall
funev=150000; % Maximum functionevaluations
IDR=1;
formatshortg
%NP = input('Enter the population size:');
NP=25; % Population size
ES=2;  % Elite size
NMP=2 ;% Number of multi population initial
NG=funev/NP;  % Maximum number of generations
DV=2; % Number of design variables
%  NG = input('Enter the no of generation :');
%  DV = input('Enter the no of Decision variable:');
for m=1:IDR % No. of Independent run
for j = 1 : DV
```

```
DV_min(j)=[-5];
DV_max(j)=[5];
end
% Initial Population Generation
for i=1:NP
for j = 1 : DV
x(i,j) = DV_min(j) + (DV_max(j) - DV_min(j))*rand(1);
end
end
for i = 1 : NP
x(i,DV + 1) = objective(x(i,:));
end
% start of generation
for k=1:NG
x=unique(x, 'rows');
[r,c]= size(x);
if r<NP % If current population size is less than NP add NP-r      %popula-
tions
for i=r+1:NP
for j = 1 : DV
x(i,j) = DV_min(j) + (DV_max(j) - DV_min(j))*rand(1);
end
end
for i = r+1:NP
x(i,DV + 1) = objective(x(i,:));
end
else
x(:,:)=x(:,:);
end
%% Replace the worst solution with elite solutions
x=sortrows(x,DV+1); % sort the solutions in ascending order
x(NP-ES+1:NP,:)= x(1:ES,:);
[f_min,ind_min] = min(x(:,DV+1));
[f_max,ind_max] = max(x(:,DV+1));
%% Devide population and modify the solutions
x1=[];
for n=1:NMP
   a=round((n-1)*NP/NMP)+1;
   b=round(NP/NMP*n);
if b>NP
for i=NP+1:b
for j=1:DV
x(i,j)= randsample(x(:,j),1);
end
x(i,DV + 1) = objective(x(i,:));
```

```
end
end
for i=a:b
for j=1:DV
    x1(i,j) = x(i,j) +rand*(x(a,j)- abs(x(i,j)))-rand*(x(b,j)- abs(x(i,
j)));
end
end
end
% Check for the bounds of decision variable
For i=1:NP
for j=1:DV
if x1(i,j)<DV_min(j)
x1(i,j)= DV_min(j);
elseif x1(i,j)>DV_max(j)
x1(i,j) = DV_max(j);
else
x1(i,j) = x1(i,j);
end
end
end
for i = 1 : NP
x1(i,DV + 1) = objective(x1(i,:));
end
for i=1:NP
if x(i,DV + 1)< x1(i,DV + 1)
x(i,DV + 1)= x(i,DV + 1);
for j=1:DV+1
x(i,j) = x(i,j);
end
else
x(i,DV + 1)= x1(i,DV + 1);
for j=1:DV+1
x(i,j) = x1(i,j);
end
end
end
[f_best(k),l2] = min(x(:,DV+1));
for j=1:DV
x_fi(j)= x(l2,j);
end
```

```
%%%%%%%self-adaptive population concept starts over here
if (f_min<f_best(k))
     NMP=NMP+1;
if (NP/NMP)>=3
        NMP=NMP;
else
     NMP=NMP-1;
end
else
if NMP>1
  NMP= NMP-1;
else
  NMP=1;
end
end
np(k)=NMP;
% disp(NMP)
%%%%%%%%%%%%%%%%% self-adaptive populations concept ends over
end
disp(['————[ ','Run No.= ',num2str(m), ' ]—————————————'])
disp(['Optimum value =',num2str(min(f_best),30)])
disp(['DV: ',num2str(x_fi,7)])
fprintf('\n \n');
 [f_min1(m),ind1(m)]= min(f_best);
 f_max1(m)= max(f_best);
end
  Best= min(f_min1);
  Worst=max(f_min1);
  Mean = mean(f_min1);
  SD  = std(f_min1);
  MFE= mean(ind1)*NP;
fprintf('\n Best=%f',Best);
fprintf('\n Worst=%f',Worst);
fprintf('\n Mean=%f',Mean);
fprintf('\n SD=%f', SD);
fprintf('\n MFE=%f', MFE);
fprintf('\n FE=%f',funev);
fprintf('\n \n');
```

A.4 SAMPE Jaya Code for Constrained Himmelblau Function

This is a complete program for solving constrained Himmelblau function using self-adaptive multi-population elitist-Jaya algorithm. The user has to create separate MATLAB files (all saved in a single folder). He or she can define the ranges, design variables and the objective function(s) as per his or her requirements.

```
%%%%%%%%%%%%%%% objective%%%%%%%%%%%%%%%%

function [ Z ] = objective( x )
% Constrained optimization
% Himmelblau function
% Detailed explanation goes here
x1=x(1);
x2=x(2);
z=(((x1^2)+x2-11)^2)+((x1+x2^2-7)^2);
g1=26-((x1-5)^2)-((x2)^2);
g2=20-(4*x1)-x2;
p1=10*((min(0,g1))^2); % penalty if constraint 1 is violated
p2=10*((min(0,g2))^2); % penalty if constraint 2 is violated
Z=z+p1+p2; % penalized objective function value
end
%%%%%%%%%Algorithm_MP_ES%%%%%%%%%%%%
clc
clearall
funev=150000; % Maximum function evaluations
IDR=1;
formatshortg
%NP = input('Enter the population size:');
NP=25; % Population size
ES=2;  % Elite size
NMP=2 ;% Number of multi population initial
NG=funev/NP;  % Maximum number of generations
DV=2; % Number of design variables
%  NG = input('Enter the no of generation :');
%  DV = input('Enter the no of Decision variable:');
for m=1:IDR % No. of Independent run
for j = 1 : DV
DV_min(j)=[-5];
DV_max(j)=[5];
end
% Initial Population Generation
for i=1:NP
for j = 1 : DV
```

```
x(i,j) = DV_min(j) + (DV_max(j) - DV_min(j))*rand(1);
end
end
for i = 1 : NP
x(i,DV + 1) = objective(x(i,:));
end
% start of generation
for k=1:NG
x=unique(x, 'rows');
[r,c]= size(x);
if r<NP % If current population size is less than NP add NP-r        %popula-
tions
fori=r+1:NP
for j = 1 : DV


x(i,j) = DV_min(j) + (DV_max(j) - DV_min(j))*rand(1);
end
end
for i = r+1:NP
x(i,DV + 1) = objective(x(i,:));
end
else
x(:,:)=x(:,:);
end
%% Replace the worst solution with elite solutions
x=sortrows(x,DV+1); % sort the solutions in ascending order
x(NP-ES+1:NP,:)= x(1:ES,:);
[f_min,ind_min] = min(x(:,DV+1));
[f_max,ind_max] = max(x(:,DV+1));
%% Devide population and modify the solutions
x1=[];
for n=1:NMP
  a=round((n-1)*NP/NMP)+1;
  b=round(NP/NMP*n);
if b>NP
for i=NP+1:b
for j=1:DV
x(i,j)= randsample(x(:,j),1);
end
x(i,DV + 1) = objective(x(i,:));
end
end
for i=a:b
```

```
for j=1:DV
     x1(i,j) = x(i,j) +rand*(x(a,j)- abs(x(i,j)))-rand*(x(b,j)- abs(x(i,
j)));
end
end
end
% Check for the bounds of decision variable
For i=1:NP
for j=1:DV
if x1(i,j)<DV_min(j)
x1(i,j)= DV_min(j);
elseif x1(i,j)>DV_max(j)
x1(i,j) = DV_max(j);
else
x1(i,j) = x1(i,j);
end
end
end
for i = 1 : NP
x1(i,DV + 1) = objective(x1(i,:));
end
for i=1:NP
if x(i,DV + 1)< x1(i,DV + 1)
x(i,DV + 1)= x(i,DV + 1);
for j=1:DV+1
x(i,j) = x(i,j);
end
else
x(i,DV + 1)= x1(i,DV + 1);
for j=1:DV+1
x(i,j) = x1(i,j);
end
end
end
  [f_best(k),l2] = min(x(:,DV+1));
for j=1:DV
x_fi(j)= x(l2,j);
end
%%%%%%% self-adaptive population concept startsover here
if (f_min<f_best(k))
     NMP=NMP+1;
if (NP/NMP)>=3
       NMP=NMP;
else
     NMP=NMP-1;
```

```
end
else
if NMP>1
  NMP= NMP-1;
else
  NMP=1;
end
end
np(k)=NMP;
% disp(NMP)
%%%%%%%%%self-adaptive populations concept ends over here%%%%%
end
disp(['————[ ','Run No.= ',num2str(m), ' ]————'])
disp(['Optimum value =',num2str(min(f_best),30)])
disp(['DV: ',num2str(x_fi,7)])
fprintf('\n \n');
 [f_min1(m),ind1(m)]= min(f_best);
 f_max1(m)= max(f_best);
end
  Best= min(f_min1);
  Worst=max(f_min1);
  Mean = mean(f_min1);
  SD  = std(f_min1);
  MFE= mean(ind1)*NP;
fprintf('\n Best=%f',Best);
fprintf('\n Worst=%f',Worst);
fprintf('\n Mean=%f',Mean);
fprintf('\n SD=%f', SD);
fprintf('\n MFE=%f', MFE);
fprintf('\n FE=%f',funev);
fprintf('\n \n');
```

A.5 MOQO Jaya Code for ECM Process

The following is the code for multi-objective quasi-oppositional (MOQO) Jaya algorithm for electro-chemical machining (ECM) process. This is a sample program for reference.

```
function [z]=moqojaya()
RUNS=0;
maxRUNS=1;
while(RUNS<maxRUNS)
clearvars -except maxRUNS RUNS;
```

```
clc;
FEs=0;
maxFEs=3750;
gen=0;
temp_cl=[];
pop=50;
sch=[0 0 1];
mini=[8 300 3];
maxi=[200 5000 21];
dim=numel(mini);
n=numel(sch);
for i=1:dim
cl(:,i)=mini(i)+(maxi(i)-mini(i))*rand(pop,1);
end
cl_new=quasi(mini,maxi,cl);
cl=[cl;cl_new];
cl(:,dim+1:dim+n)=myobj(cl);
temp_cl=cl;
cl=[];
b=[];
cl=NS(n,dim,temp_cl,sch);
cl=cl(1:pop,:);
temp_cl=[];
while(FEs<=maxFEs)
a=[]; b=[]; a= cl(:,dim+n+1)==1; b=find(cl(a,dim+n+2)==100); count=numel(b);
temp_cl=updation(cl(:,1:dim),count);
temp_cl=trimr(mini,maxi,temp_cl);
temp_cl(:,dim+1:dim+n)=myobj(temp_cl);
comb_cl=cat(1,cl(:,1:dim+n),temp_cl);
temp_cl=[];
cl=[];
cl=NS(n,dim,comb_cl,sch);
cl=cl(1:pop,:);
comb_cl=[];
FEs=FEs+pop;
temp_cl=quasi(mini,maxi,cl(:,1:dim));
temp_cl(:,dim+1:dim+n)=myobj(temp_cl);
comb_cl=cat(1,cl(:,1:dim+n),temp_cl);
temp_cl=[];
cl=[];
cl=NS(n,dim,comb_cl,sch);
cl=cl(1:pop,:);
comb_cl=[];
cl=cat(2,cl,constraint(cl(:,1:dim)));
FEs=FEs+pop;
```

```
disp(FEs);
plot3(cl(:,dim+1),cl(:,dim+2),cl(:,dim+3),'*');
 xlabel('Z1');
 ylabel('Z2');
 zlabel('Z3');
gen=gen+1;
disp(gen);
figure(1);
end
RUNS=RUNS+1;
end
end


function [z1]= NS(n,dim,cl,sch)
[pop,col]=size(cl);
x=cl(:,1:dim);
func=cl(:,dim+1:dim+n);
f=func;
srn=[1:pop];
cons=constraint(x);
tempf=f;
[p,c]=size(f);
l=1;
rank=1;
score=[];
tscore=zeros(1,2);
nd=[];
todel=[];
index=[];
t=[];
while(p>1)
for i=1:p
  win=1;
  for j=1:p
    if(i~=j)
      if(cons(i)<cons(j))
        win=win+1;
      end
      if(cons(i)>cons(j))
        break;
      end


    if(cons(i)==cons(j))
```

```
for k=1:n
 switch sch(k)
     case 0
        if(f(i,k)<f(j,k))
        score(1,k)=1;
        score(2,k)=0;
        end
        if(f(i,k)==f(j,k))
        score(1,k)=0;
        score(2,k)=0;
        end
        if(f(i,k)>f(j,k))
         score(1,k)=0;
         score(2,k)=1;
        end
     case 1
        if(f(i,k)>f(j,k))
           score(1,k)=1;
           score(2,k)=0;
        end
        if(f(i,k)==f(j,k))
           score(1,k)=0;
           score(2,k)=0;
        end
        if(f(i,k)<f(j,k))
           score(1,k)=0;
           score(2,k)=1;
        end


 end
end
 tscore(1)=sum(score(1,:));
 tscore(2)=sum(score(2,:));


 if(tscore(1)>0.0 && tscore(1)<=n)
   win=win+1;
 end
       end
      end
   end
```

```
   if(win==p)
      nd(l)=srn(i);
      todel(l)=i;
      l=l+1;
   end


   score=[];
   tscore=zeros(1,2);
end
if(numel(nd)==0)


  tempf(srn,n+1)=rank;
  rank=rank+1;
   f=[];
   srn=[];
   cons=[];
else
tempf(nd,n+1)=rank;
rank=rank+1;
f(todel,:)=[];
srn(todel)=[];
cons(todel)=[];
end
nd=[];
todel=[];
l=1;


[p,c]=size(f);
end
tempf(tempf(:,n+1)==0,n+1)=rank;
x=cat(2,x,tempf);
[t,index]=sort(tempf(:,n+1),'ascend');
x=x(index,:);
%disp(x);


for i=1:n
fmin(i)=min(x(:,dim+i));
fmax(i)=max(x(:,dim+i));
end
r=max(x(:,dim+n+1));
%disp(r);
```

```
for k=1:r
  t=[];
  index=[];
  f2=x(x(:,dim+n+1)==k,dim+1:dim+n);
[row,col]=size(f2);
if(row<2)
  x(x(:,dim+n+1)==k,dim+n+2)=100;
else


  x(x(:,dim+n+1)==k,dim+n+2)=crowdist(fmin,fmax,f2);
  x1=x(x(:,dim+n+1)==k,:);
  [t,index]=sort(x1(:,dim+n+2),'descend');
  x1=x1(index,:);
  x(x(:,dim+n+1)==k,:)=x1;
end


end
z1=x;


end


function [d]=crowdist(fmin,fmax,f)


[row,col]=size(f);
d=zeros(row,1);


for i=1:col
  [I,t]=sort(f(:,i));
  %disp(t);
  d(t(1))=100;
  d(t(row))=100;
  for j=2:row-1
    d(t(j))=d(t(j))+(f(t(j+1),i)-f(t(j-1),i))/(fmax(i)-fmin(i));
  end
end
d(d>100)=100;
end
```

```
function [z2]=updation(temp_cl,count)
[row,col]=size(temp_cl);
best=temp_cl(round(count-(count-1)*rand(1,1)),:);
worst=temp_cl(end,:);
for j=1:row
  for i=1:col
  rn=rand(1,2);
  temp_cl(j,i)=temp_cl(j,i)+rn(1)*(best(i)-abs(temp_cl(j,i)))-rn(2)*
(worst(i)-abs(temp_cl(j,i)));
  end
end
z2=temp_cl;
end


function[z4]=trimr(mini,maxi,temp)
[row,col]=size(temp);
for i=1:col
  temp(temp(:,i)<mini(i),i)=mini(i);
  temp(temp(:,i)>maxi(i),i)=maxi(i);
end
  z4=temp;
end


function [f1]=myobj(x)
[row,col]=size(x);


for i=1:row


f=x(i,1);
U=x(i,2);
V=x(i,3);


Z1=(f^0.381067)*(U^-0.372623)*(V^3.155414)*exp(-3.128926);
Z2 = (f^3.528345)*(U^0.000742)*(V^-2.52255)*exp(0.391436);
Z3=f;
```

```
f1(i,1)=Z1;
f1(i,2)=Z2;
f1(i,3)=Z3;
end
end


function[z5]=constraint(x)
[row,col]=size(x);
for i=1:row


f=x(i,1);
U=x(i,2);
V=x(i,3);


g1=(1-(f^2.133007)*(U^-1.088937)*(V^-0.351436)*exp(0.321968));
g2=((f^-0.844369)*(U^-2.526075)*(V^1.546257)*exp(12.57697))-1;
g3=1-((f^0.075213)*(U^-2.488362)*(V^0.240542)*exp(11.75651));


G1(i,1)=min(0,g1);
G2(i,1)=min(0,g2);
G3(i,1)=min(0,g3);
end
G1str=max(abs(G1));
G2str=max(abs(G2));
G3str=max(abs(G3));


if(max(abs(G1))==0)
  G1str=1;
end


if(max(abs(G2))==0)
  G2str=1;
end
```

```
if(max(abs(G3))==0)
  G3str=1;
end


z5=(abs(G1)/G1str)+(abs(G2)/G2str)+(abs(G3)/G3str);
end


function [Pnew]=quasi(mini,maxi,P)
[row,col]=size(P);
mini=repmat(mini,row,1);
maxi=repmat(maxi,row,1);
A=0.5*(mini+maxi);
B=((mini+maxi)-P);
Pnew=A+(B-A).*rand(row,col);
end
```

In addition to the above, the readers may also refer to https://sites.google.com/site/jayaalgorithm/.

Index

R. Venkata Rao, *Jaya: An Advanced Optimization Algorithm and its Engineering Applications*, https://doi.org/10.1007/978-3-319-78922-4

Printed in the United States
By Bookmasters